FUTURE TRENDS
IN TELECOMMUNICATIONS

WILEY SERIES IN COMMUNICATION AND DISTRIBUTED SYSTEMS

FUTURE TRENDS
IN TELECOMMUNICATIONS

R. J. Horrocks

Horrocks Technology
UK

R. W. A. Scarr

OFTEL
UK

JOHN WILEY & SONS

Chichester · New York · Brisbane · Toronto · Singapore

Copyright © 1993 by John Wiley & Sons Ltd,
Baffins Lane, Chichester,
West Sussex PO19 1UD, England

Other Wiley Editorial Offices

John Wiley & Sons, Inc., 605 Third Avenue,
New York, NY 10158–0012, USA

Jacaranda Wiley Ltd, G.P.O. Box 859, Brisbane,
Queenland 4001, Australia

John Wiley & Sons (Canada) Ltd, 22 Worcester Road,
Rexdale, Ontario M9W 1L1, Canada

John Wiley & Sons (SEA) Pre Ltd, 37 Jalan Pemimpin #05–04,
Block B, Union Industrial Building, Singapore 2057

British Library Cataloguing in Publication Data

A catalogue record for this book is available from the British Library

ISBN 0 471 93724 X

Typeset in 10/12 Palatino by Pure Tech Corporation, Pondicherry, India
Printed and bound in Great Britain by Bookcraft (Bath) Ltd

CONTENTS

17 DATA AND MULTISERVICE NETWORKS—UPPER LAYERS   287

18 INTELLIGENT NETWORKS   317

19 CENTREX AND VIRTUAL PRIVATE NETWORKS   327

ACKNOWLEDGEMENTS

- To Brian Evans of Tantara for information on satellite broadcasting.
- To Tony Holmes of BT and Jerry Ennis of Interconnect for information on UPT.
- To Bill Martin for certain helpful information on radio systems.
- To John Meredith of ETSI for information on certain standards.
- To Ron Samuel of BT for information on EUTELSAT.
- To Joanna Scarr for helpful advice on videotex.
- To Paul Thompson of BT and to the British Interplanetary Society for information on INTELSAT.
- To Tim Wright of ETSI TM3 and BT for information on SDH and ATM.
- To our families for their patience with the greater than normal absent mindedness and the long disappearances into the study which the preparation of this book has entailed.
- To the IEEE for permission to publish Figure 11.1, which is based on Figure 1 in the paper by J. L. Hennessy and N. L. Jouppi entitled 'Computer technology and architecture: an evolving interaction' published in *IEEE Computer*, September, 1991, 18–29.
- To BSI Standards for permission to publish Figures 10.3 and 10.5, which are based on Figure 3 of BS : 6305 : 1992. (Complete copies of the Standard can be obtained from BSI Publications, Linford Wood, Milton Keynes MK14 6LE.)
- Figures 3.4, 3.5, 3.7, 3.9, 5.1, 8.1, 8.2, 15.2, 15.4, 15.5, 15.6, 15.7 and 17.1 are based on figures in the CCITT Blue Book (IXth Plenary Assembly, Melbourne, 1988), Geneva, 1989, but have been adapted by the authors for the purpose of this book. Complete volumes of the CCITT Blue Book may be ordered from the Sales Section of the ITU General Secretariat, Place des Nations, CH-1222 Geneva 20, Switzerland.

Finally it should be made clear that the views expressed are the personal opinions of the authors and do not in any way represent the policies of any organisations with which they are connected.

PREFACE

The world of telecommunications is changing very rapidly. The old analogue public networks which have served for decades are being replaced by digital networks, and new radio services such as cellular and paging have experienced rates of growth beyond the more optimistic expectations. At the same time there are profound changes, particularly in the UK, brought about by the regulatory regime which has introduced competition in services and apparatus and by preparation for the opening up of the common market from 1992 onwards. Pervading both the technology and the regulation is the convergence of computing (or more generally information technology) with communications. The two distinct cultures are beginning to merge through the need to adopt a common approach at the interface where information from a computer system is conveyed elsewhere by a telecommunications system. Because of the diversity of approach in the past, this boundary or interface is not a precise one and standards and protocols may overlap. Rather than trying to define the boundary we have crossed it and thus the scope is somewhat wider than the pure telecommunications professional might expect. On the other hand it should make it easier for those whose background is computing to bridge the gap into telecommunications.

Because telecommunications, even on its own, is such a wide subject we have had to some extent to sacrifice depth for breadth and to concentrate on principles rather than detailed design information. We have as far as possible attempted to cover the world scene but inevitably there is bias towards the UK because this is where our experience mostly lies. However we are very conscious of the UK as a member of the European Community (EC) and in our future projections we have taken due account of pan-European factors. Nor has the North American scene been ignored and particular attention is paid to areas in which the USA, specifically, is providing the technical leadership.

All this change is surrounded by thick clouds of jargon and acronyms and it is very difficult for the uninitiated to penetrate the cloud and understand the issues. Nevertheless most of the fundamental issues and trends are relatively simple and are capable of analysis and discussion at a level which can be understood by those who have not spent their whole career in a technical field.

The objective is therefore to try to penetrate the clouds, to identify and analyse the main technical and human issues and to point the most likely way forward for the development of systems and services. The book is aimed at a wide range of readers, for example:

- telecommunications managers;
- general business managers who have a responsibility for, and whose business depends upon, good communications;
- investment analysts who need to understand technical trends in their markets;
- regulators and policy makers who need to know more about how telecommunications works;
- specialist engineers and computer scientists who wish to look more widely than their particular specialisation; and
- engineering and science graduates who are entering a field of specialisation within telecommunications for the first time.

Because we are trying to cater for such a wide range of readers, we have tried to treat each topic with a simple introduction that explains the basic terms and concepts and then gradually goes into more depth. Thus readers should not assume, if they meet something that they do not understand in an early chapter, that the the remainder of the book will be unintelligible; on the contrary they may need to skip only a page or two to be back within their depth again. For the reasons given above great depth is not feasible but we have attempted to give references to more detailed work to make follow-up easy.

We think that the value of the content may be more as a work of reference to be consulted as the unfamiliar arises rather than as something to be read from cover to cover and for that reason we have provided a comprehensive index and a separate list of the many acronyms that one is bound to encounter.

Unlike many books on future issues in technology, this work is not a collection of disparate papers which have been welded together by an editor. The structure and content have been carefully planned and close attention has been given to the cross linkages to help the reader to put together in his or her mind the complex web of interrelationships that make up telecommunications. Although it has been written by two authors, inevitably introducing some diversity of style, our method has been that one author would write those sections with which he was more familiar and the other would revise them and if necessary rewrite and add various parts. In this way we hope that the advantages of having two authors greatly exceeds the disadvantages.

The scope is mainly technology but with some reference to regulations, tariffs, numbering and other matters more directly related to management and administration. There are frequent references to costs because the costs of competing technologies are key factors in determining the shape of the future. There are also some comments on what may be called human factors because as the scope of the possible grows even wider the choice of technology or service will depend increasingly on subjective preferences and user friendliness.

The book is divided into four parts.

Part 1 is entitled Key Technologies and has 11 chapters covering topics ranging from components, via the larger elements such as transmission, switching and radio, to what might be termed concepts i.e. Open Systems Interconnection or OSI.

The book is probably unique in dealing at some length with issues concerning wiring in buildings.

Part 2 is called Networks and Services and contains 10 chapters dealing with the subject matter either from the point of view of the transmission medium, e.g. satellites and cable television, or from service concepts such as ISDN, multiservice networks and Centrex. The emphasis here is in on the immediate future, although in order to build a convincing picture it is necessary to make frequent reference to past and current practice.

Part 3 covers Administration and is divided into five chapters. Major topics are numbering, regulations and competition, and network management.

In Part 4 we conclude with a summary chapter which pulls together our principal conclusions and predictions for the future together with some future scenarios for the business and the home.

Forecasting the future is always an uncertain business and if we have stuck our heads out we hope we have not stuck them out too far. Even if all the current trends could be analysed perfectly, there is always the unpredictable new invention, the political, regulatory or the economic change or change in human taste or fashion that can alter the situation completely. But far from making analysis and forecasting a worthless pastime, they only serve to make it the more interesting.

FOREWORD

Telecommunications is an exciting topic. But it is also a topic which is of vital importance to society. Any society which wants to be competitive must have an efficiently functioning telecommunications sector. I am certain that the readers would be able to quote examples of countries which have failed to be competitive simply because they did not succeed in establishing a good telecommunications infrastructure.

Yes, indeed, exciting and important. However, the most striking feature is probably complexity. It is no exaggeration to state that telecommunications is amongst the most difficult disciplines about which to obtain a reasonable knowledge, let alone to become an expert. This is due to the fantastic development which has taken place in this sector over the last few years.

Why this recent development?

The two major reasons are most probably technology and market forces. Microelectronics and its use within Information Technology have opened the door for many new types of equipment and systems which were simply not possible to produce before. As far as the market is concerned, there is no doubt that liberalisation in Europe and other parts of the world has substantially fostered the development of the market. As far as Europe is concerned the fruitful interaction between CEPT (Conférence Européenne des Postes et Télécommunications) and CEC (Commission of the European Community) has been a real catalyst for the growth of European telecommunications. The European Telecommunications Standards Institute, ETSI, is one of the products of this interaction.

In my work in ETSI, almost daily I meet people who struggle hard to get acquainted with the complex world of telecommunications and its many facets. For those people I am convinced that this book is a 'gefundenes Fressen' as we say in my homeland or—to use the language in which this book is written–a 'treasure-trove'.

I would certainly recommend this book to everybody who is inolved in telecommunications, be they users, administrators, manufacturers, operators, buyers, standardisers, etc. It will be a valuable source of help to anyone who wishes to take advantage of the many opportunities that are available within our exciting, complex and vital sector.

Karl Heinz Rosenbrock
Director of European Telecommunications Standards Institute

ACRONYMS

AAL	ATM Adaption Layer
AC	Alternating Current
ACSE	Application Control Service Element
A/D	Analogue to Digital (conversion)
ADMD	ADministration Management Domain
ADPCM	Adaptive Differential Pulse Code Modulation
AFN	All Figure Number
AM	Amplitude Modulation
AMI	Alternate Mark Inversion
AMVSB	Amplitude Modulation with Vestigial SideBands
ANSI	American National Standards Institute
AOC-D	Advice of Charge–During call
AOC-E	Advice of Charge–at End of call
AOC-S	Advice of Charge–at call Set-up time
APB	Adaptive Prediction Backwards
APCM	Adaptive Pulse Code Modulation
APD	Avalanche Photo Diode
APF	Adaptive Prediction Forwards
APIA	Application Program Interface Association
APNSS	Analogue Private Network Signalling System
APON	Asynchronous Passive Optical Network
APPC	Advanced Program-to-Program Communication
ASE	Application Service Element
ASN	Abstract Syntax Notation
ATD	Asynchronous Time Division (multiplexing)
ATM	Asynchronous Transfer Mode
AU	Administrative Unit
AUG	Administrative Unit Group

BABT	British Approvals Board for Telecommunications
BBC	British Broadcasting Company
BCH	Bose–Chadhuri–Hocquenhem (error correction code)
BER	Bit Error Rate
BHCA	Busy Hour Call Attempts
Bi-CMOS	Bipolar Complementary Metal Oxide Semiconductor
BIDS	Broadband Integrated Distributed Star
Bi-MOS	Bipolar Metal Oxide Semiconductor
B-ISDN	Broadband Integrated Services Digital Network
BOC	Bell Operating Company
BPON	Broadband Passive Optical Network
BS	British Standard
BSC	Base Station Controller
BSI	British Standards Institute.
BT	British Telecom
BTRL	British Telecom Research Laboratories
BTS	Base Transceiver Station
CAD	Computer Aided Design
CATV	Community Antenna TeleVision
CBMS	Computer Based Message Service
CBN	Common Bonding Network
CCAF	Call Control Agent Function
CCBS	Completion of Calls to Busy Subscriber
CCF	Call Control Function
CCIR	International Radio Consultative Committee
CCITT	International Telegraph and Telephone Consultative Committee
CCR(SE)	Commitment, Concurrency and Recovery (Service Element)

CD	Call Deflection	CUG	Closed User Group
CD	Compact Disc	CVSD	Continuously Variable Slope
CDDI	Cable Distributed Data Inter-		Delta modulation
	face	CW	Call Waiting
CDM	Code Division Multiplexing		
CDMA	Code Division Multiple Access	D/A	Digital to Analogue (conversion)
CEI	Comparably Efficient Inter-	DAF	Distributed Application Frame-
	connection		work
CEN	Comité Européen de Normal-	DASS	Digital Access Signalling System
	isation	dB	deciBels
CENELEC	Comité Européen de Normal-	DBS	Direct Broadcasting by Satellite
	isation ELECtrical	DC	Direct Current
CEPT	European Conference for Posts	DCME	Digital Circuit Multiplication
	and Telecommunications		Equipment
CFB	Call Forwarding Busy	DDI	Direct Dialling In
CFDM	Companded Frequency Divi-	DDSN	Digital Derived Service Net-
	sion Multiplex		work
CFM	Companded Frequency Modul-	DECT	Digital European Cordless
	ation		Telephone
CFNR	Call Forwarding No Reply	DIS	Draft International Standard
CFU	Call Forwarding Unconditional	DIT	Directory Information Tree
C/I	Carrier to Interference (ratio)	DMSU	Digital Main Switching Unit
CLAN	Connectionless or Cordless	DNIC	Data Network Identification
	LAN		Code
CLI	Calling Line Identity	DOS	Disc Operating System
CLIP	Calling Line Identification	DP	Discussion Paper
	Presentation	DPLE	Digital Principal Local
CLIR	Calling Line Identification		Exchange
	Restriction	DPNSS	Digital Private Network Signal-
CMBS	Computer Message Bureau		ling System
	Service	DQDB	Distributed Queue Dual Bus
CMIP	Common Management Infor-	DRAM	Dynamic Random Access
	mation Protocol		Memory
CMIS	Common Management Infor-	DSRR	Digital Short Range Radio
	mation Service	DTE	Data Terminating Equipment
CMOS	Complementary Metal Oxide	DTI	Department of Trade and
	Semiconductor		Industry
C/N	Carrier to Noise (ratio)	DTMF	Dual Tone Multi-Frequency
COLP	COnnected Line identification	DTL	Diode Transistor Logic
	Presentation		
COLR	COnnected Line identification	EA	Extended Address (bit)
	Restriction	EBU	European Broadcasting Union
CONF	CONFerence (call add-on)	EC	European Community
CPU	Central Processing Unit	ECL	Emitter Coupled Logic
C/R	Command/Response (bit)	ECMA	European Computer Manufac-
CRT	Cathode Ray Tube		turers Association
CSMA/CD	Carrier Sense Multiple Access	ECT	Explicit Call Transfer
	with Collision Detection	EDI	Electronic Data Interchange
CT	Cordless Telephone	EDIM	EDI Message
CTE	Circuit Terminating Equipment	EDIN	EDI Notification
CTR	Common Technical Regulations		
CTV	Cable TeleVision		

EEPROM	Electrically Erasable Programmable Read Only Memory		GMSC	Gateway Mobile Switching Centre
EESP	End-to-End Security Protocol		GMSK	Gaussian Minimum Shift Keying
ehf	extra high frequency		GOSIP	Government OSI Procurement
EIA	Electronic Industries Association		GSC	Group Switching Centre
EIRP	Equivalent Isotropic Radiated Power		GSM	Global System for Mobile communications
ELFEXT	Equivalent Level Far End XTalk			
EM	Electronic Messaging		HCD(M)	Hybrid Companded Delta (Modulation)
E-mail	Electronic mail			
EMC	Electro-Magnetic Compatibility		HDB	High Density Binary
EN	European Norm		HDLC	High level Data Link Control
ENV	European Norm Voluntary		HDTV	High Definition TeleVision
EPHOS	EuroPean Handbook for Open Systems		hf	high frequency
			HLR	Home Location Register
EPIRB	Emergency Position Indicating Radio Beacon		HOLD	call HOLD
			Hz	hertz
EPROM	Erasable Programmable Read Only Memory		IBCN	Integrated Broadband Communications Network
ERMES	European Radio MEssage System		IBS	INTELSAT Business Service
			IC	Integrated Circuit
ERP	Effective Radiated Power		ICI	Interface Control Information
ESS	Electronic Switching System		IDA	Integrated Digital Access
ETACS	Extended Total Access Communications System		IDR	Intermediate Data Rate
			IEC	Inter-Exchange Carrier
ETS	European Telecommunication Standard		IEC	International Electrotechnical Commission
ETSI	European Telecommunications Standards Institute		IEE	Institution of Electrical Engineers (UK)
FAS	Flexible Access System		IEEE	Institution of Electrical and Electronic Engineers (USA)
Fax	Facsimile			
FCC	Federal Communications Commission		I-ETS	Interim European Technical Standard
FDDI	Fibre Distributed Data Interface		IF	Intermediate Frequency
FDM	Frequency Division Multiplex		IHS	Interactive Home System
FDMA	Frequency Division Multiple Access		IIR	Infinite Impulse Response
			IKBS	Intelligent Knowledge Based System
FET	Field Effect Transistor			
FEXT	Far End XTalk (crosstalk)		IMO	International Marine Organisation
FIR	Finite Impulse Response			
FM	Frequency Modulation		INMARSAT	INternational MARitime SATellite (organisation)
FPH	FreePHone			
FSK	Frequency Shift Keying		INTELSAT	INternational TELecommunication SATellite (organisation)
FTAM	File Access Transfer and Management			
FTTC	Fibre To The Curb (kerb)		IPMS	Inter-Personal Message Service
FTTH	Fibre To The Home		IPNS	Inter-PBX Networking Specification
GAP	Analysis and Forecasting Group		IPS (ips)	Impulses Per Second
GMDSS	Global Maritime Distress and Safety System		IS	International Standard
			ISDN	Integrated Services Digital Network

ISO	International Standards Organisation	MTP	Message Transfer Part
		MTS	Message Transfer Service
ISPBX	Integrated Services Private Branch eXchange	Mux	Multiplex or Multiplexer
ISUP	ISDN User Part	NEMA	National Electrical Manufacturers Association (USA)
IT	Information Technology		
ITA	Independent Television Authority	NET	European Telecommunication Standard (Norme Européene de Télécommunications)
ITU	International Telecommunications Union		
		NEXT	Near End XTalk (crosstalk)
		N-ISDN	Narrow-band ISDN
JEEC	Joint ETSI-ECMA Committee	nm	nanometres
JTC	Joint Technical Committee	NMOS	N-type Metal Oxide Semiconductor
JTM	Job Transfer and Manipulation		
		NNG	National Numbering Group
LAN	Local Area Network	NPI	Numbering Plan Indicator
LAP	Link Access Protocol	NSAP	Network Service Access Point
LATA	Local Access and Transport Area	NT	Network Termination
		NTE	Network Terminal Equipment
LEC	Local Exchange Carrier	NTP	Network Termination Point
LED	Light Emitting Diode	NTTA	Network Terminating and Test Apparatus
lf	low frequency		
LLC	Link Layer Control		
LSI	Large Scale Integration	OCR	Optical Character Reader
LTP	Long Term Prediction	ODA	Office Document Architecture
		ODIF	Office Documentation Interchange Format
MAC	Multiplexed Analogue Component		
		ODP	Open Distributed Processing
MAN	Metropolitan Area Network	ODSA	Open Distributed Systems Architecture
MAP	Manufacturing Automation Protocol		
		OFTEL	OFfice of TELecommunications
MAP	Mobile Application Part	OMAP	Operations and Maintenance Applications Part
MAU	Multi-station Access Unit		
MCI	Microwave Communication Inc.	ONA	Open Network Architecture
MCID	Malicious Call IDentification	ONP	Open Network Provision
mf	medium frequency	O/R	Originator/Recipient
MHS	Message Handling System	OSI	Open Systems Interconnection
MMC	Meet-Me Conference		
MMS	Manufacturing Message Service	PABX	Private Automatic Branch EXchange
MMS	Modified Monitoring Code		
MOS	Metal Oxide Semiconductor	PAD	Packet Assembler/Disassembler
MOU	Memorandum Of Understanding	PBX	Private Branch Exchange
		PC	Personal Computer
		PC	Private Circuit
MQW	Multi-Quantum Well	PCI	Protocol Control Information
MSC	Management Service Control	PCM	Pulse Code Modulation
MSC	Mobile Switching Centre	PCN	Personal Communications Network
MSI	Medium Scale Integration		
MSK	Minimum Shift Keying	PDH	Plesiochronous Digital Hierarchy
MSN	Multiple Subscriber Number	PDS	Planned Domestic Leases
MSRN	Mobile Station Roaming Number	PDU	Protocol Data Unit
MTA	Message Transfer Agent	pels	picture elements

PHB	Photochemical Hole Burning	RTL	Resistor Transistor Logic
p-i-n	p-type–intrinsic–n-type	RTS(E)	Remote Transfer Service (Element)
PIN	Personal Identification Number		
PMOS	P-type Metal Oxide Semiconductor	Rx	Receiver
PMR	Private Mobile Radio	SAPI	Service Access Point Identifier
POCSAG	Post Office Code Standardisation Advisory Group	SAT	Supervisory Audio Tones
		SCCP	Signal Connection and Control Part
POEEJ	Post Office Electrical Engineers Journal	SCF	Service Control Function
POH	Path OverHead	SCP	Service Control Point
POTS	Plain Ordinary Telephone Service	SCPC	Single Channel Per Carrier
		SDF	Service Data Function
PRMD	PRivate Management Domain	SDH	Synchronous Digital Hierarchy
PS	Packet Switching	SDIF	Standard Document Interchange Format
PS	Paging Service		
ps	Picosecond	SDU	Service Data Unit
PSK	Phase Shift Keying	SEED	Self ElEctrooptic Device
PSPDN	Packet Switched Public Data Network	SG	Study Group
		SGML	Standard Generalised Mark-up Language
PSS	Packet Switching Service		
PSTN	Public Switched Telephone Network	SHF (shf)	Super High Frequency
		SI	Speech Interpolation
PTNX	Private Network Telecommunication eXchange	SIB	Service Independent Building block
PTO	Public Telecommunications Operator	SMAF	Service Management Access Function
PTT	Post Telegraph and Telephone (a general international term for a PTO)	SMATV	Satellite Master Antenna TeleVision
		SMF	Service Management Function
		S/N	Signal to Noise (ratio)
PVC	Poly-Vinyl Chloride	SNA	Systems Network Architecture
		SNACF	Sub-Network Access Control Function
QDU (qdu)	Quantising Distortion Unit		
QOS	Quality Of Service	SNDCF	Sub-Network Dependent Convergent Function
QPSK	Quaternary Phase Shift Keying		
QPSX	Queued Packet Synchronous switch	SNICF	Sub-Network Independent Convergent Function
		SOGT	Senior Officials Group on Telecommunications
RACE	R & D in Advanced Communications technologies in Europe		
		SOH	Section OverHead
RAM	Random Access Memory	SONET	Synchronous Optical NETwork
RBOC	Regional Bell Operating Company	SOS	Silicon On Sapphire
		SPADE	SCPC Access with Demand Assignment
RF (rf)	Radio Frequency	SPC	Stored Program Control
ROM	Read Only Memory	SQL	Standard Query Language
ROS(E)	Remote Operations Service (Element)	SRAM	Static Random Access Memory
		SRF	Specialised Resource Function
RPC	Remote Procedure Call	SSB	Single SideBand
RPE	Regular Pulse Excitation	SSBSC	Single SideBand Suppressed Carrier
RPI	Retail Price Index		

SSF	Service Switching Function		TWT	Travelling Wave Tube
SSTDMA	Switched Satellite Time Division Multiple Access		Tx	Transmitter
STD	Subscriber Trunk Dialling		UA	User Agent
STD	Synchronous Time Division		UHF (uhf)	Ultra High Frequency
STM	Synchronous Transfer Mode		UL	Underwriters Laboratory (Inc., USA)
STP	Screened Twisted Pair			
STS	Space–Time–Space (switching)		UMTS	Universal Mobile Telecommunications System
SUB	SUB-addressing			
Sub Mux	Sub Multiplexer		UPT	Universal Personal Telecommunications
TA	Terminal Adapter			
TACS	Total Access Communications System		UTP	Unscreened Twisted Pair
			UUS	User–User Signalling
TASI	Time Assigned Speech Interpolation		VADS	Value Added and Data Services
TC	Technical Committee		VC	Virtual Container
TCAP	Transaction Capabilities Application Part		VCI	Virtual Channel Identifier
			VHF (vhf)	Very High Frequency
TCP/IP	Transmission Control Protocol/ Internet Protocol		vlf	very low frequency
			VLR	Vehicle Location Register
TDM	Time Division Multiplex		VLSI	Very Large Scale Integration
TDMA	Time Division Multiple Access		VPC	Virtual Private Circuit
TE	Terminal Equipment		VPI	Virtual Path Identifier
TEI	Terminal Equipment Identifier		VPN	Virtual Private Network
TIA	Telecommunication Industries Association		VSAT	Very Small Aperture Terminal
			VT	Virtual Terminal
TJF	Test Jack Frame		VTAM	Virtual Terminal Access and Management
TM	Terminal Management			
TNIC	Transit-Network Identification Code		VTP	Virtual Terminal Protocol
TO	Telecommunication Organisation		Wal	Walsh
TON	Type Of Number		WAN	Wide Area Network
TOP	Technical Office Protocol		WARC	World Administrative Radio Conference.
TP	Terminal Portability			
TP	Transaction Processing		WATS	Wide Area Transmission Service
TPON	Telephony over Passive Optical Network		WDM	Wavelength Division Multiplex
			WIMP	Windows, Icons, Mouse and Pointers
TSC	Trunk System Controller			
TST	Time–Space–Time (switching)		WLAN	Wireless LAN
TTL	Transistor Transistor Logic		WORM	Write Once Read Many
TU	Tributary Unit		WSF	Work Station Function
TUG	Tributary Unit Group		WSI	Wafer Scale Integration
TUP	Telephony User Part			
TV	TeleVision		μm	micrometre
TWA	Two-Way Alternate			
TWS	Two-Way Simultaneous		3PTY	Three Party Service

Part 1

KEY TECHNOLOGIES

OVERVIEW

New developments in telecommunication systems and services are strongly influenced by developments in base technologies. Base technologies in their turn depend on developments in theoretical physics and fundamental research into materials and manufacturing processes. Very often this research finds its first application in new components and this is the starting point here. In Chapter 1 we consider advances in VLSI and memories, technologies that are having a major impact in the field of telecommunications generally and are particularly relevant in the context of Chapters 6,8,9 and 11. These four chapters deal with coders, radio, terminal equipment and processors respectively and here the impact of VLSI and memory become very apparent in complex equipment realisations. The other major component technology, optical, is covered in Chapter 2 and its impact is mainly on transmission as covered in Chapter 3, but has yet to exert much influence on switching as covered in Chapter 4. Chapters 3 and 4 are partly of an introductory nature, but the synchronous digital hierarchy is treated in Chapter 3 and other aspects of the topics they cover are enlarged upon later in the context of networks.

Other subjects covered are signalling in Chapter 5 with the emphasis on the newer types of signalling system; line codes and the coding of speech and video in Chapter 6 (line codes because of their increasing importance in the context of local transmission networks) and building wiring in Chapter 10 because of its increasing economic importance and technical complexity.

An understanding of the principles of OSI are fundamental to an understanding of networks and services and Chapter 8 provides an introduction to this subject. An increasing amount of telecommunication apparatus contains stored programs, and software design which has aspects in common with VLSI design is treated in Chapter 11.

Some relatively basic knowledge is assumed on the part of the reader, for example familiarity with what a transistor is and what computer memories are and do, but it is hoped that even a totally non-technical reader can obtain some benefit from the content. We shall during the course of Parts 2 and 3 refer back to the various topics in Part 1 and some readers may prefer to regard at least some of the chapters in Part 1 as reference material to be consulted as and when necessary. On the other hand to obtain a more fundamental understanding of what is possible by building up from the component level (the authors' vision of what might be built is inevitably circumscribed) a reading of these chapters is to be commended.

1 VLSI AND MEMORY

1.1 INTRODUCTION

Semiconductors have played the key role in technological advances since the mid 1950s and here we concentrate on one particular aspect of integrated circuits, i.e. VLSI (Very Large Scale Integration), because this is probably the key means of obtaining highly complex and 'intelligent' behaviour from the equipments, particularly mobile equipments, of the future.

Memories of all types too are essential to realising complex systems and provide the means of storing data or information or perhaps even knowledge. A memory is an essential part of a computer and computing in the broadest sense of the term has permeated throughout telecommunications.

1.2 VERY LARGE SCALE INTEGRATION (VLSI)

1.2.1 Integrated circuits in general

Integrated circuits, or microchips as they are commonly known, were developed in the late 1960s and the advances in their performance and the reduction in the cost per function have been the central feature of the electronics industry for the last two decades. While discrete semiconductor devices are made from germanium and silicon, silicon is a much better material than germanium for integrated circuits because it is easy to provide surface passivation by growing a protective oxide layer on silicon. Figure 1.1 shows how a planar epitaxial bipolar transistor and metal oxide semiconductor field effect transistor are made from silicon using the oxide layer as a mask. This oxide also acts as an insulating layer on which interconnecting metallic tracks can be deposited. Gallium arsenide, which is intrinsically faster than silicon, can be also be produced with its surface protected by an oxide but is in general a much more difficult material than silicon to process (and hence more expensive) and finds application only where the very highest speeds are essential.

Integrated circuits are fabricated on a substrate which is a single die (or chip) cut from a wafer of semiconductor material. The circuit elements are made from circumscribed layers or regions of n- and p-type semiconductor material. The n- and

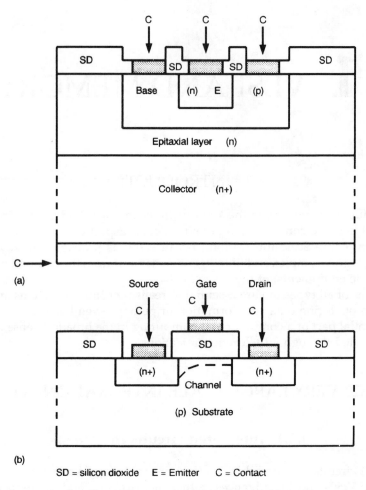

Fig. 1.1 (a) Structure of a silicon planar epitaxial transistor; (b) structure of a metal oxide transistor

p- refer to the dopants that have been added to the semiconductor material either during the wafer fabrication process (e.g. as a complete layer of material in a process known as epitaxy) or subsequently and selectively by diffusion through a mask. The n- and p-layers so formed make transistors, diodes, resistors and one plate of capacitors. The second plate of capacitors and the conductors joining the circuit elements together are formed by metallic tracks which run over the surface oxide; the oxide provides both insulation and the capacitor dielectric. (Alternatively capacitors can be realised as reverse biased p–n junctions or diodes: such capacitors are however voltage sensitive.) Conductors are made from aluminium and gold evaporated through a mask. Inductors are not really practicable as discrete components and are simulated using capacitors and active elements.

There are varying degrees of integration leading to a series of mnemonics which in ascending order of complexity are as follows:

- MSI—Medium Scale Integration. Several tens of components per chip.

- LSI—Large Scale Integration. Several hundred to several tens of thousand components per chip.

- VLSI—Very Large Scale Integration. Over 100 000 components per chip to several million.

- WSI—Wafer Scale Integration. A number (10–100) of VLSI circuits interconnected on the same wafer.

VLSI is current state-of-the art technology and is a major enabling technology for future telecommunications developments. WSI is some way from being an established production possibility and is something of an unknown quantity in the current context. The main emphasis from here on is on VLSI but much of the terminology is common to all levels of integration.

A measure of the technological advance in going from MSI to VLSI is the feature size of the devices that form the circuit elements. Feature sizes of these circuit elements are expressed in microns (10^{-6} metres), currently VLSI using 1–5 micron circuits are readily available and the trend is towards sub-micron technology, e.g. 0.5 micron in a few years' time and 0.2 micron by the mid-1990s. Reduction in feature size obviously means more devices per chip and because chip manufacturing methods are continually being improved, the size of chip that can be produced without defects to give an adequate yield of working circuits is increasing. The combination of smaller feature size and a larger chip means that the number of devices, or specifically logic gates, per packaged chip is increasing very rapidly with time and because the cost of the finished product is strongly dependent on the cost of the package, the cost per logic function is decreasing rapidly with time.

1.2.2 Processes and circuit types

There are two main types of transistor which can be fabricated in discrete or integrated circuit form; the bipolar transistor and the field effect transistor (FET). As a general rule for integrated circuits, bipolar transistors are used in analogue circuits, high power circuits and logic circuits where the fastest switching times are required. FET transistors are used in logic circuits of high complexity and where low power dissipation is an advantage.

Assemblages of devices are given various designations such as MOS and TTL which may refer to the process by which they are made or to the type of circuit that results. It is not necessary to give an exhaustive list of the nomenclature and it should suffice to explain briefly the commoner terms:

MOS technologies

- MOS stands for Metal Oxide Semiconductor and the key components of an MOS circuit are FETs with an oxide or nitride insulated gate electrode. (Rather than a reverse-biased p–n junction.) There are two versions of MOS, NMOS and PMOS (NMOS being the more common) where N and P stand for n-channel and p-channel transistors respectively. MOS transistors are very simple and can therefore be realised at a relatively high packing density.

- CMOS stands for Complementary Metal Oxide Semiconductor and CMOS circuits contain both p-channel and n-channel insulated gate field effect transistors. With this technology logic circuits can be designed which only draw current during the time that switching actually takes place, thus minimising power consumption. CMOS circuits involve many more fabrication steps than MOS and for a given level of performance their development tends to lag behind that of NMOS.

- SOS stands for Silicon On Sapphire, which is a technology used in military applications where high resistance to nuclear radiation is required. As the name implies the silicon chip is attached to an insulating sapphire substrate.

- Bi-(C)MOS stands for Bipolar-(C)MOS and is an emerging technology that allows a mixture of (Complementary) MOS and Bipolar devices on the same chip. This is of particular interest in signal processing applications where an analogue input and output are often required for the main digital processing function.

MOS memories

There are two major categories of semiconductor memory—Read Only Memory (ROM) and Random Access Memory (RAM). Whereas ROM has a fixed content and is used for permanently resident programs and fixed data tables, RAM can have its content changed under program control and is used for the transient program and data sector of a computer's address space. ROM uses a variety of technologies which are discussed under a separate heading below, but MOS is by far the most widely used technology for RAM.

The two major circuit divisions of RAM are Static and Dynamic (SRAM and DRAM) respectively. In SRAM each memory cell contains transistors which are self-latching into the 1 or 0 state but with DRAM each cell contains one or more transistors and a capacitor with the charge stored on the capacitor being indicative of a 1 or 0 state. Because the charge on the capacitor leaks away it is necessary to refresh DRAM periodically, hence its title. For both SRAM and DRAM memory content is destroyed if the power supply is switched off and sometimes battery back-up is provided to save information during a mains supply failure. A greater memory cell density can be achieved with DRAM and hence it is cheaper and much more widely used. SRAM is faster and tends to be used in special applications and as a small high speed 'cache' memory in certain computer architectures.

Because of the very large market for DRAM, DRAM technology tends to be very much the leading edge both in terms of feature size and in the general sophistication of its processing. The object is to achieve an ever increasing number of bits per chip; progress has moved over the years from 4 kbit in 1975 with a four-fold step size about every five years, through 64 kbit in 1981, to 4 Mbit in 1989 and is still continuing into the 1990's. (Although the pace may be expected to slow before before long.)

Bipolar logic types

Under the bipolar heading there are a number of logic circuit types most of which are largely of historical interest, e.g. Resistor Transistor Logic (RTL) and Diode Transistor Logic (DTL). The two remaining main contenders are:

- TTL which stands for Transistor-Transistor-Logic. TTL is reasonably fast and uses less silicon area per gate than alternative types of logic and is the bread and butter means of realising bipolar logic circuits.

- ECL which stands for Emitter Coupled Logic and is used where the highest speed of operation is required. It is faster than TTL and uses more power than TTL, but less power than TTL would use were TTL to operate at the same speed.

Read only memories

There is potentially a very wide range of technologies available for read only memories; in its simplest form (and one sometimes used for prototyping) a ROM can be realised as a pegboard matrix with a peg in position for a 1 and the absence of a peg for 0. Categories of ROM are briefly outlined below, methods or realisation are only examples. (Reprogrammable read only memory is usually called Programmable Read Only Memory or PROM.)

- Program Once ROM

    —Factory programmed by mask during manufacture.

    —User programmed. For example a diode matrix in which the programming involves burning out diodes to form the pattern of 1s and 0s.

- Reprogrammable PROM or Erasable Programmable Read Only Memory (EPROM)

    —Reprogramming is a special operation which is generally done off line, most current EPROMs are erased by ultra violet light and reprogrammed electrically but devices are becoming available (EEPROMS) in which both steps are electrical ones. The number of times an EPROM may be reprogrammed may be limited.

1.2.3 Wafer Scale Integration (WSI)

WSI is based on the interconnection of chips on the wafer itself, thus reducing the length of connections, the number of individual packages and the number of bonds

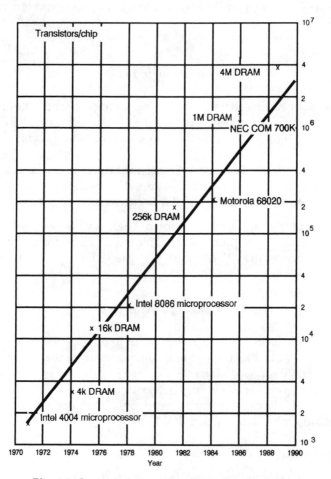

Fig. 1.2 Increase in integrated circuit complexity

joining lands on the IC to pins on the package. Short interconnections improve speed of operation and the reduced number of bonds is a significant factor in increasing reliability.

The difficulty with WSI is with the yield of integrated circuits during manufacture. Because of localised defects in the silicon wafer, yields are usually well below 100% and the probability that all the circuits connected together on a wafer will work is very low. The solution is to build into the wafer sufficient redundancy and perhaps to include a test system and an automatic configuration controller on the wafer as well. In any case a means must be found, either automatically or manually, to configure a sufficient set of working circuits to make the wafer viable. This is easier if the circuits are of a repetitive nature, and hence memory applications are likely to be the first to be realised. A 180–200 Mbit replacement for a disk memory is one potential product being developed in the UK and Japan.

1.2.4 Physical limits and future developments

Is there a limit to the increasing performance of integrated circuits and if so when is growth going to stop? There are theoretical thermodynamic and quantum limits which suggest that speed cannot be increased by more than three to four orders of magnitude above that currently achieved by the fastest silicon devices. Limits on speed combined with circuit component density are set by the practical problem of heat removal but it possible to design circuits to minimise dissipation and to circumvent this problem as least to some extent. There are limits on device size but it would be rash to predict what these might be except to say that the limit of a few times the inter-atomic distance is within the bounds of possibility. Already in the dimension perpendicular to the chip surface it is possible to fabricate layers not much more than one atom thick to make what are called multi-quantum-well (see Figure 2.7) and heterojunction devices. The resolution of photolithography using X-rays might be thought to set a limitation in the other dimension but the ability to fabricate devices using electron beam technology could well extend the frontiers further still. Another possibility, still very much in the research stage, is to grow devices in layers in the vertical direction to make what are called 3D VLSI circuits. Cooling is certainly likely to be a serious limitation to such an approach. Figure 1.2 shows how the complexity of integrated circuits in terms of the number of transistors per chip has increased over the 20 year period from 1970.

For the moment all that can be said is that we are a long way from both theoretical and practical limits and progress will continue to be made albeit perhaps less

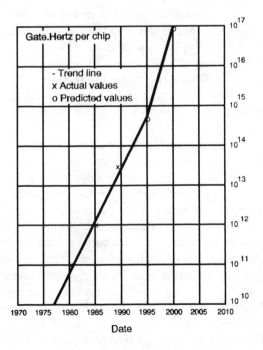

Fig. 1.3 Increase in chip performance since 1975

rapidly well, into the next century. A suitable measure (applicable strictly to logic circuits but a good trend indicator) is one that combines the factors of increasing speed (expressed as computer clock rate), decreasing feature size and increasing chip size. This measure is expressed as gate.herz/chip and as Figure 1.3 shows performance has improved by a factor of about three every year since 1975. Figure 1.3 contains a point that is a prediction for the year 2000 (Gosling, 1989) which in view of past history would seem to be on the optimistic side. Nevertheless progress is likely to continue at about the current rate for some time to come.

1.3 MASS MEMORY TECHNOLOGY

1.3.1 Introduction

There is an increasing tendency for the information conveyed by telecommunication systems to be stored electronically at both ends of the connection and in some cases to be stored in transit as well and we shall be dealing with the use of storage or memory widely throughout the rest of the book. Electronic storage of information at both ends of a connection is essential when machines are communicating, but memory technology has become involved in the human communication process too because word processors, personal computers and transaction terminals (which contain memory) frequently act as an interface between the human being and the telecommunication system.

A steady reduction over a number of years in the cost of memory and in some cases its access time (i.e. the time taken to extract information) are the main factors that are enabling increased application, and as we have shown there is reason to believe that these reductions will continue for many years to come. It will remain the case that at any point in time the faster the memory of a given type the more expensive it is. Memories will continue to be organised hierarchically with the cheaper (generally) and slower storage media at the top of the hierarchy as depicted in Figure 1.4.

Currently semiconductor memories of various types as outlined in the previous section form the bottom or 'immediate access' layer of the hierarchy, being the fastest with access times of a few tens of nanoseconds up to about a microsecond but having limited capacity in the region of tens of kilobytes up to several megabytes. Unmounted costs in 1990 are of the order 0.0006 pence/bit for DRAM and 0.0016 pence/bit for SRAM. Semiconductor memories are volatile (unlike all the other items in the memory hierarchy) and battery back-up has to be provided if their content is to be preserved when the power supply is switched off.

At the middle layer of the hierarchy revolving or disk memories with capacities in the region of hundreds of kilobytes to gigabytes take over as 'backing store'. Disk memories fall into two broad categories those with a rigid recording media and 'floppies'. With floppies the recording media can be removed; some systems with rigid recording media have removable platens but others (e.g. 'Winchesters') are fixed and sealed.

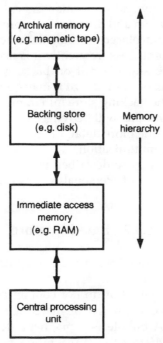

Fig. 1.4 Computer memory hierarchy

Rigid disk memories have access times for indexed data (i.e. where the address of the information on the disk is known) of the ten millisecond order but for un-indexed data which involves searching large areas of the disk access times will be considerably longer, e.g. seconds to tens of seconds. Once found data can be transferred at a rate of the megabit/s order. Capacity is normally well in excess of a megabyte and costs range from about 0.002 pence/bit to 0.0001 pence/bit (including drive). Floppy disks revolve more slowly and have a lower bit density giving access times and transfer rates some two orders of magnitude slower. Capacity ranges from a few hundred kilobytes to a few megabytes. Media costs are of the range 0.002–0.000 05 pence/bit and the drive cost of the £50–500 order.

The third or top layer in the hierarchy, which is sometimes known as an 'archival store' has in the past usually been realised by magnetic tape but there is an increasing tendency for floppy disks to take over the role, often in cases where a fixed media disk provides the second layer in the hierarchy. There are non-electronic means of archival storage (which are not considered further here) such as microfilm but these cannot be readily read electronically and are best suited to human examination. An important distinction between archival storage and the lower layers is that the lower layers must have a rewrite capability whereas archival storage is so rarely changed that technologies that permit only a single write operation are acceptable if the write once media is much cheaper or has a longer shelf life. Media cost will generally be of the order 0.000 002 pence/bit. Tape drives (which are rewritable) will entail costs ranging from a few hundred to many thousands of pounds depending on capacity and performance.

When programs need to be run and data has to be processed the relevant material is taken off backing store and placed in immediate access memory where it will overwrite the previous content. Less frequently it is necessary to transfer information from the archival store and place it on backing store as well. Because the immediate access memory is of limited size it may be necessary to transfer blocks of program and data from the backing store into immediate access memory during the running of a program and this will be reflected in a long running time.

We go on to consider in a little more detail the emerging technologies for backing and archival storage. Telecommunications applications in which memory hierarchies are relevant are such as those described in Chapter 21 on directories and in Sction 25.5.4 under the heading of 'Personal Numbering'.

1.3.2 Backing store

There are two potential competitors to the magnetic backing stores; bubble memories and optical memories.

Bubble memories are magnetic but do not contain any moving parts. They are made from single crystal layers of magnetic material, usually garnet, deposited on a non-magnetic substrate. A bubble is a magnetic domain with its direction of magnetisation opposite to that of the surrounding material. Bubbles are moved through the magnetic material in a similar manner to bits in an electronic shift register, bits are read out serially at an output port and then re-entered again at the input port thus forming a ring. For economic and practical reasons only a limited number of such rings can exist in parallel and hence access time on average is relatively long. Access time depends on the size of the ring, ranging for various designs from under 1 ms up to about 40 ms; once the data has been located it can be read out serially at a rate of the order 1 to 2 Mbit/s. Memory size ranges from typically about 0.5 megabytes up to a few tens of megabytes. Bubble memories are usually mounted on printed circuit boards but take up considerably more board space than semiconductor memories of comparable capacity. Their main attraction is that they are non-volatile (unlike semiconductors) and easily rewritable and yet can be mounted on printed circuit boards. Bubble memories are however considerably more expensive than large disk memories and seem likely to remain so and they therefore find application in environments, such as military ones, where a rotating device would not withstand vibration and shock. With the increasing sophistication of civil mobile communications it seems possible that a larger market for bubbles memories might develop for use in vehicles. Costs per bit are of the same order as non-volatile immediate access memory but the serial rather than parallel nature of bubble memory access makes it slow and therefore an unsuitable alternative to semiconductor memory in most applications.

The need for backing store memories to withstand a very large number of read–write cycles has had an inhibiting effect on their realisation in any of the many possible optical forms. Today it is not clear whether optical memories will supersede magnetic memories or not. Currently recording densities for both magnetic and optical media can approach one micron (10^{-6} m) with track spacings not very much greater than that. The major limitation on magnetic memories are the read head and air gap dimensions, but with optical memories, because read out is by a light beam, dimensional constraints are less severe, which gives them a significant advantage. The answer may well be a compromise in which magneto-optic films are used; the write process being magnetic and the read process being an optical one based on the property that the polarisation of the light depends on the direction of magnetisation of the bit being read. An all optical memory has the inherent advantage of not demanding close head spacing leading more readily to the use of a removable recording medium. Thus the search for novel optical media goes on; a number of approaches too numerous to discuss here are being pursued but for further details see for example (Bowhuis *et al.*, 1985). Whatever the technology, backing stores are getting larger and will exceed 1 gigabyte in capacity per disk. From the user's point of view backing store can be regarded to some extent as a 'black box' which is getting cheaper in real terms at roughly 5%/annum.

1.3.3 Archival storage

There is little doubt that for archival storage, optical recording is a strong competitor to magnetic storage. The current technology is called Write Once Read Many (WORM). The write operation uses a laser which heats and changes the surface of the disk in an irreversible way commonly by mechanical deformation. Disks are made of plastic, are very cheap and can be handled with little fear of damage, their capacity is of the gigabyte order and one disk can replace a dozen or so tape reels. The archival storage life is predicted to be very long, e.g. 20 years at least, which gives a distinct advantage compared with magnetic tape which has to be rewritten every year or two if high confidence in the maintenance of bit integrity is to be insured. Disk drive systems are still relatively expensive, costing tens of thousand of pounds, and are therefore only justified when large amounts of storage are required, but no doubt prices will fall in due course as market volumes increase. Mechanical deformation is only one of many technologies being investigated and claims have been made for a technique called 'Photochemical Hole Burning' (PHB) in which the line spectra of the material is changed resulting in very much greater information densities (6×10^{10} bits/cm^2 have been claimed) compared with what is currently realised (e.g. 5×10^7 bits/cm^2) by more conventional means. Very low temperatures and X-ray wavelengths are involved so that PHB may be a long way from being a marketable approach.

The WORM technology is similar to that used for making masters for compact disks (CDs) and hence lends itself to the mass production of software for distribution, for example for operating system updates.

1.4 OPTICAL INTERFACES TO MEMORY

When the memory storage medium is an optical one as with the WORM technology, then an optical interface is a necessity, but optics is well suited (as is explained

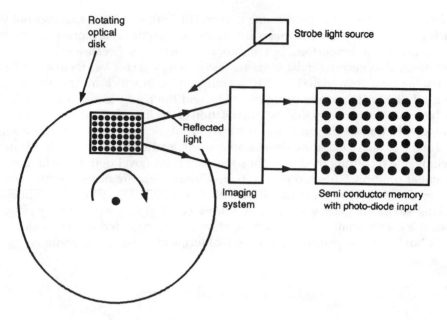

Fig. 1.5 Parallel transfer of data from optical disk to semiconductor memory

in the next chapter) to highly parallel modes of operation. Consideration is being given to reading sectors of optical disks in parallel directly into a block of semiconductor memory. Each memory cell of the semiconductor memory is connected to a photo-diode which acts as a read-in device, and a sector of the revolving optical memory disk is imaged directly on to the photo-diodes using a very short pulse of light (see Figure 1.5). By this means a whole sector is transferred in parallel in one operation and hence very high data transfer rates between backing store and immediate access memory can be realised.

A similar technique can be used to transfer information en bloc between semiconductor memories. The read-in process is as above and the write process is achieved by placing a liquid crystal layer over the block of memory cells, each one of which is joined to an electrode which forms a back-electrode of the liquid crystal layer. If a 1 is written into a memory cell the liquid crystal reflects light and if a 0 it does not, hence the content of one memory can be transferred to another. This could have applications in processor arrays and perhaps in switching.

1.5 SUMMARY

There is no doubt that VLSI is *a*, if not *the*, current key technology for telecommunications which impacts strongly on logic and immediate access memory functions. Magnetic materials remain the main media for backing stores but one may confidently expect that optical storage will be increasingly important in an archival role at least.

1.6 BIBLIOGRAPHY

G. Bouwhuis, J. Braat, A. Huijser, J. Pasman, G. Van Rosmalen and K. Schouhamer Immink (1985) *Principles of Optical Recording*, Adam Hilger.

W. Gosling. (1989) Telecommunications set free—the era of personal communication. *2nd IEE national conference on telecommunications, York*, April, pp 6–9 (CPN-300).

2 OPTICAL COMPONENTS

2.1 FIBRE OPTICS

2.1.1 Introduction

Electronics has a competitor called optics, and optics has already achieved an established position in line transmission, and we need to understand how that position will advance further and what other roles optics might play in switching and computing. The role of optics in transmission is treated in Section 2.1 and switching and computing in Section 2.2.

The development of communications by light transmission along fibres is probably the most significant technological development in communications in the last twenty years. Even though a substantial part of the public networks in the United Kingdom and many other countries use optical fibres, there is still significant scope for further development because the inherent bandwidth, or traffic-carrying capacity, of a fibre is some three orders of magnitude greater than coaxial cable and some five orders of magnitude greater than that of a copper pair. However, only a fraction (0.1%) of optical fibre's inherent capacity is currently being exploited.

2.1.2 Transmission windows or bands

The operating point of an optical system is normally expressed in terms of a wavelength measured in nanometres (nm) or microns (micrometres, μm) and rarely in terms of the frequency which is of the order 10^{14} Hz. Optical fibre communication makes use of transmission windows or bands where for a particular wavelength the loss, or attenuation, in the optically transparent material from which the fibre is made is relatively low. Figure 2.1 shows the two lowest loss windows for silica based glass, nominally at 1300 and 1500 nm (1.3 μm and 1.5 μm) and well into the infra-red. These are the windows used for long-distance transmission but there is also a window just into the infra-red, at about 850 nm; a frequency at which sources tend to be cheaper but which is only suitable for use at shorter distances because of the relatively high attenuation.

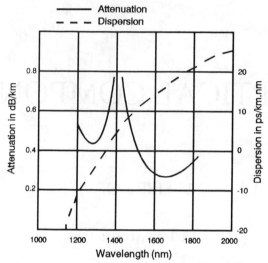

Fig. 2.1 Transmission windows in silicon fibre

The choice of window is influenced by loss, dispersion and the availability of the components to provide the light source at the transmitting end and the receiver at the receiving end. These topics are discussed below.

2.1.3 Loss and dispersion

Optical attenuation or loss is influenced by the choice of base material (e.g. silica), its purity and regularity of structure, and the amounts of various dopants that are added to tailor the refractive index to meet specific performance objectives. Typical attenuation figures in decibels/km (dB) at various wavelengths are :

Wavelength (nm)	Attenuation
850	1.2–4.0
1300	0.2–0.5
1550	0.1–0.3

Fibres using fluorozirconate glasses are capable in theory of giving a loss about a factor of ten lower than silica but at a longer wavelength in the 2100–2600 nm region. At the time of writing these were very much at the research stage and what follows assumes the use of silica fibre unless otherwise stated. The wavelength at which long haul transmission systems operate has always been dictated by fibre transmission behaviour, and as was the case at 1300 and 1550 nm, it will be necessary to develop transmitters and receivers in the 2000 nm band before a practical system can be realised.

Modal dispersion (which is discussed further in Section 2.1.4) is due to multi-path propagation through the optical wave guide, and material dispersion, which is relatively a much smaller effect, occurs because the various frequency components of the signal itself are propagated with different velocities. Material dispersion (or more strictly chromatic dispersion) is usually expressed in units of picoseconds per nanometre per kilometre, where the nanometres refer to the spectral bandwidth of the transmitted signal and the kilometres refer to the transmission distance. Material dispersion goes through zero at a wavelength that depends on the design of the fibre, and one of the objects of fibre design is to make the zero material dispersion wavelength correspond to one of the windows, e.g. in the vicinity of 1550 nm (see Figure 2.1).

Attenuation, and to a lesser extent dispersion, determines the maximum distance over which an optical transmission can work without amplification or regeneration. Currently optical systems use regenerators or repeaters that are spaced typically between 10 and 100 km apart. At each repeater the optical signal, which is normally in a digital form, is detected, amplified and reshaped electronically and then relaunched optically. Improvements in optical components should allow the spacing between repeaters to be increased in the future, leading to significant cost savings.

2.1.4 Modes of transmission

Light is an electromagnetic wave and an optical fibre is a dielectric wave guide of circular cross-section and so the transmission through the wave guide can be expressed by Maxwell's equations, which also form the theoretical basis for radio. Light is confined to the fibre by making the dielectric constant of the glass greater at the centre of the fibre than it is at the circumference. Essentially there are three forms of transmission:

1. *Multimode*. Multimode fibres have a core diameter typically of 50 μm, or about 50 wavelengths, and a step index profile (see Figure 2.2a). Light propagates along a multitude of paths, being reflected by total internal reflection from the wall, or cladding, of the fibre. The different path lengths result in modal dispersion because of time differences for the light reaching the receiving end of the fibre.

2. *Monomode*. Monomode fibres (Figure 2.2b) have a core diameter of typically 3–6 μm, and because the core diameter approaches the wavelength of the transmitted light, that light propagates in a single mode, hence the name monomode. Monomode fibres are consequently subject only to material dispersion. Monomode fibres are slightly more expensive to manufacture than multimode fibres, but more significantly the smaller core dimension makes the jointing of fibres and the connection of transmitting and receiving devices difficult. Multimode fibre is therefore still used for some less demanding applications.

3. *Graded and Complex Index Profiles*. Graded index fibres have a refractive index profile which varies in a complex fashion with distance from the centre of the

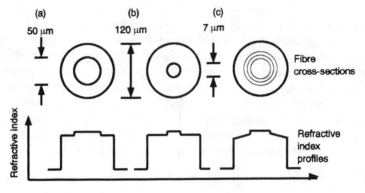

Fig. 2.2 Fibre cross sections and refractive index profiles; (a) multimode; (b) monomode; (c) graded index

fibre (see Figure 2.2c). In multimode operation, because the velocity of light in the fibre depends directly on the refractive index, the refractive index profile can be chosen to compensate for path length differences. Dispersion performance is not as good as for monomode fibres and graded index fibres have not so far been adopted for long distance transmission use and seem unlikely to be used to any great extent in the future.

Monomode fibres can have complex index profiles that are tailored to make the zero dispersion point correspond to the desired operating wavelength.

2.1.5 Light sources

The two principal types of device used as light sources are light emitting diodes (LEDs) and semiconductor lasers. LEDs have a wide spectral spread and are used

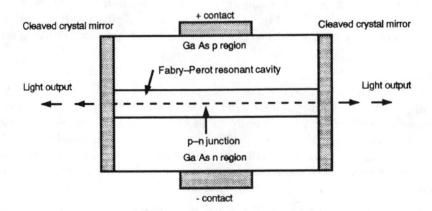

Fig. 2.3 Basic structure of a laser

only in very non-critical applications. Lasers differ from LEDs in having the dimensions of the light-emitting region of the semiconductor junction fabricated to form an optically resonant cavity, thus making their spectral response much narrower. Figure 2.3 shows a cross section through a basic laser structure but in practice laser structures are a great deal more complex than this. Ideally one would like a spectral response consisting of a single frequency or line and much of current laser research and development is directed to that objective. The distributed feedback laser, or dfb laser, is a comparatively recent step along that road.

The semiconductor materials used for laser construction depend on the operating wavelength. Typically these are as follows (where Ga = gallium, Al = aluminium, As = arsenide, In = indium, P = phosphide):

Wavelength (nm)	Type of source	Material
850	LED/Laser	GaAIAs on GaAs
1300	Laser	InGaAsP on InP
1500	Laser	InGaAsP on InP

2.1.6 Receivers

Receivers use avalanche photo detectors (APDs) or p–i–n photodiodes. There are two modes of operation: (a) as a direct detector or (b) in a heterodyne or homodyne (see further in Section 7.4.4) mode with a local oscillator. The latter, (b), results in some 20 dB improvement in signal to noise ratio at the expense of a considerable increase in receiver complexity.

Semiconductor materials used in the fabrication of detectors are commonly:

Wavelength (nm)	Type of detector	Material
850	APD	Silicon
1300/1500	APD	Germanium
1300/1500	p–i–n	Gallium indium arsenide

but a number of other materials are under investigation at the longer wavelengths.

In order to achieve a low cost, high performance receiver design, the detector and the pre-amplifier that follows it should be mounted in the same package and very close together; ideally they should be fabricated as one integrated circuit. Although detector materials are not suitable for the fabrication of integrated circuits it is possible by matching the thermal expansion characteristics of GaInAs to an indium phosphide substrate to fabricate an FET transistor on the substrate and obtain an integrated solution. The design, however, involves compromises and it is difficult to achieve a performance that is better than the best combination of discrete components.

With WDM (see next section), a separate receiver is needed for every wavelength carried in the multiplex and the integrated circuit approach then becomes particularly attractive because it is possible to mount all the detectors and their associated circuitry on the same substrate.

2.1.7 Making the best use of the signal carrying capability of optical fibre

Wavelength division multiplexing (WDM)

Ideally the spectrum of the signal should be determined by the information it is required to convey but as we have seen above the spectrum is also affected by the spectral spread of the laser that acts as the light source. From Figure 2.1 it may be seen that the windows available to transmit the signals are wide and it should therefore be possible to send signals simultaneously on different wavelengths both within and between windows. This technique is called wavelength division multi-plexing. Minimum wavelength spacing is continually decreasing as laser designs are improved. Spectral spacing is expressed either in nanometres or frequency and it should be noted that 1 nm corresponds to 300 GHz. Currently various demon-strations and trials have shown spacings ranging from 15 nm down to 300 MHz but optical filters as well as highly stable laser sources are necessary at the closer of these spacings.

The ability to send a multiplicity of signals over the same fibre without cross-modulation depends on the linearity of the optical material. Fortunately silica is highly linear at low power levels but with very high order multiplexing the peak power level could reach a point where the effects of intermodulation products produced by the non-linearity are a limitation because they give rise to cross-talk between channels, a limitation that can be avoided by decreasing the distance between repeaters.

Coherent operation

Coherent operation depends on the light wave carrier being a good approximation to a stable single frequency a situation that we take for granted in the context of radio frequency transmission. With coherent operation the receiver contains a local oscillator and can select the signal it wants to receive by choosing the appropriate local oscillator frequency. Superheterodyne and homodyne (zero intermediate fre-quency, see Section 7.4.4) techniques are well known to radio engineers and are beginning to be applied in optical communication. In Section 2.1.6 above we have seen how receiver performance is improved by such an arrangement. Thus coher-ent transmission is an important adjunct to WDM, allowing improved receiver selectivity and sensitivity.

Coherent operation allows one wavelength within a WDM multiplex to be divided into a number of separate channels by the use of sub-carriers, still further increasing the scope for increasing the overall capacity of the fibre.

Transmission capacities

At the time of writing, fibre systems with a transmission capacity of 565 Mbit/s using a single light source are already in operation in the public network. Experiments and trials have shown that rates in excess of 2 Gbit/s are quite feasible. With the addition of wavelength division multiplexing, coherent operation and further improvements in optical sources, detectors and filters, overall rates of 10–100 Gbit/s should be obtained within the next five years (see further in Section 2.1.9).

However, the important consideration is that these improvements in system performance can in the main be achieved using existing fibre designs. Thus telecommunication operators can upgrade their systems without further fibre and installation costs.

2.1.8 Optical amplification

Currently all signal regeneration in long haul systems involves repeaters that convert the optical signal back to the electrical domain, amplify and equalise it and then relaunch it optically. There are a number of potential means of optical amplification of which the most promising are (a) laser-like devices biased so that they amplify but do not oscillate and (b) non-linear fibres. The latter, which can be classed as parametric amplifiers, depend on having a high power optical 'pump' which drives the fibre material into a non-linear region so that some of the pump energy is converted into signal energy. The pump works at a shorter wavelength than the signal and hence can be prevented by attenuation from propagating through the fibre. While conventional silica fibres can work in this mode, they require very high pump powers. The alternative is to use special purpose fibres doped with rare earths, such as erbium, which results in increased material non-linearity and hence allows operation at a lower pump power. Gains in excess of 30 dB with adequate noise margins have been quite widely reported with doped fibres.

The laser-like devices, because they need no pump, look to be the simpler approach in the medium term at least and their use in submerged repeaters for submarine cables is being given serious consideration. Here one of their attractions is the resultant reduction in component count and hence increased reliability. Another advantage is that whereas an electrical repeater design is dependent on signal-characteristics, an optical amplifier is broadband and the signal carrying capacity of the system can be increased without modifications having to be made to the optical repeater. A gain of 15 dB and a bandwidth of 15 nm is achievable.

For the distribution of television and other broadband signals over optical fibres in the local area the use of erbium-doped fibre amplifiers is favoured on the ground of their superior linearity as an amplifier compared to the laser-like devices. They are linear enough to give acceptable levels of cross coupling or crosstalk when a multiplicity of signals are carried on the same fibre. Figure 2.4 shows the configuration for an erbium doped fibre amplifier. Such amplifiers are capable of giving outputs of the 100 mW order with a gain of 20–30 dB and simultaneous operation at at least 10 wavelengths.

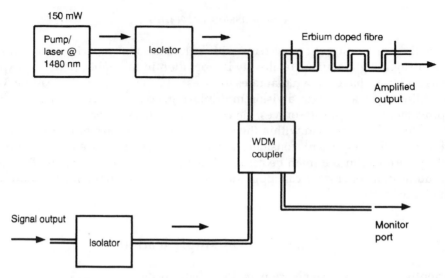

Fig. 2.4 Fibre amplifier configuration

2.1.9   Non-linear fibres and solitons

Optically transparent materials generally show very linear behaviour which is fortunate in terms of preventing cross-modulation terms in WDM and FDM transmission systems. However at very high power densities non-linear effects always become significant, but sometimes these levels approach those at which the material is in danger of being permanently damaged due to local over heating. Very small non-linear effects if they accumulate systematically over very long transmission distances can, however, be very significant, and of particular interest in this context is the 'soliton'.

A soliton is a pulse travelling along a fibre in such a manner that the non-linear behaviour can be matched to the pulse power and width so that instead of the pulse getting broader as it travels along the fibre it gets narrower. The non-linearity gives rise to 'self-phase modulation' and if the pulse is launched within specified limits of power and width it will propagate (if the effects of attenuation are negligible) so that it never changes shape, resulting in what is called propagation in the 'fundamental mode'. In essence the pulse narrowing effects of the non-linearity are exactly balance by the pulse broadening effects of dispersion. In higher order modes (corresponding to higher launch energies) the pulse can split into two and oscillation in behaviour along the fibre usually takes place. At energy levels very much in excess of those required for the fundamental soliton, very short pulses (e.g. 0.2 ps) can be produced in relatively short lengths of fibre (tens of metres) from an input pulse of, say, 30 times the width. These very narrow pulses are, however, situated on a broad pedestal pulse many times wider; this pedestal pulse can however be removed by making its polarisation orthogonal to that of the narrow pulse.

Early in 1992 AT&T claimed a breakthrough in long distance transmission using soliton light pulse by sending 5 Gbit/s signals over a distance of 15 000 km and 10 Gbit/s signals over a distance of 11 000 km. This means that the Atlantic or Pacific can be crossed without any repeaters. What is particular surprising is that the 10 Gbit/s transmission was achieved by wavelength division multiplexing, implying that the soliton interaction can take place at two wavelengths simultaneously.

The soliton effect can also be used for switching, see Section 2.2.2.

2.1.10 Conclusions

Optical fibre technology is well established in the 850, 1300 and 1500 nm bands and is likely to be further developed at longer wavelengths. We may expect to see the unused potential of installed monomode fibre in the main transmission network exploited as the demand for wideband services increases and in later chapters, particularly Chapter 16, we shall see how optical fibre can be used in local area networks and LANs.

2.2 OPTICAL SWITCHING (AND COMPUTING)

2.2.1 Introduction

The emphasis is on optical switching but because much of the technology shares common ground with optical computing, the latter is implicitly included. As pointed out above optical devices are not strongly non-linear, there is no known optical device that is anything like as non-linear as a semiconductor p–n junction. (That is assuming that non-linearity is measured in terms of how little energy is required to switch from one side of a non-linear region to the other.) One of the major problems with optical switching is therefore excessive power dissipation at high switching rates. Because optical beams can propagate in free space with no mutual interference and very little frequency distortion, there is great potential for highly parallel systems of interconnection. With electrical interconnection, on the other hand, straight line point-to-point connections on a printed circuit board are not generally practicable and cross-overs and indirect routes are a potential source of delay, frequency distortion and interference. To reduce the number of wires, interconnection tends to take place using buses of limited bandwidth making for a more serial mode of operation and one that can result in processing bottlenecks. This propensity towards parallelism in optics is of major importance but is probably of more immediate interest in computing than it is in switching. Electronic and mechanical switches of necessity already exhibit high orders of parallelism and direct optical analogues of electrical space switches are possible. However, in the longer term there is scope to make optical switch blocks with even higher orders of parallelism.

A distinction is necessary between the basic switching elements, the switch cross points, and assemblies of such elements into, for example, the switching unit or switch block of a telephone exchange. The overall design of switch blocks, whether the switches work on the time division or space division principle (see Section 4.3) is technology independent; given a specific switch architecture (e.g. time–space–time), it does not matter to the first approximation at least whether the switching elements are electronic, optical or a hybrid of both. We shall therefore concentrate here on the switching elements or cross points. It is always possible however that an optical approach could lead to a radically new switch architecture.

In the electrical domain very few if any practical switch blocks have been realized using frequency division techniques, although there is nothing against this in principle; in the optical domain there is interest in switches that use WDM and FDM as the means of selecting channels to switch and the implications of this are given further consideration.

2.2.2 Switch elements

Potentially there are three types of switching element (or cross point) that need to be considered:

1. elements working wholly in the optical domain;

2. elements working with optical input and electrical output; and

3. elements working wholly in the electrical domain.

The main concern here is with (1), (2) is of some interest as a fall-back situation, but (3) is considered inappropriate in this context.

There are numerous physical processes on which an all optical switch might be based and some of the more relevant of these are itemised below.

Mechanical/electromechanical

A light beam is moved to point in some specific direction by moving a transmissive or reflective body (e.g. a tilted mirror).

Electro-optic

The properties of optical transmission or reflection are changed under the influence of an applied electric field. There are no moving parts. Usually the property changed is the refractive index of the transmission medium but changes in absorption or polarisation are also possibilities.

Magneto-optic

As electro-optic above, but under the influence of an applied magnetic field.

Acousto-optic

Refractive index changes in crystalline solids can be brought about by mechanical stress. Acoustic energy having a wavelength in the optical medium of the same order of the wavelength of the optical signal is used to set up a standing wave stress pattern that form an optical grating which steers the light in the required direction.

Photo-optic

Control light of one wavelength can be used to change the optical properties of the transmission medium so that the signal light (normally of a longer wavelength) is switched. Such changes will normally be to the refractive index of the material, exceptionally to its absorptivity. A variant of this approach is to make the signal itself do the switching; which either depends on the switching element being photo-optically bistable or on some other non-linear behaviour, e.g. the soliton as discussed above.

To treat all these possibilities comprehensively is beyond the scope of this book and a few examples follow of some of the approaches that have received fairly widespread attention in the literature.

Electromechanical switches

There are two main categories of switching element:

1. an element in which incoming optical fibre ends are moved mechanically and aligned with outgoing fibre ends which are either butted or directed via a fixed optical system;

2. an element in which fixed fibre ends are connected by means of moving lenses, mirrors or prisms.

There are a number of products commercially available in the second category. Typically these might be 10×10 matrices with crosstalk levels better than 65 dB and an insertion loss of 2–4 dB. Switching times are of the millisecond order.

Electro-optic

Perhaps the best-known electro-optic device is the directional coupler illustrated in Figure 2.5. Optical waveguides are constructed in a semiconductor material (e.g. GaAs). There are four ports and the two switch states are straight through and crossed over. The controlling signal is a DC voltage applied to the control electrode which changes the dielectric constant of the waveguide material. Crosstalk levels with careful design can be as low as 25–30 dB and the insertion loss is several dB.

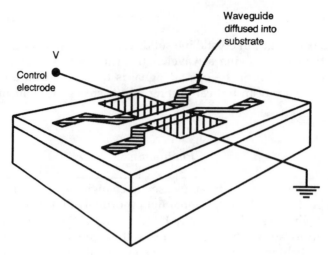

Fig. 2.5 Electro-optic four port switch

Switching speeds of 100 ps have been reported and 30 ps appears feasible, so clearly these elements are suited to a TDM mode of operation. The device needs to be several tens of wavelengths long, so it is difficult for it to compete in terms of size with electronic crosspoints based on a micron semiconductor device geometry. As far as is known directional couplers of this kind are not in widespread use in switching systems.

A rather less ambitious approach is illustrated in Figure 2.6, but at the expense of reverting to the electrical domain. Photo devices (e.g. diodes) are used as crosspoints and are biased so as not to act in a photo-sensitive mode when a cross point is required to be open. When closed a cross point photo device produces an electrical replica of the signal which can be used to modulate an optical source on the outgoing side. A switch block based on this principle has been used as a switch for TV video signals in studios in Canada. On/off ratios of 60–80 dB are claimed but no information on switching speed is available.

For WDM switching one approach is to use an optical grating which is electrically tunable, e.g. by applying an electric field that changes the refractive index of the grating material. The frequency selectivity of the grating is proportional to its length and for high selectivity, or narrow bandwidths, large grating sizes are required.

An MQW (see next section) electrically addressed, optically two-terminal device called a SEED (Self ElEctrooptic Device) was used by AT&T in 1990 for a demonstration optical TDM switch. Its performance, particularly with respect to switching speed, was not competitive but there is presumably scope for improvement.

Photo-optic

A three-terminal element has a lot to commend it as switch or logic device, and the type of device illustrated in Figure 2.7 takes this form. The upper right-hand

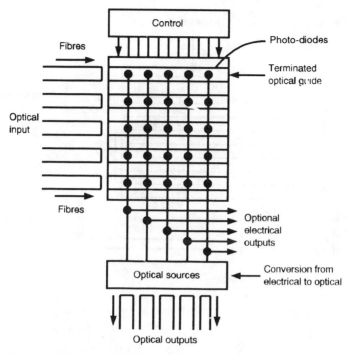

Fig. 2.6 Hybrid optical–electronic switch

portion of the device is a photo transistor which is coupled to a Multi-Quantum-Well (MQW) modulator which forms the lower part. Pump power (light) is turned on and off at the output under the control of the signal light. Thus we have the optical analogue of the transistor. While all the external signals are optical the internal operation is partly electronic and the device may be considered as an electro-optic hybrid. It is early days to make a prophecy about what its future will be; there are considerable problems in making MQW modulators with high on/off ratios.

Solitons of sub-picosecond pulse width can make effective use of the non-linear properties of optical fibres in switching applications by virtue of their very high peak power level. The arrangement shown in Figure 2.8 is capable of providing switching. There are two fibre paths each having different velocity dispersion characteristics. A pulse launched at the input is split and arrives at the coupler (which is similar to the one shown in Figure 2.5 but without a control electrode) at the same time by both paths but with a phase difference that depends on its amplitude. The pulses combine and depending on the net phase of the resultant pulse divide in specific proportions between the output ports of the coupler. As the input pulse increases in amplitude output is switched from one port to the other repeatedly and for certain levels all, or the great majority, of the output is directed to one port. This type of device is still very much in the research phase and a long way from becoming a practical system component. In the form described it is a

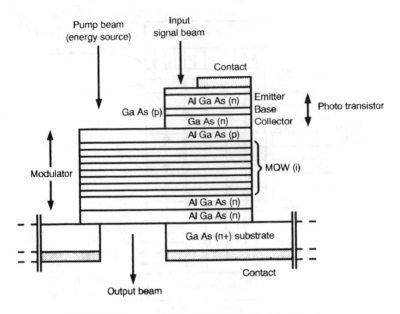

Fig. 2.7 Three-terminal optical switching device

two-terminal switch and past experience with two-terminal switching elements (e.g. gas tubes and tunnel diodes) is not encouraging. This limitation is less serious in other applications, e.g. sub picosecond pulse generation to provide the clock for an optical computing or switching system.

2.2.3 Conclusions

The current conclusion in the context of the European RACE project (see Section 16.3.5) is that electrical rather than optical elements will form the cross points in the next generation of switching systems for all but some rather highly specialised applications (such as cross connects and protection switching, see Section 4.2). Electronic switch blocks are capable of meeting the switching rate requirement, are cheaper and dissipate less power. From consideration of the physics of switch operation there are reasons for believing that, for the ultimate in switching speed

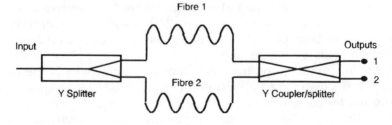

Fig. 2.8 Outline representation of a soliton switch

(sub-picosecond), optical devices have the advantage. However, from a switching (as opposed to a computing and control) system point of view, while a need for higher speeds is there it is not a pressing one.

2.3 BIBLIOGRAPHY

P. Cochrane. (1984) Future trends in telecommunication transmission—a personal view. *IEE Proceedings* **131**, Pt F, No 7, Dec, 669–683.

N. J Doran and D. Wood. (1987) Soliton processing elements for all-optical logic. *J. Optical Soc. Amer., Part B*, **4** (11), 1843–1846.

ECOC-88 (1988) *14th IEE European conference on optical communication*. Brighton Sept. (CPN-292) (Contains a number of relevant papers).

N. W Horne. (1988) Changing light. *IEE Computing & Control Division: Chairman's Address*, IEE paper 6294E (C2, E13) Oct. pp 1–8.

J. E. Midwinter. (1984) Optical fibre communications, present and future. *Proc. Royal Soc. London*, **A392**, 247,277.

L. F. Mollenaur and R. H Stolen. (1982) Solitons in optical fibre. *Laser Focus* **18**, April 193–198.

3 TRANSMISSION SYSTEMS

3.1 TRANSMISSION MEDIA

The transmission system is the means for conveying signals between two points, or, in the case of radio broadcasts, between one and many points. The transmission medium may be copper wire, normally in the form of bundles of pairs of twisted wires, or coaxial cables, or optical fibre cables (coaxial and optical cables can carry much higher bandwidths than twisted pairs) or radio. The transmission system includes both the transmission medium and the means of converting the signal into an appropriate form for the medium.

Optical fibre and radio technologies are described in more detail in Chapters 2 and 7 respectively. So far as wire or cable communications are concerned, it is important to note that the signals may be processed in some way to make them more suitable for transmission. In particular in the digital case, where timing information has to be transmitted even if the signal is not changing, special coding schemes called line codes are used (see Chapter 6).

The choice of the telecommunications medium will depend on the requirements of the system, the environment in which it is to be used, and the costs of the different technologies. Radio is inherently well suited to broadcasting services, mobile services, and temporary communication where a link has to be set up quickly. The quantity of radio communications is limited by the availability of the electromagnetic spectrum; over use of this resource will lead to interference. Line communication is less suitable for broadcasting and unsuitable for mobiles but is not spectrum limited in capacity in the same way that radio is because more cables can always be provided.

In addition to the transmission medium there are two important topics that are generally considered to be part of transmission, and these are multiplexing and configuration control, which are treated next, but before doing so it is important to make the distinction between a circuit and a channel.

The strict definition of a circuit is a number of elements connected together for the purpose of carrying a current, but by common usage the term is also used to designate an aggregation of paths or a specific part of a complete path. Here we shall generally use the term to mean the totality of a connection or a number of connections carrying traffic. A circuit may be single or multi-user.

A channel is defined as a means of communication between a source and a receiver and hence is one-way only. When we talk of channels we are talking about one-way communication but it is usually implied that there is a complementary channel carrying information in the contrary direction. Here, unless specifically stated to the contrary a channel will always be assumed to be carrying information in one direction relating only to one user.

3.2 MULTIPLEXING

Communications media can be shared between a multiplicity of users by a process called multiplexing and multiplexing is normally considered to be a transmission rather than a switching function. The two main forms of multiplexing are frequency division multiplexing (FDM) and time division multiplexing (TDM). Both types of multiplexing are illustrated in Figures 3.1a, b.

Frequency division multiplexing

In Frequency Division Multiplexing (FDM), the individual channels are transmitted on carriers of a different frequency and the receiver has to be tuned to the carrier of the signal which it wants to receive. The radio spectrum is in effect frequency division multiplexed, and a domestic radio is an example of a frequency division demultiplexer. Analogue line systems use FDM but the gradual evolution towards digital transmission means that the majority of multiplexers are currently TDM. Systems based on frequency division where the transmission medium contains fewer channels than the number of users are called Frequency Division Multiple Access (FDMA). (See below under statistical multiplexers for the principles involved.)

Time division multiplexing

With time division multiplexing (TDM) the bit streams of different information-bearing channels are interleaved with the individual bit lengths being correspondingly shortened. At each stage of time division multiplexing, the ensemble of groups of bits from each channel is called a frame, and a flag, which is a special pre-defined pattern of bits, is inserted to indicate the beginning of the frame; by means of the flag the receiver can work out which bits belong to which channel. Successive stages of time division multiplexing can be used and the hierarchy of multiplexes recommended by CCITT is shown in Figure 3.2.

The type of time division multiplexing described above relates to circuit switching and is synchronous. In packet switching the individual packets, which may be of different lengths and which contain information to identify them, pass along a channel one after the other and packets relating to different calls will be intermingled in an undefined order. This type of time division multiplexing is described as asynchronous.

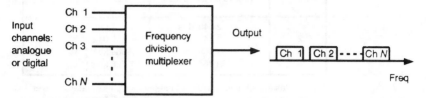

(a) Frequency division multiplexing

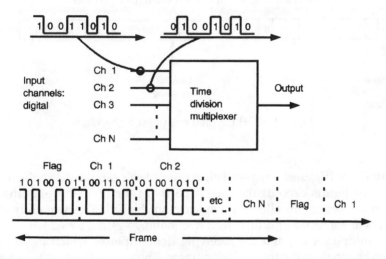

Note: Each input channel is frequency converted by mixing with a 'carrier' whose frequency determines the channel's position in the multiplexed signal; it is then filtered to remove one sideband and the other sidebands of each input are combined to form the output.

(b) Time division multiplexing

Notes
(1) Each input channel is read into a buffer store. Each output frame is formed by transmitting a flag followed by a fixed number of bits from each buffer in succession. After the bits from the buffer for channel N have been transmitted, a subsequent frame is sent.
(2) The duration of each output channel is $1/(N+1)$ of the duration of each input bit.
(3) Normally TDM is used only for digital signals but an exception is the multiplexed analogue component (MAC) system for colour television, where the analogue waveform is compressed in time for transmission.

Fig. 3.1 Frequency and time division multiplexing

Time division multiplexers

There are several different types of time division multiplexer (Figure 3.3).

- *Simple fixed format multiplexers* In a simple multiplexer the relationship between the inputs and the output is fixed (Fig. 3.3a) but in a more sophisticated multiplexer it may be varied.

Digital hierarchy level	European (based on 2 048 kb/s)	North American (based on 1 544 kb/s)
1	2 048 kb/s	1 544 kb/s
2	8 448 kb/s	6 312 kb/s
3	34 368 kb/s	32 064 kb/s ¦ 44 736 kb/s
4	139 264 kb/s	97 728 kb/s ¦

Note: Rates at higher levels are not integer multiples of rates at lower values because additional bits are included to give a framing signal and scope for justification (rate adjustment) if the inputs are not synchronised.

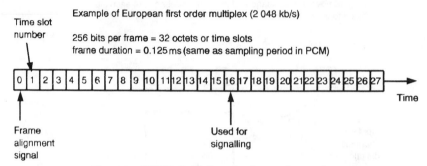

Example of European first order multiplex (2 048 kb/s)

Time slot number

256 bits per frame = 32 octets or time slots
frame duration = 0.125 ms (same as sampling period in PCM)

Frame alignment signal

Used for signalling

Fig. 3.2 CCITT digital multiplexing hierarchy

- *Statistical multiplexers* In a statistical multiplexer (Figure 3.3b), the total capacity of the inputs exceeds the total capacity of the link, and only those inputs which are active are allocated capacity on the link. Consequently the multiplexers at each end of a link need a signalling system by which the transmitting multiplexer can tell the receiving demultiplexer which inputs are using which channels in the transmission frame. These signalling protocols are normally proprietary. Time division systems that share a transmission medium based on this principle are often called Time Division Multiple Access (TDMA) systems.

  The simplest form of statistical multiplexer assigns a channel for the duration of a call but the technique can be carried a stage further for speech channels by assigning a channel only when the party at the sending end is talking. This technique is called Speech Interpolation (SI) and is often referred to as TASI for Time Assigned Speech Interpolation. TASI is widely used for long distance international telephony where transmission costs are very high.

- *Drop and insert multiplexers* Another more complex type of multiplexer which is used commonly in private networks is a drop and insert multiplexer (also called an add/drop multiplexer). This type of multiplexer is used to remove one or more channels from a multiplexed transmission system or to add one or more channels to vacant slots in a multiplexed system. These multiplexers are commonly used to extract or insert 64 kbit/s channels in a 2 Mbit/s transmission. A drop and insert multiplexer has two distinct parts (see Figure 3.3c).

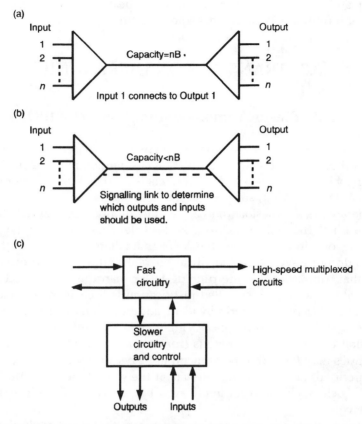

Fig. 3.3 Multiplexer types: (a) simple multiplexer; (b) statistical multiplexer; (c) drop and insert multiplexer

The part which is inserted into the multiplexed transmission receives the multiplexed signal into a buffer and transmits the outgoing signal from the buffer. The bits of the channel to be extracted or inserted are transferred from or to the appropriate parts of the buffer. This part of the multiplexer uses high speed digital components, the performance of which matches the rate of the multiplexed signal. The second part of the multiplexer handles the interfacing between the extracted channels and the low rate outputs and inputs and uses lower speed and hence cheaper components.

Drop and insert multiplexers can be used to traverse a number of levels in the multiplex hierarchy in one operation provided that the position of the channels in the multiplexed signal is explicit. If it is not explicit because the hierarchy is not synchronous (e.g. because various tributaries are not completely synchronised and need to be time justified) then they cannot be so used and each level must be demultiplexed one at a time. The combination of drop and insert techniques and synchronous multiplex hierarchies (see below) will become increasingly important in the future.

Drop and insert multiplexers can normally be controlled and reconfigured by command from a network management centre.

3.3 TRANSMISSION HIERARCHIES

3.3.1 Plesiochronous digital hierarchy (PDH)

Digital transmission in today's wide area networks is based on either of the two well-established CCITT transmission hierarchies defined in G.702 (see Figure 3.2). The hierarchy based on 2.048 Mbit/s systems is used in Europe whereas the hierarchy based on 1.544 Mbit/s is used in the USA and Japan.

These hierarchies are plesiochronous, i.e. they do not run at exactly the same bit rate and are not therefore synchronous. Each level in the hierarchy allows the multiplexing of tributaries from the level immediately below, but because the tributaries do not have exactly identical bit rates, additional bit capacity is provided so that tributaries that are running faster than the nominal bit rate can be accommodated and a process called justification is then used to handle the disparity in rates. This process works by using additional bits, called justification bits, to indicate whether other bits are carrying real data or are spare.

In addition to being plesiochronous (running at similar but not identical rates) the tributaries may be said to be asynchronous because their frame structures are not synchronised, i.e. not all frames start at the same time. At 2.048 Mbit/s and 1.544 Mbit/s channels are octet interleaved but above 2.048 Mbit/s channels are bit interleaved.

The asynchronism and bit interleaving mean that to demultiplex, say, one 64 kbit/s channel from say a 34.368 Mbit/s bearer it is necessary to take all the traffic down to 2.048 Mbit/s, demultiplex the required channel and re-multiplex the remaining channels again. This is a complex and expensive procedure and a synchronous system in which any number of 64 kbit/s channels, either individually or in combination to provide a broadband circuit, could be extracted directly at any level in the hierarchy has obvious attractions particularly for a multi-service network.

3.4 SYNCHRONOUS DIGITAL HIERARCHY

3.4.1 Introduction

Work on a synchronous transmission system began in the USA in the early 1980s and led to the development of the SONET (Synchronous Optical NETwork) standards. SONET is based on a transmission rate of 51.48 Mb/s which can not carry the European rate of 139.264 Mbit/s and whose frame structure can not accommodate the 2.048 Mbit/s rate signal.

In 1988, agreement was reached in CCITT on a set of three recommendations, G.707 to G.709, for an international Synchronous Digital Hierarchy (SDH) which can accommodate both the European and US hierarchies and is not limited to a specific technology (SONET is designed only for optical transmission). The term synchronous transmission mode (STM) describes the nature of the structure. The following information is based on the May 1990 versions of G.707 to 709 which are different in a number of places from the 1988 versions.

The SDH multiplexing structure is complicated by the need to embrace a wide variety of sub 155 Mbit/s existing and future multiplexing structures from both the European and North American plesiochronous hierarchies. The basic principle on which SDH works is to include within the multiplex structure a sort of header, called a section overhead, which contains an index in the form of a set of pointers to what is called the payload, i.e. the information carried by the multiplex. These pointers enable the contents of the multiplex to be removed and inserted independently, whatever the level in multiplex hierarchy, whenever the multiplexed data goes through the process of regeneration.

3.4.2 The SDH hierarchy

The basic rate of the SDH is 155.520 Mbit/s. This rate is called STM-1, where STM stands for Synchronous Transport Module, and at higher rates it is called STM-*N*, where the rate is *N* × the basic rate. To date only the levels STM-1 (155.520 Mbit/s), STM-4 (622.080 Mbit/s) and STM-16 (2488.320 Mbit/s) have been defined.

It is important to understand that there is no direct multiplexing between STM levels, i.e. four STM-1s are not multiplexed together to make an STM-4. This is because the STM exists only in a single section (i.e. between network node interfaces), it does not pass through a node. Within each section, *N* administration unit groups (see below) are multiplexed together, using byte interleaving, and a section overhead is added to make an STM-*N*. This arrangement is shown in Figure 3.4.

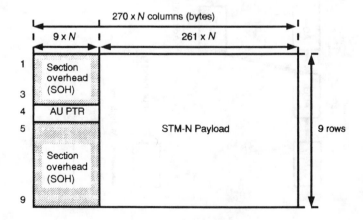

Fig. 3.4 STM-*N* frame structure

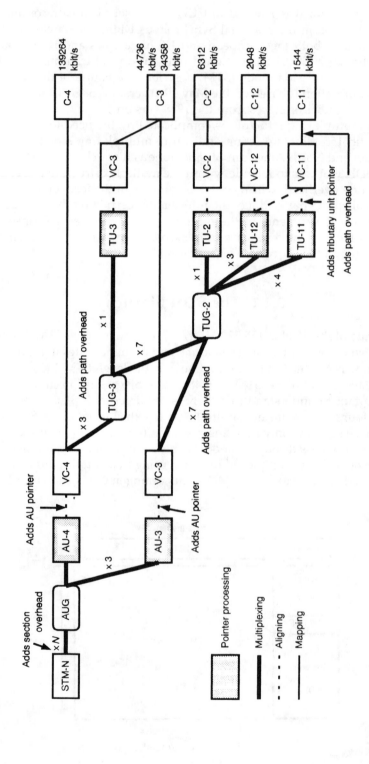

Fig. 3.5 Options within CCITT for multiplexing from the PDH to the SDH

The frame duration is 125 microseconds which matches that of the PDH, and the STM-1 frame contains 2430 bytes or 19 440 bits.

3.4.3 The STM structure

Figure 3.5 shows the way in which different combinations of signals at the G.702 hierarchy rates may be multiplexed together into the STM-N frame.

The unit of capacity is the Container C_{ij} which carries a particular signal rate from the G.702 hierarchy. i = the level of the signal in the G.702 hierarchy, and j = 1 for the lower value or 2 for the higher value of signal bit rate. (Note: For i = 1 or 2, j = 2 for Europe but for i = 3, j = 1 for Europe.) A Path OverHead (POH) for management, including performance monitoring, of the end-to-end path through the SDH is added to the container and the ensemble is called a Virtual Container (VC_{ij}).

The SDH can interwork with the PDH at any rate in the hierarchy, but PDH signals will not be synchronised with respect to the SDH, and therefore a justification system must be used. This is achieved by placing the virtual container in a Tributary Unit (TU_{ij}) which contains a pointer which defines the phase (i.e. the relative position in the frame) of the virtual container with respect to the path overhead at the next layer. An ensemble of tributary units is called a Tributary Unit Group (TUG_i). The details of the justification process are complex and an explanation is not attempted here.

At layers 3 and 4, virtual containers and tributary unit groups are placed in an Administration Unit (AU) together with a pointer to align their phase with respect to the STM-N. One or more AUs are byte interleaved to make an Administrative Unit Group (AUG).

All pointers are referenced by chaining or linking from the AU pointer that is carried in a fixed place within the STM-N frame so that individual containers can be readily extracted without demultiplexing the whole frame layer by layer (as in the PDH). Thus drop and insert multiplexers can be used.

Figure 3.6 shows the ways defined in ETS 300 147 that Europe is expected to choose to construct a STM-N frame. These ways are a subset of those in G.709.

3.4.4 Example of multiplexing a 2.048 Mbit/s signal into the STM-1 frame

Figure 3.7 shows the STM-1 frame which is repeated every 125 µs. The frame is conveniently represented as an array of octets consisting of 270 columns and nine rows. (The octets can be conveniently thought of as being eight bits in the third dimension perpendicular to the plane of the diagram.) The order of transmission is left to right then top to bottom. The first nine columns are the section overhead and so far have been only partially assigned by CCITT for functions such as frame alignment and link management.

Figure 3.7 shows a tributary in the form of a 2.048 Mbit/s link. It is required to multiplex this tributary into an STM-1 frame running at 155.520 Mbit/s. The tributary consists of 32 8 kbit/s channels so there are 32 octets to insert into the multi-

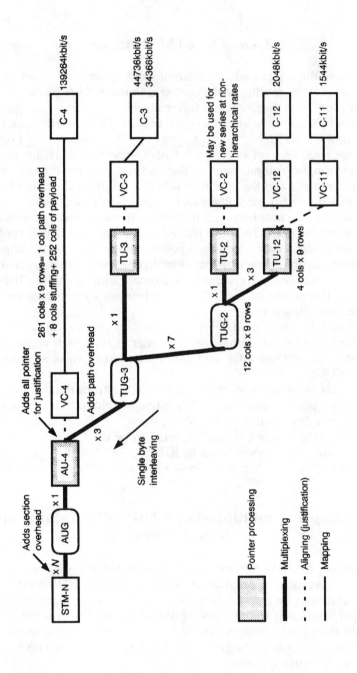

Fig. 3.6 European version of multiplexing paths into the SDH

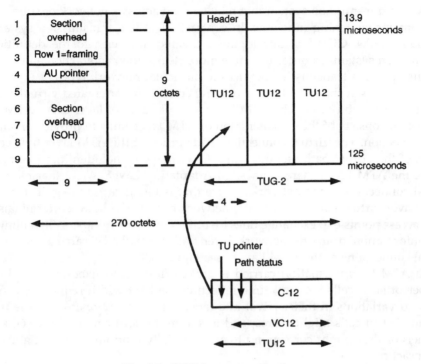

Fig. 3.7 SDH frame organisation

plex every 125 μs. These 32 octets are represented as being in a container which is assigned the number C-12. Four more octets are added as a header which is split into the path overhead (POH) which contains status information and the tributary unit (TU) pointer which acts as the identity of the originator. When the POH is added the container is called a (virtual container) VC-12 and with the TU added it becomes a TU-12.

The TU-12 is inserted in the STM-1 frame as shown in Figure 3.7; note that the content of the tributary is confined to specific columns and this means that it is evenly distributed in time throughout the frame period, and, because the first column contains the pointer, the identity of the source can be recognised before disassembly takes place. In fact the TU-12 does not have to have a fixed position in the frame. This is important, as discussed above, if the source is not precisely synchronised to the transmission clock.

3.5 ASYNCHRONOUS TRANSFER MODE (ATM)

Within CCITT and the RACE Project, considerable attention has been given to a transmission structure called Asynchronous Transfer Mode (ATM) as a major element in the development of broadband ISDN. ATM is relevant both to both switching and transmission and so it is covered more fully in Section 16.3.5 Only the transmission aspects are covered here.

ATM is a transmission system that may be used across a network or subnetwork that contains switching functions. In contrast, SDH is a transmission system for individual links. CCITT is developing a layered approach to the definition of transmission systems in order to define more clearly their interaction and the way that they provide transmission services to particular network types.

ATM operates at the ATM layer providing connection-orientated virtual connections between two points. The ATM layer is a sub-layer of layer 2. An ATM layer connection consists of the concatenation of ATM layer links to provide an end-to-end connection. The virtual connection is identified at the ATM layer by a combination of the virtual path identifier and the virtual channel identifier (see below). Above the ATM layer, there is an ATM Adaptation Layer (AAL) that relates the virtual connection to the network service that is being carried (e.g. it relates the ATM Layer virtual connection to the network connection between terminals network access points e.g. exchange lines identified by E.164 numbers. Signalling and user information from the network layer will normally be carried on separate virtual connections at the ATM Layer. (see Figure 3.8).

With ATM the information carried is broken down into fixed-length cells, the number of such cells per unit time reflecting the bandwidth requirement of the user, and variations in their arrival rate reflecting the bursty nature of the traffic. The number of cells might correspond to sub-multiples or multiples of 64 kbit/s channels or they might correspond to an essentially random process, e.g. packets in transaction processing.

Each cell of 53 octets is divided into a header of 5 octets and an information field of 48 octets; the primary purpose of the header is to identify the connection number for the sequence of cells that constitute a virtual channel for a specific call.

A number of virtual paths multiplexed at the ATM layer may be connected into the same physical layer with each path being identified by a Virtual Path Identifier (VPI) of 8 bits at the user-network interface and 12 bits at the network-node interface. A number of channels may exist within each path with each being identified by a Virtual Channel Identifier (VCI) of 16 bits. The header also contains

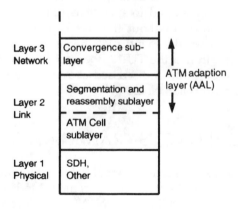

Fig. 3.8 Relationship between ATM sub-layers and OSI layers

8	7	6	5	4	3	2	1	Bit / Octet
GFC				VPI				1
VPI				VCI				2
VCI								3
VCI				PT			CLP	4
HEC								5

CLP   Cell loss priority
GFC   Generic flow control
PT   Payload type
HEC   Header error control
VPI   Virtual path identifier
VCI   Virtual channel identifier

Fig. 3.9 ATM header structure at user–network interface

header error control, generic flow control, cell loss priority and payload type fields, (see Figure 3.9).

The SDH may provide the physical layer for the transmission of ATM cells. Because the SDH container capacity may not be an integer multiple of the ATM cell length, cells are allowed to cross container boundaries. In order to provide security against false cell delineation and the possibility of the information field replicating the STM-N frame alignment word, the cell information field (but not the header) is scrambled before mapping into the container.

In 1989, CCITT agreed to define two physical interfaces to ATM, one based on SDH, and the other a version Asynchronous Time Division (ATD) consisting of an unframed concatenation of ATM cells. At the user-network interface the rate in both cases is 155.520 Mbit/s.

Only the first of these interfaces which involves the mapping of ATM cells into the VC4 has been defined to date in G.709. In addition, ANSI has defined an ATM frame structure for SONET at 51.840 Mbit/s which is similar in concept to G.709 but, of course, is incompatible with it.

3.6 BIBLIOGRAPHY

Bell Telephone Laboratories (1970/71) *Transmission Systems for Communication* (4th edn).
M. T. Hills and B.G. Evans (1973) *Transmission Systems*, George Allen & Unwin, London.
H. Inose (1979) *An Introduction to Digital Integrated Communication Systems*, Peter Peregrinus (for IEE).

4 SWITCHING SYSTEMS

4.1 INTRODUCTION

Switching system designs are influenced by the type of network in which they operate (e.g. a circuit or packet switching network) and the function that they have to perform in that network, i.e. to make a per transaction connection or a more permanent connection. We start by considering the functional requirements and then devote Sections 4.3 and 4.4 to circuit and packet switching respectively. In Section 4.5 we introduce some of the possibilities for switch architectures in an ATM environment. The emphasis is on switching in public networks although the same technology and principles apply to switching in private networks. We end by considering wireless PBXs, which is one concept that is not applicable to the public network.

4.2 SWITCHING FUNCTIONS

In the past the primary function of a switching network has been to set up connections between users for the duration of a call and then to release the call to allow the paths through the network to be used for another call. The more permanent connections, for example that of a particular customer to a particular line circuit at his local exchange, have been made on a distribution frame which involves physical changes to the wiring if for any reason a rearrangement of hardware resources is necessary. The tendency is for the situation to evolve into one in which distribution frames are replaced by cross-connects, which are switches under central control and which can effect circuit re-arrangement in a time which may well not be critical on a microsecond or millisecond time scale. One function of cross-connects is to provide 'route switching' if for example a transmission route fails or becomes overloaded.

Cross-connects are likely to be realised as circuit switches using space switching concepts (see below) and as such do not demand any particularly novel technology. However, there is an exception for optical transmission where carrying out the cross-connect function in the optical domain saves a conversion–reconversion process and this may well be one of the first applications of optical switching in telecommunication networks. (See Section 2.2 for material on optical switching.)

The advantage of cross-connects over distribution frames and patching is that changes can be controlled centrally and record keeping can be fully automated as an integral part of the process.

4.3 CIRCUIT SWITCHES

Circuit switches switch circuits from one of many inputs to one of many outputs and in a normal exchange, the configuration of the switch will be controlled by signalling associated with the calls that are being switched.

Historically there are two main forms of switch control. In a step-by-step switch, the setting up of the call takes place sequentially from stage to stage in the switch with each stage responding in turn to the signalling information (principally the addressing information which with telephony is the dialled digits). The other form is called common control, where the control parts of the switch examine all the signalling or addressing information and then set up the path through the switch.

Common control can take a number of forms ranging from register control, which is usually realised as an assembly of electromechanical components and can only handle one call at a time, through special-purpose electronic units which can handle a substantial percentage of calls simultaneously, to a computer or (stored program) control in which all the calls are handled centrally.

In analogue switches where the inputs to the switch are normally individual analogue circuits, the switches are space switches which provide physical paths between the input and output ports of the switch for the duration of a call. However, for digital circuits where the inputs may be time division multiplexed together for transmission, there are usually three or more stages, perhaps one of space switching and two of time switching, which rearrange in space and time the channels or time slots used by the individual circuits.

A modern digital switch consists essentially of two parts—a digital switch matrix and a processor. The switch matrix undertakes the time and space switching of the calls under the control of the processor which handles the signalling. Both the switching matrix and the processor are normally modular in design so that they can be enlarged without difficulty, and all the main parts of both systems are duplicated or replicated to ensure high reliability. The technology of the processor is normally rather older than that of contemporary commercial computing equipment because switches have to be designed to be highly reliable and so have to use very well-proven designs and construction.

Digital switches operate under Stored Program Control (SPC), which means that the operation of the switch is controlled by the software programs and data which are stored in the processor memory. The great advantage of this arrangement is that it is much easier to make changes to the operation of the switch than it was for electromechanical switches. This particularly applies to customers' lines where the relationship between line, or equipment, number and directory number can be rearranged by alterations to software tables instead of making wiring changes on the Main Distribution Frame (MDF).

With 2 Mbit/s digital systems the 30 channels from the customer to the exchange are time division multiplexed and by storing the content of each channel at the

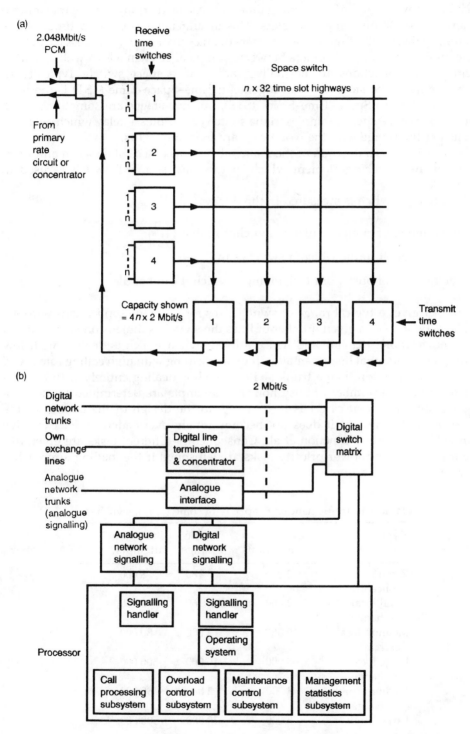

Fig. 4.1 (a) Time–space–time switching matrix; (b) block diagram of a digital switch.

exchange, any one of the 30 incoming time slots can be rearranged to be transmitted in any one of 30 outgoing time slots. This arrangement is called a time switch. Obviously there will be more than 30 users of an exchange so it is necessary to have a means of connecting customers between groups of 30 channels as well as within them. Such connections are normally made with a space switch. Typically an exchange has three stages of rearrangement; Time–Space–Time (TST) is the most common followed Space–Time–Space (STS), but five stages are sometimes used. The storage of the speech samples in the switch constitutes a delay which averages 125 μs per time stage with the worst case approaching 250 μs.

Figure 4.1 shows a diagram of the main subsystems of a digital switch and more detail of the switching element which in this case is a TST switch as used in System X.

The size of a switch is measured in three ways:

- the number of circuits which it switches in lines;

- the volume of traffic which it switches in Erlangs;

- the number of busy hour call attempts which it can handle.

Table 4.1 gives a typical range of switch sizes for different applications within a public network. For a given number of lines the switch is dimensioned in terms of the average traffic per line and hence a local switch which is dealing with low calling rate residential lines is smaller than one dealing with high calling rate PABX lines and smaller still than a trunk switch which is dealing entirely with concentrated traffic. The number of busy hour call attempts are determined by the processing capacity. Great care has to be given to the design of the control of the processor to ensure that it does not become totally overloaded under excessive demands and cease to function at all. Consequently different tasks are given different priorities and lower priority tasks are jettisoned if the loading approaches predetermined limits.

Table 4.1 Traffic handling capacity of some typical switches

Type of call	Circuit terminations	Traffic Erlangs	Busy hour attempts
Remote concentrator	2 000	160	8 000
Small local exchange	2 000	160	8 000
Medium local exchange	10 000	2 000	80 000
Large local exchange	60 000	10 000	500 000
Medium trunk exchange	8 000	2 000	80 000
Large trunk exchange	85 000	20 000	500 000

Although circuit switches may contain some temporary storage to enable the information to be transferred from one time slot to another traffic is not queued but there is delay as discussed above. When a large number of digital switches are connected in tandem, delay can build up and if echo control is to be avoided such delays can be quite significant factors in the design of a network which will normally be such as to ensure that calls do not need to traverse more than a certain number of switches.

4.4 PACKET SWITCHES

Packet switches also switch information between input circuits and output circuits but unlike circuit switches they need to examine each packet to find out which call it relates to or which is its destination before they can take a routing decision. Early packet switch designs had a typical computer architecture and were controlled wholly by software. With such designs it proved to be very difficult to achieve high levels of traffic handling capacity and they were used only in those data networks where the traffic levels were comparatively low. With the increase in data traffic generally and the inclusion of packetised speech on some data networks the reading of packet headers and the consequential routing is carried out by special-purpose hardware or dedicated microprocessors.

Unlike circuit switches, packet switches contain significant storage to queue the packets which are to be transmitted onwards. The length of the queues vary with the traffic loading, giving a variable delay characteristic and a finite loss of packets.

4.5 FUTURE SWITCHING SYSTEMS

We have considered in Section 3.5 how circuit switched and packet switched information can be combined onto a single transmission bearer given a suitable form of multiplexing arrangement. The issue here is how switching will develop and to what extent the different traffic types that are multiplexed together for transmission will be handled in the same switch.

The widespread acceptance of ATM would put the major emphasis on packet switching and the switching of ATM cells is a topic which is absorbing a lot of research and development effort. In a mixed environment of ATM and circuit switching the current consensus seems to be that switches dealing with both types of traffic should treat them separately and switch designs are partitioned accordingly with the circuit switch design being a conventional digital one. (In the view of one of the authors such partitioning is a non-essential constraint.) The novelty lies in some of the ATM cell or packet switch designs which we discuss briefly below. These designs are all likely to be implemented electronically in the first instance at least. Various optical designs are being considered (see Section 2.2.2) but except perhaps for the rather specialised cross-connect applications, as mentioned above, they are currently not competitive in terms of performance or cost.

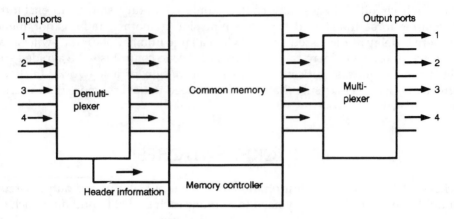

Fig. 4.2 Shared memory switch

Shared memory switches

The simplest type of switch is a pure time slot rearranger, equivalent in principle to the 30 time slot arrangement discussed in Section 4.3 except that it deals with a lot more than 30 customers; we need in fact to consider traffic of the order of several millions of packets per second order. If all the incoming packets are written into a common block of memory by the input channels (Figure 4.2) they can be read out again by the appropriate output channel. With very fast memory and dynamic partitioning of that memory to prevent access clashes, throughputs corresponding to the traffic carried by an STM-16 transmission system (i.e a bit rate of about 2.4 Gbit/s) appear to be feasible.

Broadcast switches

Figure 4.3 shows one outgoing link of the broadcast switch. The incoming traffic is demultiplexed down to, say N, streams at the STM-1 level and is then broadcast to N equipment modules of which Figure 4.3 is one. By looking at the packet headers, each of the N modules selects (or filters) only the traffic that is addressed to it and ignores the rest (for this reason switches of this type are also called 'knockout' switches). The filtering process concentrates the traffic, after which it is reordered or rephased by the shifter and put into packet buffers that form an output queue. The packet buffers are emptied as capacity becomes available on the outgoing link. The performance of each module depends only slightly on the number of other modules and hence it is possible to design a switch that is capable of being expanded in size to meet increasing demand (i.e. one substantially independent of N).

The term 'broadcast' applies to the internal structure of the switch rather than what is perhaps the more familiar usage where traffic is externally broadcast to

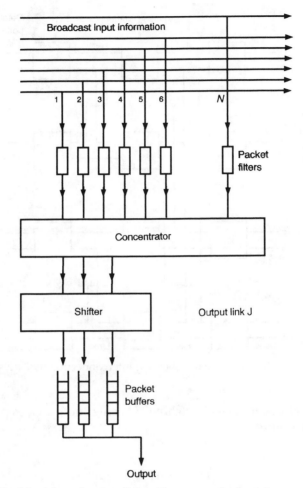

Fig. 4.3 A broadcast switch with a concentration stage

multiple destinations; however, the broadcast switch is quite capable of providing the external broadcast facility as well.

Multi-stage self-steering switches

Multi-stage switching networks of various types (e.g. a perfect shuffle in the form of a Batcher–Banyan network) are capable of connecting any input to any output. Each stage in the switch reads the packet header and selects a path to a subsequent stage that is potentially capable of completing the connection. If no path is available, the packet may have to be stored until one becomes free. Switches consisting of three-stage networks and folded three-stage networks are quite common. This type of switch design is sensitive to traffic levels and is not as easy to expand as a broadcast switch. Figure 4.4 gives an example of a Banyan network which is

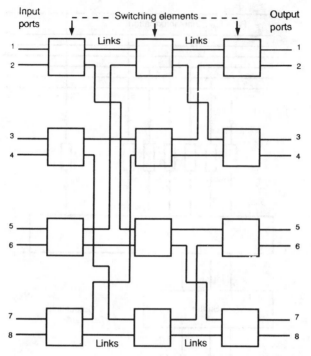

Fig. 4.4 Three-stage Banyan network showing interconnection by 2×2 switching elements.

constructed entirely of binary switching elements and provides paths between any input and any output.

The packet loss problem

Unlike circuit switching networks, packet switching networks have a finite probability of losing traffic because of the finite size of their internal queues. In the ATM environment a packet loss target of 1 in 10^9 has been set. In view of the unknown properties of much of the traffic that an ATM network is going to have to carry this is likely to provide a very considerable problem. Video coders (see Section 6.3.8) are very much of an unknown quantity because they can produce packets at high and variable rates. A key factor in the choice of a switch design may well be its peak load packet loss and delay characteristic and much remains to be established on this subject.

4.6 WIRELESS PBXs

Several designs of wireless PBX were announced in 1991/92. The PBX's cross-office switch hardware is of conventional design but the hardwired in-house cabling is partially or completely replaced by radio connections. The radio standards that are potentially available for this application are discussed in Chapter 12 and the choice is likely to depend on the country where the equipment is operated. The choice in Europe is between CT2, CT3, DECT and GSM derivatives. The advantages claimed for wireless PBXs include:

- Mobility of terminals—A person no longer needs to be associated with a specific point in a building.

- Staff location. It is easier to contact people outside their offices.

- Reduced investment in building wiring.

- No wiring re-arrangement/maintenance costs on the extension side of the PBX.

Disadvantages include:

- Worries about security/confidentiality although various security schemes can mitigate this problem.

- Poor reception in parts of some buildings.

- Services other than speech may demand a performance that is not readily achieved over a radio link.

For voice services at least there are significant advantages in the wireless approach; it remains to be seen how well it works out in practice.

4.7 CONCLUSIONS

Switches are becoming cheaper and smaller as VLSI technology is introduced and as that technology becomes faster. It is quite within the bounds of feasibility that a switch that with a Strowger design once filled the floor of a building can be realised in a single circuit board. The size of the exchange is then determined largely by the wiring coming into the building and the line cards that interface to that wiring. With the increasing use of optical transmission in the local area that too will decrease in volume.

4.8 BIBLIOGRAPHY

R. Barnett and S. Maynard Smith (1988) *Packet Switched Networks: Theory and Practice,* Sigma Press.

G. Oliver (1980) Architecture of System X—Part 3 local exchanges. *POEEJ*, **73**, 27–35.

S. Poulton (1989) *Packet Switching and X.25 Networks,* Pitman.

J. Tippler (1979) Architecture of System X—Part 1. An introduction to the System X family. *POEEJ*, **72**, 138–141.

N.J. Vanner (1979) Architecture of System X—Part 2. The digital trunk exchange. *POEEJ*, **72**, 142–148.

5 SIGNALLING SYSTEMS

5.1 INTRODUCTION

This chapter describes the main digital signalling systems currently in use or under development for public and private networks. With the approach of digital networks, in the 1970s, CCITT began development of a digital common channel signalling system called CCITT No. 7. This system was designed in a modular and layered manner and therefore is capable of being enhanced to provide whatever new features are needed for networks and services to develop.

Unfortunately the development of CCITT No. 7 has been rather slow because it has taken a long time to reach final agreement on some of the standards. This has driven some network operators who wished to introduce digital networks with ISDN features as early as possible to implement their own versions of CCITT No. 7 in advance of the standards finally being agreed. Thus there will be a period of realignment once the international standards are complete.

CCITT No. 7 is designed for public networks, but in parallel access signalling systems for ISDN, and private signalling systems have been developed. This chapter covers all these three areas.

All the digital signalling systems described here are common channel systems which means that the signalling information is carried on a separate logical or physical channel from the user's information and therefore signalling interactions can take place independently and in parallel with the user's communications. This independence greatly increases the range of features that can be provided.

5.2 CCITT SIGNALLING SYSTEM NO. 7

5.2.1 Introduction

CCITT Signalling System No. 7 is a digital common channel signalling system for use with stored program control exchanges. Initially CCITT No. 7 was specified in the context of international inter-exchange signalling, but it is also widely used, with national variations, in a number of countries. For example Systems X and Y use it in the UK, and System 12 uses it in some other European countries. It operates over 64 kbit/s links using addressed (or labelled) messages. Although not designed

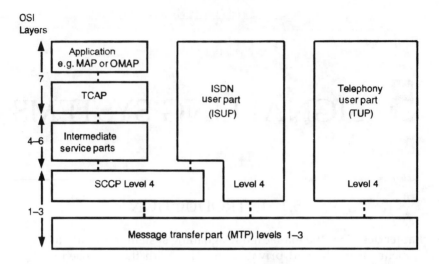

Fig. 5.1 Structure of CCITT signalling system No. 7

primarily for ISDN, CCITT No. 7 will provide the basis for signalling in connection with ISDN services.

CCITT No. 7 is a methodically structured signalling system which provides a wide range of functions. The main parts of the signalling system have already been defined sufficiently for it to be used by a number of public network operators for basic call progression functions; however, work is continuing on some of the details, especially those concerned with supplementary services, with conformance testing and with some of the more sophisticated higher layer features. The definition of the system may never be complete if CCITT No. 7 continues to be used as the basis for new services.

Signalling takes place between signalling points within a network. These points may be part of an exchange or an operations centre or a database (see Chapter 18 on Intelligent Networks), and any network node may have more than one signalling point associated with it. The signalling points are identified by unique point codes.

Figure 5.1 shows the main structure which is based in general terms on the OSI reference model. Note that in CCITT No. 7 the structure is broken down into levels, whereas the ISO model breaks it into layers. Because CCITT started the definition of CCITT No. 7 for telephony well before the promulgation of the OSI model (which is for data services), CCITT No. 7's levels have only an approximate correspondence to OSI layers. Levels 1 and 2 correspond in general terms to layers 1 and 2 while levels 3 and 4 combined correspond roughly to layer 3. The use of CCITT No. 7 in conjunction with data services has highlighted the incompatibility between the two approaches.

Table 5.1 lists the European standards currently published or nearing completion. The standards that relate to access to ISDN by terminals are shown in Table 5.2.

Table 5.1 Public network signalling standards

Number	Title	Status
ETS 300 008	Message Transfer Part to support international connections	Published
ETS 300 009	Signalling Connection Control Part (connectionless) to support international service	Published
ETS 300 100	Routing in support of ISUP Version 1	1992
ETS 300 121	Application of ISUP for international connections	due 1992
ETS 300 134	Transactions Capabilities Part	due 1992

Table 5.2 Access standards

Number	Title	Status
ETS 300 046–1	Primary rate access Safety and Protection— General	Published
ETS 300 046–2	Primary rate access Safety and Protection— Interface Ia-safety	Published
ETS 300 046–3	Primary rate access Safety and Protection— Interface Ia-protection	Published
ETS 300 046–4	Primary rate access Safety and Protection— Interface Ib-safety	Published
ETS 300 046–5	Primary rate access Safety and Protection— Interface Ib-protection	Published
ETS 300 047–1	Basic rate access Safey and Protection— General	Published
ETS 300 047–2	Basic rate access Safey and Protection— Interface Ia-safety	Published
ETS 300 047–3	Basic rate access Safey and Protection— Interface Ia-protection	Published
ETS 300 047–4	Basic rate access Safey and Protection— Interface Ib-safety	Published
ETS 300 047–5	Basic rate access Safey and Protection— Interface Ib-protection	Published
ETS 300 077	Attachment requirements for terminal adaptors at S/T reference point	due 1992
ETS 300 102–1	User–network interface layer 3: Basic call control	Published
ETS 300 102–2	User–network interface layer 3: Basic call control Specification Description Language Diagrams	Published
ETS 300 011	Primary rate access user–network interface	Published
ETS 300 012	Basic rate access user–network interface Layer 1 specification and test principles	Published
ETS 300 125	User–network data link layer specification	Published
ETS 300 153	Attachment requirements for basic rate access	Published
ETS 300 156	Attachment requirements for primary rate access	Published

5.2.2 Message transfer part

The Message Transfer Part (MTP) is common for all applications and has the function of transferring messages across the network.

Level 1 is concerned with the physical transmission of data across a link which is normally time slot 16 on a 2 Mbit/s transmission system, although other time slots and even analogue circuits with modems can be used.

Level 2 is concerned with the transmission of messages across a single link with a satisfactorily low level of errors. Messages are transferred in signal units which are of three types:

- message signal units;

- fill-in signal units (when no messages are to be sent);

- link status signal units (which are used to control the link).

These signal units are delimited by 8-bit flags and contain sequence numbers and codes for error detection (not correction). All message signal units are normally subject to positive and negative acknowledgement and errors are corrected by retransmission.

Level 3 is divided into two parts: the signalling message handling part which carries out the examination and routing of individual messages, and the signalling network management part which manages links, routes and traffic and effects the reconfiguration which may be necessary in the event of congestion or failure.

At layers above the message transfer part there are a number of alternative user parts which are application dependent.

5.2.3 Signalling traffic and reliability

The use of a common channel signalling between public exchanges enables the signalling traffic associated with several hundred speech circuits to be carried on one 64 kbit/s channel. Clearly the risk of failure of such a channel constitutes a serious exposure and it is necessary to provide a number of safeguards. Among these are:

- (a) The use or error detection (by means of a cyclic redundancy check code) on each channel followed by retransmission in the event of detected errors.

- (b) Switching to another 64 kbit/s link (channel) if the error rate exceeds a predetermined level or the link fails completely. In the no-failure state there is always more than one link in service, links are lightly loaded and traffic is shared between them so that in the event of a single link failure there is ample capacity on the remaining link(s) to take the reallocated signalling traffic. Change-over should take place without loss of calls.

- (c) Switching over to a new signalling route in the event that all the links on an existing route fail simultaneously (e.g. because a cable is severed).

This need to provide high availability adds significantly to the hardware cost and software complexity of the No. 7 signalling system, tending to inhibit its use in applications where a high availability is not required.

5.2.4 Telephony user part

The Telephony User Part (TUP) handles the call set up and supplementary services for ordinary telephony. The features available in the international TUP were very limited and so British Telecom developed their own National User Part for use within the UK. Subsequently at the request of the European Commission, CEPT has produced an improved version with more features called TUP+.

5.2.5 ISDN user part

The ISDN User Part (ISUP) handles the provision of ISDN bearer and supplementary services. The basic connection types include speech and unrestricted 64 kbit/s and it is possible to alternate in mid-call between these types.

5.2.6 Signalling connection and control part

The Signalling Connection and Control Part (SCCP) adapts the message transfer part in order to provide a service for transferring data which conforms to the network layer of the OSI model. Such a service contains more functionality than the TP particularly in the area of addressing where it uses globally defined addresses, which give much more flexibility than the message repertoire of the MTP. It also offers both connection-orientated and connectionless data transfer and allows data to be exchanged between points which are not connected by telephony circuits. The SCCP will normally be used for applications such as:

* updating of vehicle location registers in mobile radio networks;

* interrogation of data bases by call processors in intelligent networks, (although there are several other possibilities here);

* transfer of network management information—particularly that relating to the performance of the signalling system itself;

* user-to-user signalling that is not related to calls and which is handled by the ISUP or the more specialised application parts.

Although considerable progress had been made on the definition of the SCCP by the end of the 1985–1988 study period, some further work is required before it can be implemented extensively.

5.2.7 Transactions capabilities application part and intermediate service part

The Transactions Capabilities Applications Part (TCAP) and Intermediate Service Part (ISP) use the services of the SCCP to provide a high-level transaction protocol based largely on English words for the use of applications such as

- mobile network control;
- operations and maintenance;
- enhanced supplementary and value-added services;
- customer-to-customer data transfer.

5.2.8 Mobile application part

The Mobile Application Part (MAP) uses the services of TCAP to handle the messages and data which are exchanged between the mobile switching centres and home and visiting location registers in a digital cellular radio system such as GSM.

5.2.9 Operations and maintenance applications part

The Operations and Maintenance Applications Part (OMAP) contains procedures for controlling, supervising and testing the signalling network, and carrying management commands for the main network. OMAP uses the services of TCAP for its operation.

5.2.10 The future for CCITT No. 7

CCITT Signalling System No. 7 has already established itself as the most important signalling system for digital public networks and it is being used as the basis for some private network signalling systems. The basic parts of CCITT No. 7 necessary for bearer services and some supplementary services have already been defined in sufficient detail for implementation, although work on the more complex supplementary services will be continued. Therefore the general development and improvement of the main parts of CCITT No. 7 is likely to continue for several years. In the areas of the SCCP, TCAP and the other application parts, CCITT No. 7 is less advanced although implementations are taking place. These areas are likely to see significant further development to meet the needs of intelligent networks especially microcellular radio systems.

The sophistication and comprehensiveness of CCITT No. 7, as well as the enormous investment which has gone into it, will ensure that it remains the most important public network digital signalling system for at least two decades, and it is very likely that it will be adapted to handle any changes in technology such as the introduction of fast packet switching exchanges and the use of SDH (see Section 1.3.4) for transmission.

5.3 DIGITAL SUBSCRIBER SIGNALLING SYSTEMS FOR ACCESS TO ISDN

CCITT began the development of a digital subscriber signalling system for ISDN access, signalling between the subscriber's terminal equipment and the public exchange, in parallel with the development of CCITT No. 7. Like CCITT No. 7, the development has been slower than hoped, and several countries, including the UK, France, Germany, USA and Japan have all developed their own incompatible variants, in advance of the standards being finally agreed within CCITT, in order to offer digital services earlier than would otherwise have been the case.

The European Commission has attached great importance to the introduction of ISDN to improve services within Europe and to initiate a common market in terminal apparatus. Consequently the Commission persuaded the CEPT members to sign a memorandum in 1989, for the introduction of ISDN using a common access standard by the end of 1993. At the same time, ETSI has been developing a set of European standards for ISDN access, some of which will be used for apparatus approval.

The standards to be used for approval purposes in Europe are shown in Figure 5.2. The top level approval standards are:

- basic rate access: NET3 (ETS 300 104 and ETS 300 153);

- primary rate access: NET5 (ETS 300 156).

Both these standards will be converted into Common Technical Regulations (CTRs) during 1992/3 for apparatus approval under the Directive 91/263/EEC (see Section 23.5.4).

5.4 COMMON CHANNEL SIGNALLING SYSTEMS FOR ISDN IN PRIVATE NETWORKS

5.4.1 Introduction

In the early 1980s the need for common channel signalling systems for the private network sector was realised and it was thought that CCITT No. 7 was both too elaborate and not (at that time) sufficiently defined in the TUP (see Section 5.2.4) to serve the purpose. The need for a different approach was recognised and action was taken in the UK in advance of other countries since the UK had regulations that were more favourable to private networks than those of most other countries,

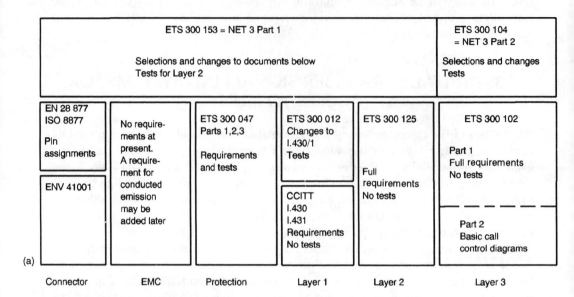

(a)

NB: Safety is handled separately under the low voltage directive (EN 60950 & EN 41003 apply)
    EMC is handled separately under the EMC directive

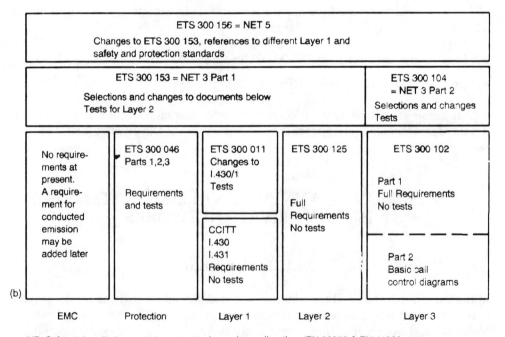

(b)

NB: Safety is handled separately under the low voltage directive (EN 60950 & EN 41003 apply)
    EMC is handled separately under the EMC directive

Fig. 5.2 Structure of ISDN NETs and related standards

to develop an open, vendor-independent standard called DPNSS for private networks. The hope was that the UK work would be adopted as the basis of European and International standards but this has not come fully to fruition.

In 1985, following the publication of the 1984 CCITT Recommendations, CEPT began work on adapting ISDN standards for use between PBXs in private networks. DPNSS was not accepted as the basis for this new standard because the adoption of a UK standard was unacceptable politically and because the protocol structure and encoding were too different from what CCITT had recommended for ISDN, although the information flows of DPNSS have been followed. The detailed technical work on the new European standard, which became known as Q-SIG, passed to the European Computer Manufacturers Association (ECMA), and CEPT's interest passed eventually to ETSI after a confused period during which the various and somewhat conflicting interests of ETSI, CENELEC and ECMA had to be resolved.

In July 1990, the ISDN PBX Networking Specification (IPNS) forum was established by Alcatel, Siemens, GPT, SAT and Telenorma to accelerate the development of Q-SIG and to ensure interworking capability with DPNSS. Since its formation, several other manufacturers including Ericsson, MatraCom, Northern Telecom and Philips have joined IPNS. The particular attention given to DPNSS is a recognition that it is an open standard in widespread use internationally and supported by many manufacturers. Although other manufacturers have their own proprietary standards such as Cornet (Siemens) and ABC (Alcatel) information on their specifications is not readily available and their use is not nearly as widespread as DPNSS.

5.4.2 DPNSS

DPNSS was developed initially by British Telecom, GPT and Mitel for use in the UK Government's telecommunications network. The first issue of the specification (BTNR 188) was published in 1983 and defined basic call and a small number of supplementary services. The specification has been enhanced continually since then and over 40 supplementary services have now been defined. The standard is supported by more than 12 manufacturers and more than 3000 DPNSS links are in use throughout the world. In addition to the specification, conformance testing procedures were published in 1990 and a committee meets regularly to review proposals for changes in the light of experience in implementing the standard.

The current standards are:

- BTNR 188: Digital Private Networking Signalling System No. 1;

- BTNR 188-T:Digital Private Networking Signalling System No. 1: Testing Schedule;

- BTNR 189: Interworking between DPNSS1 and other Signalling Systems.

The DPNSS working party is currently working on defining the interworking arrangements with:

- Q-SIG;

- ISDN access;

- ISDN extension interface to data processing equipment.

The essential features of DPNSS are as follows:

- At layer 1, DPNSS can either use time slot 16 or a physically separate transmission bearer including modems operating over analogue circuits at 9.6 kbit/s. The acronym APNSS is sometimes used in the case of an analogue bearer. There are no back-up signalling links.

- At layer 2, DPNSS uses compelled signalling, i.e. each frame is transmitted repeatedly until a positive acknowledgement is received, whereas with the I-Series retransmission occurs only if a positive acknowledgement is not received within a preset period. DPNSS provides 30 virtual circuits for data in addition to the signalling associated with the 30 main traffic channels.

- At layer 3, DPNSS has a basic set of messages for more than 40 supplementary services. In addition individual manufacturers are allowed to extend the message repertoire to provide the facilities necessary to support proprietary features on their own PABX and private network products. Thus DPNSS has a very extensive message set.

5.4.3 Q-SIG

Q-SIG is named after the Q reference point defined in ENV 41004 (see Section 5.4.4). The basic call capabilities of Q-SIG are defined in the following two standards which were published by ECMA in February 1991:

- ECMA 141: Data link layer protocol for signalling channel between two private telecommunications network exchanges;

- ECMA 143: Layer 3 protocol between exchanges of private telecommunications networks for the control of circuit switched calls.

There is no layer 1 standard because the private network exchanges will use the network services (leased or switched circuits) offered by the public networks. Work on supplementary services is well in hand but at the time of writing (January 1992) no standards on these services had been published.

ISO has agreed to adopt the ECMA work on Q-SIG as the basis for worldwide standardisation.

5.4.4 Other relevant standards work

There are three standards that have been prepared by CENELEC that define the framework for private network standardisation:

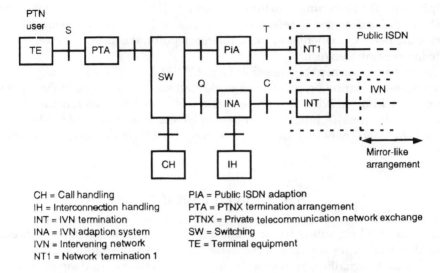

CH = Call handling
IH = Interconnection handling
INT = IVN termination
INA = IVN adaption system
IVN = Intervening network
NT1 = Network termination 1

PIA = Public ISDN adaption
PTA = PTNX termination arrangement
PTNX = Private telecommunication network exchange
SW = Switching
TE = Terminal equipment

Fig. 5.3 Reference configuration in ENV 41004

- ENV 41004: Reference configuration for connectivity relations of private network telecommunications exchanges;

- ENV 41006: Scenarios for interconnections between exchanges of private telecommunications networks;

- ENV 41007-1: Definition of terms in private telecommunications networks Part 1: General terms.

The reference configuration defined in ENV 41004 is shown in Figure 5.3, which introduces and defines the Q and C reference points. The node of a private network that provides switching and call handling functions is called a private telecommunication network exchange (PTNX) which may in practice be an ISDN PBX or equipment that is part of a public network (e.g. Centrex).

Interconnections between PTNXs are provided by an intervening network which may be a public or privately provided facility. The intervening network may take various forms from a dedicated transmission facility at one extreme to a switched public ISDN service at the other.

A distinction is made between calls and connections, with calls being associations of two or more end users (people or processes) and connections being associations of telecommunications facilities for the transfer of information between two or more points in a network. This means that a number of calls may take place sequentially across a given connection. For example, two PTNXs may establish a connection across an intervening network and use the connection to carry a number of calls, one after the other.

The C reference point is the boundary between the intervening network and the intervening network adaptation functional grouping, whereas the Q reference point is the boundary between the intervening network adaptation functional grouping and the switching functional grouping. Thus the signalling between the PTNXs is defined at the Q reference point in a way that is independent of the intervening network, whereas at the C reference point, it is dependent on the form of the intervening network.

Following an agreement between ETSI and CENELEC, ETSI is now fully responsible for standardisation on private networks and no further work in this area will take place within CENELEC. Future work will take place within the Business Telecommunications Technical Committee of ETSI and within ECMA. These activities will be coordinated by the ISDN Management Coordination Committee of ETSI.

5.5 CONCLUSIONS

Common channel signalling has so many advantages that its use from terminal to terminal as well as for the core of the network is highly desirable. CCITT No. 7 is too elaborate a signalling system to be in private networks and simpler systems based on compatible principles are necessary. We have seen how delays in the international standardisation of CCITT No. 7 have provoked the development of several different national standards, but these standards will be superseded rapidly as soon as the international standards are fully established.

The transition to the international standards will take place sooner in the case of the access standards than in the case of DPNSS, because it will be several years before the international standard Q-SIG includes all the features of DPNSS.

Although this chapter has tended to focus on differences between the various digital common channel signalling systems, they are all capable of working together over public and private networks to provide end-to-end ISDN connections, together with the more common supplementary services.

5.6 BIBLIOGRAPHY

K. Fretten and C. Davies (1988) CCITT Signalling System No 7: Overview. *Br. Telecommunications Engineering*, **7**, 4–6.

M. Ozdamur, M. Trought and P. Jones (1988) Experience gained from the implementation of DASS and DPNSS. *Private switching systems and networks, IEE international conference*, June, pp 45–50 (CPN 288).

6 CODES AND CODERS

6.1 INTRODUCTION

Three aspects of coding are considered:

1. the codes used to represent binary information for line transmission;
2. algorithms used in analogue to digital (A/D) conversion and digital to analogue (D/A) conversion;
3. coder designs for A/D and D/A conversion with particular reference to reduced bit rate coding.

The need for a digital representation of analogue waveforms arises mainly in the context of speech and video signals and hence is where most of the emphasis lies but the coding of redundant data patterns is also briefly treated

6.2 CODING OF BINARY INFORMATION FOR LINE TRANSMISSION

The simplest representation of binary information is by a specific voltage to represent the 1 (or mark) state and zero voltage to represent the 0 or (space) state and we call this pure binary. Pure binary is a perfectly satisfactory arrangement in terminal or other compact pieces of apparatus but is often unsatisfactory for transmitting digital information over distances in excess of a few tens of metres. There are a considerable number of what are called line codes for use in digital transmission systems and they are designed with one or more of the requirements that are discussed below in mind:

1. The need to have no direct current content in the code because the transmission system involves transformers or capacitors that will not pass DC. The solution is to balance the code so that the mean voltage is zero. The simplest form of balancing is balanced binary in which the voltage representing a space is the complement of the voltage representing a mark, but a better solution is to use what is called alternate mark inverse (AMI) in which alternate 1s are of alternate sign and balanced about zero and a zero is 0. AMI is a pseudo-ternary code;

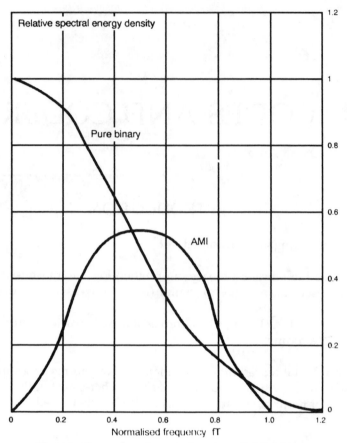

Fig. 6.1 Examples of spectral energy distributions

pseudo because although it has three levels the information transmitted is
binary coded.

2. The need to have a strong frequency component at the basic bit or clock rate so
    that the clock can be easily extracted on reception and synchronism maintained.
    (Transmitting the clock as a separate signal is not usually considered to be
    economic.) The solution is to prevent long runs of bits with the same sign by
    replacing bit with 'violation' pulse every time a run reaches the limit length.
    The most widely used code of this type is a modified version of AMI called
    HDB3 (high density binary with a maximum of three 0s). The violation pulse
    is distinguished from a 1 because it violates the alternate mark inverse rule.
    Codes such as HDB3 are called non-alphabetic codes.

3. The need to make the most economic use of the channel bandwidth in the
    presence of some specific level of noise. We have already indicated that the
    number of levels in a channel can be more than two and ternary encoding can
    be used to increase channel capacity. Such ternary codes are called alphabetic

codes. An example of a ternary code is 4B3T where four binary bits are replaced by 3 ternary levels (-1, 0, $+1$). There are various techniques for balancing ternary codes but a method of mode changing when the out of balance sum reaches a set threshold is perhaps the most common.

4. A possible need to provide some form of error detection or perhaps even correction. Violation rules provide one potential form of error detection and redundant ternary codes another.

5. The need to minimise the pulse distortion that arises from the variation of attenuation with frequency and the need to minimise crosstalk. For both these requirements reduction of signal bandwidth is desirable. Compared with pure binary most of the codes mentioned above and listed in the table below show energy concentrated in a relatively narrow spectrum. Figure 6.1 compares the spectrum of AMI with that of pure binary.

Table 6.1 summarises the characteristics of some of the better known line codes. Line codes are becoming increasingly important because of their use for digital transmission in the local area and private networks, for example with ISDN transmission between the exchange and the customer (see Chapter 15 and Section 10.4.1).

Table 6.1 Properties of line codes

Name of code	Balance unbal	Binary/ ternary	Non-alpha/ alpha	Spectrum	Timing extraction
Pure binary	unbal	bin		poor	poor
Balanced binary	bal	bin		poor	poor
AMI	bal	tern	non	good	fair
ISDN S/T	bal	tern	non	fair	good
Wal 1 (Manchester)	bal	bin		fair	good
Wal 2	bal	bin		fair	good
Miller	bal	bin		narrow DC component	poor
HDB3	bal	tern	non	good	good
B6Z6	bal	tern	non	good	good
2B2T (PST)	bal	tern	alpha	fair/good	good
3B2T	bal	tern	alpha	fair/good	good

6.3 DIGITAL CODING OF ANALOGUE WAVEFORMS

A/D and D/A conversion has two major applications in telecommunications, coding for transmission and coding for processing. For transmission the emphasis is on sampling at the Nyquist rate and hence minimising channel bandwidth. For processing (e.g. digital filtering, see Chapter 9.3) the emphasis may be more on coder simplicity and sampling may be at many times the Nyquist rate which is commonly referred to as over-sampling. (The Nyquist rate is twice the reciprocal of the bandwidth of the analogue signal that is to be converted to a digital one.) Because much of the early work on coders was done with speech coding as the objective, the history of coders is very much a history of speech coding and speech will be used as a vehicle to explain many of the various types of coding algorithm although the concepts are applicable in some case to the coding of video signals as well.

The equipment combination of a coder and a decoder as one unit is called a codec.

6.3.1 Pulse code modulation—the need for lower bit rates

Speech waveforms are commonly digitally encoded as Pulse Code Modulation (PCM) at a transmission rate of 64 kbit/s (see the G.711 to G.716 series of CCITT recommendations). The bandwidth required to transmit 64 kbit/s is very much in excess of the 3.4 kHz needed to transmit speech in analogue form. For many years research has been going on to achieve satisfactory coding of speech using a lower bit rate and while some lower bit rate systems have been realised in a military context they have proved uneconomic in the civilian field. VLSI is changing all that but paradoxically optical communication has in the meanwhile reduced the cost of bandwidth to the point where it is even more difficult to justify extra cost in the coding process. However, there are a number of applications where bandwidth remains a scarce and expensive resource and, as we shall see in Section 12.5, elaborate speech coding schemes are justifiable.

PCM at 64 kbit/s works by sampling the analogue speech waveform every 125 μs, i.e. 8000 times a second, and converting the amplitude of the sample into an 8 bit code. The 8 bit code allows 128 positive and 128 negative levels, and the value of the sample is represented by the code that corresponds to the level nearest to the value of the sample. This process is called 'quantisation' and the error introduced in choosing the nearest level is called quantisation noise or quantisation distortion. The Nyquist rate for a 3.4 kbit/s channel is 6.8 kbit/s but sampling at 8 kbit/s makes the design of filters in the decoder much easier.

6.3.2 Redundancy in speech production

The amplitude of speech signals from the softest sound of the quietest talker to the loudest sound from the loudest talker covers a range of some 50 dB. Coding such a large range with equally spaced levels would need 12–13 bits/sample, but fortunately the human ear has a logarithmic response and it is possible to code the speech waveform with 256 logarithmically spaced levels (8 bits) to give a result that is subjectively acceptable. The resultant coding laws are A-law as used in Europe and elsewhere and μ-law as used in North America.

To convey the information relevant to speech amplitude with fewer bits involves taking into account the statistical properties of speech and perhaps also moving away from a waveform, or time domain, encoding approach towards one of analysis and synthesis which will in general involve both time and frequency domains. To take an extreme and perhaps somewhat ridiculous example, if the input speech were fed into a phoneme* recogniser which recognised up to 64 phonemes at a production rate of say 20/s, then speech could be conveyed at 120 bits/s and reproduced on a speech synthesiser. Of course such speech would sound totally unnatural and talker identity would be lost; furthermore it would be difficult to convey other sounds such as tones and background noise. In practice therefore there is a limit, probably corresponding to a bit rate of the order 1 kbit/s, below which it is impractical to go. Nevertheless the gap between 64 kbit/s and 1 kbit/s is large and there are a number of ways in which it might be filled.

Listed below are four ways in which the statistical properties of speech amplitude and spectrum are constrained in human speech production. We need to understand these constraints before we go on to examine how to exploit them in various coder designs.

1. The voice production mechanism is such that at frequencies above about 500–1000 Hz the long-term average of the speech amplitude falls off at about 12 dB per octave.
2. Speech sounds are distinguished from each other according to the position of lips, teeth, tongue and soft palate (vellum). The position of these articulators at any time determines the spectrum of the speech sound at that time and because articulatory movements take place in times of several tens of milliseconds, the spectrum is relatively stationary over periods of between, say, 10 and 50 ms.
3. For voiced sounds, e.g. vowels, for which the larynx vibrates, the vocal tract is excited with what approximates to a saw-tooth waveform. The frequency of excitation, which is lower for men than for women, conveys little phonemic information in English but is important for other reasons. This frequency is relatively stable over periods of several tens of milliseconds in general and hence can be predicted with good accuracy.
4. Because the amplitude of speech sounds correlates well with their spectra and because the larynx output does not change rapidly, the RMS amplitude of speech sounds, particularly vowel sounds is relatively stable too, typically over a 10–5 ms period. Potentially use can be made of these long periods of relative stability to reduce the data or information rate on the speech channel.

Codecs can be categorised in terms of whether they work in the amplitude domain, or the time domain or both when they are called hybrid.

6.3.3 Redundancy in pictures

The bandwidth of a video transmission can be reduced by making use of one or more of the following properties of typical pictures and human appreciation of them:

---

* Phonemes are the basic phonetic elements of voice production, corresponding approximately to letters of the alphabet in written material.

1. The human eye is more concerned with edge information than it is with large expanses of relatively constant amplitude and colour. Thus less bits can be used to code flats than are needed for edges.
2. In many pictures the frame content and often the focus of interest as well is static for periods of seconds or more. By transmitting only the difference between one frame and the next large savings in bandwidth can be achieved. For this reason many picture transmission systems have what is known as a frame store which stores the content of one picture frame.
3. When the picture is moving *en bloc*, e.g. because the video camera is tracking, motion estimation allows much of the frame store information to be used in a translated form and only the new scene content need be transmitted. Rapid changes in scene can be updated over more than one frame.
4. It is not always vital to transmit rapid movement in a picture accurately and hence a low frame transmission rate can be tolerated. (At the receiver the same frame may need to be repeated several times to avoid flicker.)

One implication of the above is that it could be useful for video services to have variable bit rate transmission (and switching) so that major changes of scene can be assigned a high bit rate that falls to a low bit rate once the scene becomes relatively static, the overall objective being to keep the picture quality constant. In Section 16.3 we discuss the type of network where it is possible to vary the bit rate on demand.

6.3.4 Redundancy in text

Certain character strings have a much higher probability in most languages than others, for example in English 'the' or 'que' has a much higher probability of occurrence than 'ghn' or 'zjt'. The data rate needed to send plain textual information can therefore be reduced by making use of the statistical properties of written languages.

6.3.5 Time domain coding

Delta modulation

Rather than sample the waveform at the minimum (or Nyquist) rate determined by the bandwidth of the signal, it is possible to sample at a higher rate but reduce the number of coding levels correspondingly. In the extreme the number of coding levels reduces to one and the resultant coding scheme is then called delta modulation. Delta modulation is therefore a one bit differential coding scheme running at a sampling rate equal to the transmission rate; a rate typically in the range 16–32 kHz for speech. With delta modulation the coder contains a replica of the decoder and sends a one or a zero depending upon whether the current input level is above

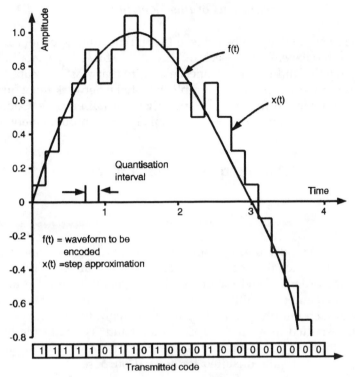

Fig. 6.2 Delta modulation—waveform encoding

or below the decoder's output (see Figure 6.2). The decoder treats a one as an increase of unit step size and a zero as a decrease of unit step size. Coders containing replicas of the decoder and making decisions depending on a comparison between decoder output and coder input are called predictive coders and delta modulation is the simplest form of predictive coding.

There are a number of elaborations of simple delta modulation of which CVSD and HCDM are possibly the best known. With CVSD (continuously variable slope delta modulation) the significance of the step size depends on past history and step size increases with increasing signal level as averaged over a defined period, e.g. 5 ms. The system known as HCDM (hybrid companded delta modulation) has two adaptive processes with different time constants, one of about 1 ms and the other of about 5–10 ms.

Currently there is little interest in simple delta modulation schemes for speech transmission but delta modulation codecs are quite widely used because of their simplicity for A/D and D/A conversion with digital signal processing VLSI chips. The codec can be built on the chip with perhaps a pair of external resistors and capacitors. Signals are often grossly over-sampled; speech may be sampled at 2.5 Mbit/s, say, to obtain the accuracy necessary for subsequent digital processing.

Limitations of time domain coding

The bit rate reduction that can be achieved with time domain coders is mainly but not entirely at the expense of speech quality. A 16 kbit/s delta modulation coder of whatever type would produce a speech quality that was noticeably lower than that which is normally expected in a public telephone network but a quality which is more readily tolerated for military and other specialised uses. Coders, such as hybrid coders (below), are therefore preferable for use on public networks.

6.3.6 Frequency domain coding

The vocoder

A really low bit rate coder, dating from the 1950s but improved considerably since, is the vocoder, in which the speech is separated by a number of filters into bands (typically of the order 12) and the energy in each band is sampled at a relatively low rate, encoded and transmitted. The speech is reconstituted at the receiving end by recovering pitch information and exciting a bank of filters with energy according to the transmitted level in the appropriate band. Typically the bit rate for a vocoder is in the range 3–6 kbit/s. The speech is, however, not at all natural, of low intelligibility and quite unsuited for general-purpose use.

A very different approach to frequency domain coding is being developed in the context of Digital Audio Broadcasting (DAB) to give audio reproduction of CD quality. Here the problem is multipath propagation arising from reflections which, with high bit or symbol rate, can result in destructive interference between successive bits. The solution is to send information at a much lower rate on, say, 20 frequency divided channels. For a net bit rate of 128 kbit/s this would mean 6.4 kbit/s per channel giving symbol period long enough for destructive interference not to be problem in the great majority of situations.

6.3.7 Hybrid and predictive coding—speech

Sub-band coders

There is a class of coders called sub-band coders which have something in common with vocoders in that the speech band is separated by filters. Unlike the vocoder the energy in the bands is not coded but the waveforms are. There are fewer bands, typically four, and there is a dynamic allocation of coding bits between bands. Because quantisation noise is confined to its own band, noise from a band containing a high amplitude signal does not mask a low level signal in another band.

There is some interest in speech of a higher quality than is available when the bandwidth is limited to an upper frequency of 3.4 kHz. CCITT Recommendation

Table 6.2 Modes of operation for a coder to G.722.

Mode	Audio coding bit rate (kHz)	Auxiliary channel bit rate (kHz)	Comment
1	64	0	7 kHz speech
2	56	8	7 kHz speech
3	48	16	7 kHz speech
0	64	0	3.4 kHz speech

G.722 covers a system of speech coding with a 7 kHz bandwidth. The speech is first coded with a sampling rate of 16 kHz and 14 bit PCM with uniformly spaced coding levels; this is followed by a combined 2-band sub-band and an ADPCM coder (see below for ADPCM coding). The coder has four modes of operation, as shown in Table 6.2.

Mode 0 is to enable the kHz coder to interwork with existing 64 kbit/s, 3.4 kHz coders. In modes 2 and 3 the coding levels in the lower sub-band are sacrificed to provide an additional data channel.

Sub-band coding in also proposed as part of the DAB coding scheme discussed in Section 6.3.6.

Linear predictive coders

The advent of VLSI makes codecs of very high complexity and considerable computing power possible. The number of design variations is very large indeed and it is only practicable to describe briefly two designs, one of which should see service shortly and the other in the mid 1990s.

Adaptive Differential Pulse Code Modulation (ADPCM) codecs are an example of a class of linear predictive waveform coders called APB (Adaptive Predictor Backwards). As the name implies they only use past samples for adaptively predicting the current sample and are distinguished from APF (Adaptive Predictor Forward) coders which by storing and hence delaying samples are able to appear to predict using future samples. CCITT has made a recommendation for a 3 kbit/s ADPCM (APB) coder aimed primarily at use on expensive long distance transmission routes (e.g. submarine cables or satellites) where a 2:1 saving in bandwidth is well worthwhile. The coder algorithm and its interfaces are specified by CCITT; the coder is realised as a VLSI design and there are likely to be a number of competitive commercial products. Further description of the ADPCM design is to be found below in the somewhat more general context of the GSM coder.

The GSM (Global System for Mobile communications) codec is intended for the next generation pan-European mobile radio network (see Section 12.5.2) . The first version is intended to operate at a speech coding rate of 13 kbit/s, with a half-rate version at 6.5 kbit/s to follow (see further below). Its class is APF (Adaptive Predictor Forward) and it is called a Regular Pulse Excitation–Long Term Prediction–Linear predictive coder or RPE–LTP for short. A block diagram of the coder is shown in Figure 6.3 and the decoder in Figure 6.4.

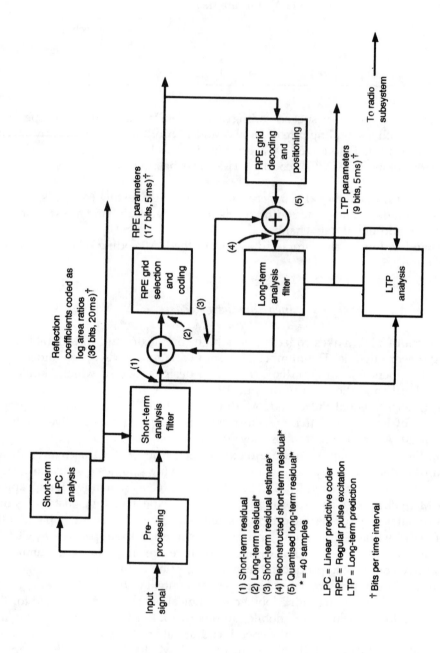

Reflection
coefficients coded as
log area ratios
(36 bits, 20ms)[†]

RPE parameters
(17 bits, 5ms)[†]

LTP parameters
(9 bits, 5ms)[†]

To radio
subsystem

(1) Short-term residual
(2) Long-term residual*
(3) Short-term residual estimate*
(4) Reconstructed short-term residual*
(5) Quantised long-term residual*
* = 40 samples

LPC = Linear predictive coder
RPE = Regular pulse excitation
LTP = Long-term prediction

[†] Bits per time interval

Fig. 6.3 Simplified diagram of the linear predictive (RPE–LTP) encoder

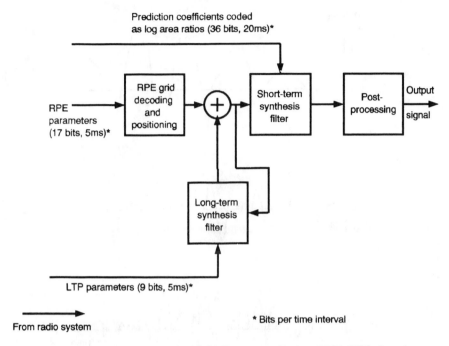

Fig. 6.4 Simplified diagram of the linear predictive (RPE–LTP) decoder

The operation of the RPE–LTP coder is best understood by reference to Figures 6.3 and 6.5. Consider a voiced sound (e.g. a vowel) with a waveform such as (a) in Figure 6.5. The preprocessing block of the GSM coder removes any DC component from the speech waveform and pre-emphasises it (i.e. compensates for the fall off in the speech spectrum). Analysis is carried out using 160 speech samples at an 8 kHz sampling rate (implying a delay of at least 20 ms in the coder and decoder). The so-called short term spectral analysis simulates the configuration of the vocal tract and results in eight parameters called 'log area ratios' or more generally 'reflection coefficients', which emulate the human vocal tract in the form of eight cylindrical sections. The short-term analysis filter is an inverse filter so its output corresponds to the larynx output in human terms, which, as Figure 6.5(b) shows, is a series of short repetitive pulses of varying amplitude. These pulses provide the input to the next stage of the coder. 20 ms is rather a long time interval over which to report the vocal tract configuration and linear interpolation is used from the position reported for the previous 20 ms.

The ADPCM 32 kbit/s coder is similar in principle but rather than predicting the spectral cross section of the vocal tract, it predicts the resonances (poles) and anti-resonances (zeros) of it, but note because the ADPCM design is APB, rather than APF, it does not delay the speech by more than a millisecond or two before sending it.

It would be possible to waveform encode the output at (1) in Fig. 6.3 and send it together with the reflection coefficients to the decoder at the other end of the connection, and in the case of the ADPCM design this is what is done. But in

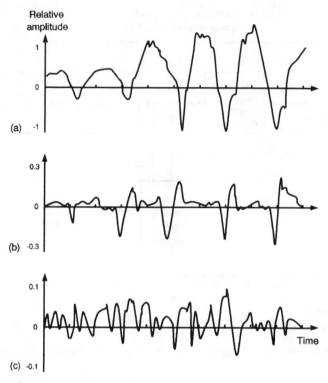

Fig. 6.5 Coder waveforms; (a) speech input; (b) prediction error—no pitch extraction; (c) prediction error—with pitch extraction

addition to having the advantage of hindsight, the GSM coder is of a more elaborate design and also predicts the excitation waveform. The block marked LTP does this and another inverse filter removes the predictable part of the excitation waveform leaving a noise like residue as in Figure 6.5(c). The parameters of the LTP are sent to the decoder as amplitude and phase (or lag) terms. The residual waveform is adaptive pulse code modulation (APCM) encoded and sent as the RPE parameter. The excitation part of the coder works with a 5 ms wide moving window and its parameters are determined by cross correlation of the current window's samples with samples from earlier windows. For unvoiced sounds such as the fricatives f and s, the LTP analysis will extract low frequency resonances from the sounds, rather than repetitive features, making the RPE waveform more like white noise than the spectrally shaped form that it has before processing. The coder needs to contain a decoder so that it can evaluate the effect of its own predictions, thus the decoder proper is largely a sub-set of the coder.

A further development of linear predictive coders goes by the name of CELP (Code-book Excited Linear Predictive), pronounced 'kelp'. The reflection coefficients mention above are not independent and a further bandwidth reduction can be achieved if, instead of sending them individually, they are treated as a set and the set is looked-up in a code-book. From the code-book search the best match is found and the code representing this entry is sent. The design for the half rate GSM codec (see above) is the subject of competition and the current front runner (in early 1992 no design had met the specification) is a CELP coder.

One type of CELP coder that has received fairly wide acceptance is a Motorola design for Vector Sum Excited Linear Prediction (VSELP) which run at 4.8 or 8 kbit/s. Vector-excited has the same meaning as code or code-book-excited. Rather than having a set of unstructured code in the code-book, VSELP uses a set or sets of what are called basis vectors. The 4.8 kbit/s version has a set of 11 basis vectors and the 8 kbit/s version 2 sets of basis vectors. Each code vector is constructed as a linear combination (sum) of the 11 or 7 basis vectors. For the 8 kbit/s coder, where there are two code-books, the vectors are chosen by minimising by successive approximations the combined error from both code-books. The vector-sum approach is more resistent to bit errors than a simpler unstructured code-book.

All the signal processing operations that are performed in the ADPCM and GSM codecs are on digitised signals and hence the filters in them are adaptive digital filters as described in Section 9.3. The inverse filters that are also an essential part of equalisers are the subject of Section 9.4.2.

CCITT Recommendation G.763 is concerned with Digital Circuit Multiplication Equipment (DCME): this equipment is used on long-distance transmission circuits where bandwidth is at a premium and where circuits are shared on a statistical basis by means of digital speech interpolation (SI), see Section 3.2. During peaks in talker activity, 'freeze out' (talk spurts lost as a consequence of a high traffic loading) can be prevented if circuits can be temporarily assigned a lower bit rate. An ADPCM system can be made to adapt by switching between bit rates of 16, 24, 32 and 40 kbit/s according to traffic circumstances. The method of coding is covered by CCITT recommendation G.726 (replacing G.721 and G.723) which is extended in recommendation G.727 to cover transcoding between 64 kbit/s and packetised speech operating according to the packetised voice protocol (PVP) of G.764.

The International Marine Satellite Organisation (INMARSAT) plans to use an adaptive predictive coding algorithm (APC) working at 16 kHz for satellite maritime communication.

6.3.8 Hybrid and predictive coding—pictures

Rather than attempt a general survey of picture coding we will concentrate on the CCITT codec requirements contained in Recommendations H.261. The 1990 'blue book' version of these recommendations was far from complete and a fuller version was to have been published in 1991 but was not available at the time of writing, hence it is not possible to be precise about all the aspects of the coding algorithm. The basic philosophy is to specify only what is essential in terms of a general structure and transmission interface leaving scope for innovation in areas where this is possible. In the 1990 version codecs work in multiples of 384 kbit/s ($n \times 384$ kbit/s) or multiples of 64 kbit/s ($m \times 64$ kbit/s) but in the latter version the n and the m are replaced by $p \times 64$ kbit/s. Sound uses a separate 64 kbit/s channel which can be shared with data according to recommendation G.722 (see Section 6.3.6). (There are commercially available codecs that will operate at any data rate between 64 kbit/s and 2 Mbit/s with flexible allocation of bandwidth between sound and vision.)

The picture is represented by

- a frame rate of approximately 30 frames/s, not interlaced;
- 288 lines/frame of luminance and 144 lines/frame of two-colour difference component;
- 352 picture elements (pels)/ line for luminance and 176 pels/line for colour difference interlaced in both dimensions;
- picture aspect ratio 4 : 3.

Without making use of redundancy a bit rate of about 3.8 Mbit/s is needed to transmit pictures having the parameters listed above. Working at much lower rates means that full use must be made of picture redundancy and that rapid changes of scene will be subject to delay before they are completely updated.

The picture is divided into 1584 blocks of 8 pels by 8 lines of luminance information plus 4 pels by 4 lines of chrominance information. There are two main modes of operation per block; INTER in which the difference signal is sent and INTRA which represents a major change of scene. In both modes blocks are segmented into transmitted and non-transmitted.

The coding algorithm is represented by Figure 6.6. The INTER mode has the switch in the lower position and the INTRA mode has the switch in the upper position as determined by the block marked comparator. Common to both modes are the blocks called transform and quantiser and their inverses. The transform block executes a discrete cosine transform in two dimensions of size 8×8. The discrete cosine transform is an effective means of coding when adjacent samples are highly correlated as they are with most pictures. The output of the transform block is quantised for transmission and the coder performs the inverse operation so that it has a signal identical to that of the decoder for reference. The filter when switched into circuit performs a smoothing operation by computing a weighted average of adjacent pels within a block. This reduces noise and patterning on flat areas of the picture. The block marked threshold determines whether the current block is sufficiently different to be sent at all. The picture memory optionally contains motion compensation, see Section 6.3.3, which is expressed in terms of horizontal and vertical components indicating the displacement of the picture from its last position.

In the INTER mode the current pel is compared with the corresponding pel from the last frame and the difference is transmitted. On decoding the difference signal is added to the corresponding pel in the last frame and stored. Operation in the INTRA mode involves transmitting absolute pels and on reception adding the difference between the received + filtered pels and the received pels to the received pels to give the stored pels. Hence the stored pels constitute a filtered version of the received pels.

6.3.9 Textual data compression

CCITT Recommendation V.42bis describes a data compression scheme for use over telephone networks. The data reduction technique is based on what is sometimes called 'codebook' coding (which can also be used for speech). For textual informa-

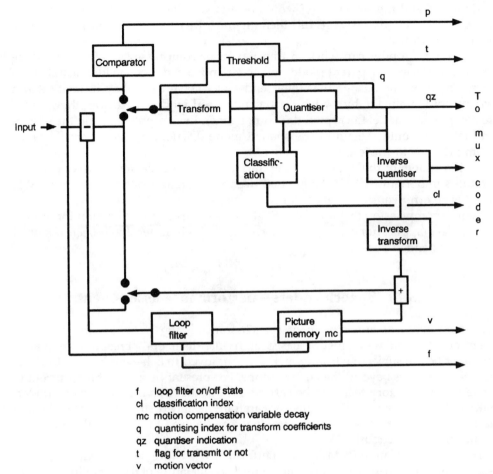

f   loop filter on/off state
cl   classification index
mc motion compensation variable decay
q   quantising index for transform coefficients
qz   quantiser indication
t   flag for transmit or not
v   motion vector

Fig. 6.6 Video coding algorithm—block diagram

tion the algorithm is called BTLZ which is a British Telecom variant of an algorithm published by A. Lempel and J. Ziv in 1978. A dictionary of strings (i.e. the code-book) is maintained dynamically as a tree of strings, see example shown in Figure 6.7. An input string will be matched to the lowest possible leaf on the tree and the appropriate code for that leaf transmitted. In the example shown in Figure 6.7 if the word 'they' were sent then the code for 'the' would be sent followed by the

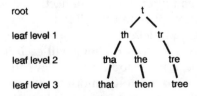

Fig. 6.7 Tree structure for BLTZ encoding algorithm

code for 'y' and the string 'they' would be added at level 3 to both end's diction-aries. If the dictionaries were full then an entry that had not been used recently would be discarded.

The coder/decoder can work in transparent or compressed mode. Initially the system works in transparent mode, and coder and decoder create identical diction-aries based on the strings that they send and receive until they estimate that a saving could be made by switching to compressed mode whereupon they switch to compressed mode. Encrypted data is not generally compressible and the system in that case would remain in transparent mode. Dictionary sizes typically range from 512 to 2048 entries.

With a system of this type the coder and decoder must be kept perfectly in step and it is important that V.42bis is used over an error-controlled connection. (V.42 itself is an error control protocol for modems.)

Typical compression factors with ASCII text is three or four to one but for limited vocabulary applications compression factors as high as ten have been obtained over an extended period.

6.3.10 Speech coders—performance assessment

The assessment of any system of speech transmission in which the speech is degraded in some way must be based on human listeners' perception of quality. To describe the subjective testing methods employed to achieve consistent subjec-tive test results is beyond the current scope. For present purposes it is sufficient to state that satisfactory tests can be carried out but that this is an expensive under-taking. Based on a limited number of carefully designed subjective tests, the aim is always to find an objective measure than can be used to quantify the quality of a specific speech circuit.

The relatively simple PCM coder's performance is expressed in terms of quanti-sation noise (i.e. errors produced due to the discrete nature of the sampling process) and when a number of PCM coder–decoder pairs are connected in tandem, i.e. the signal returns to analogue several times between one end of the transmission path and the other, further quantisation noise is introduced at every coder. It has been established from subjective tests that the quantisation noise adds when codecs are operated in tandem. The units used are called quantisation distortion units, (qdus) for short, and a scale has been chosen such that the quantisation noise introduced by one 8 bit A- or μ-law PCM codec is one qdu. Thus four PCM codecs in tandem would introduce 4 qdus. In subjective terms, 4 qdus represent a detectable but not very significant degradation. CCITT has recommended a limit of 14 qdus as the worst case for an end-to-end international connection, and while this does repres-ent a significant degradation, it is still possible to carry on a conversation over a circuit with a much higher qdu level.

The more elaborate coders discussed above will each introduce their own char-acteristic type of noise and it is by no means obvious that the noise that one kind of coder introduces can be added to the noise produced by an entirely different type of coder. However, there is some evidence that the noise from one coder type expressed in terms of qdus can be added to the noise produced by another coder type similarly expressed with fair accuracy. In order to express, for example, the GSM coder's performance in qdus it is necessary to carry out subjective tests and to equate the result obtained to the number of PCM systems in tandem that give the same subjective score. On this basis the GSM codec rates 7–8 qdus while the ADPCM codec rates 3–3.5 qdus. Sinc 8 qdus is a very significant fraction of the overall 14 qdus recommended by CCITT, and the implications of this on overall network design is a subject that will need to be considered further by international standardisation bodies. There is scope for a completely fresh approach to the treatment of quantisation distortion.

6.4 SUMMARY

The need to transmit digital information over relatively long distances in the local area of the public network and in private networks has led to the introduction of a number of line codes by various administrations and companies to solve the problem. CCITT has in fact recognised six such codes in the context of the ISDN digital section (see Recommendation G.961). This diversity is partly a reflection of different system requirements in terms of protocols, distance and cost and the situation is still fluid as far as convergence on a specific code choice is concerned.

In the field of mobile and video communication the sophistication of codecs is remarkable and it would be rash to predict when it will reach a limit; however, wide practical field experience with these new devices is limited and a period of retrenchment may be called for before even more advanced ideas are allowed to develop.

Picture coding is receiving increasing attention and it too is becoming remarkably complex. An area that has still to be fully explored is that of very high quality audio coding, i.e. with a 20 kHz bandwidth, most of the principles have been established in the context of very low bit rate coders but it remains to be seen which of these works best for the very highest quality reproduction.

6.5 BIBLIOGRAPHY

F. Bigi and M. Decina (1988) CCITT standardisation work on speech processing. *Telecommunications J.*, **55** (11), 748–753.

R. Dettmer (1990) The big squeeze—a CELP coder. *IEE Review*, **30** (2), 55–58.

I. A. Gerson and M.A. Jasuik (1989) *Vector sum excited linear prediction (VSELP), IEEE Workshop on Speech Coding for Telecommunications*, September, pp 66–68.

N. S. Jayant and P. Noll (1984) *Digital Coding of Waveforms. Principles and Application to Speech and Video.* Prentice-Hall.

N. S. Jayant (1990) High-quality coding of telephone speech and wideband audio. *IEEE Communications Magazine* **28** (1), pp 10–20.

R. Pietroiusti (1988) Speech coding of waveforms at 16 kbit/s and below. *Telecommunications J.*, **55** (11) 765–9.

H. Price (1992) CD by radio—digital audio broadcasting. *IEE Review*, **38** (4), 131–5.

M. Taka (1988) ADPCM coding of speech signals. *Telecommunications J.*, **55** (11), 754–757.

7 RADIO

7.1 INTRODUCTION

With line communications undergoing revolutionary changes through the development of optical fibres, it is easy to ignore the major developments which are taking place in radio communications. Yet in terms of public awareness of new services the introduction of new mobile radio services would be perceived by many as the more important development.

In this chapter, we will begin by explaining some of the fundamental features of radio communications and then consider some of the main developments in radio technology. The main developments in the service or system areas which are changing most rapidly and are of greatest significance for the future, namely mobile services and satellite systems are discussed in Chapters 12 and 13 respectively.

7.2 THE RADIO SPECTRUM

Radio is an electromagnetic wave which propagates at the speed of light in a vacuum. Radio communications work through the transmission and reception of a carrier signal, which is modulated to make it carry the information that is to be communicated. The frequency of the carrier can vary at present over the range of 10 kHz to 100 GHz, which can be thought of as the extent of the currently usable radio spectrum. Radio transmitters are allocated discrete frequencies and receivers are selective (i.e. they reject by filtering all frequencies other than the wanted ones) so that the spectrum can be shared by a large number of users. This sharing of the radio spectrum is termed frequency multiplexing.

Use of the radio spectrum is divided by international agreement through the CCIR, which is part of the International Telecommunications Union (ITU), into bands which are designated for specified services. The reason for grouping identical services into bands is that the total possible usage of a given band of spectrum is greatest if the services using it have similar transmitter powers and receiver sensitivities. The division of services into bands also means that industry is able to concentrate on developing specific ranges of those components, the characteristics of which vary with frequency. Telecommunications is not the only user of the spectrum. Heavy use is also made by radar and other navigation systems, and there is some use by astronomy and other scientific services.

The propagation of radio waves depends on their frequency. In the band 30–300 kHz the Earth's lower atmosphere bends radio waves back towards the Earth producing 'sky waves' which can provide very long distance, even tending towards global, communications. However the total quantity of communications which can take place simultaneously is very low because the total amount of spectrum over which this phenomenon occurs (approximately 270 kHz) is low. In the band 3–30 MHz the ionosphere, an ionised layer of the Earth's atmosphere, reflects radio waves, and in the band 30–60 MHz the ionosphere scatters radio waves, and both effects help to extend the propagation distance beyond the line of sight even to give global coverage. From about 60 MHz through to about 1 GHz the troposphere, another layer of the Earth's atmosphere, will scatter radio waves, and tropospheric scattering can be used for longer-distance communications, although the angle of the transmitter beam relative to the troposphere is critical. Above about 1 GHz all propagation is effectively line of sight only, although in an urban environment reflections off buildings and other objects can make communications possible where there is no direct line of sight.

The other two factors that affect propagation are absorption of the radio wave energy by atoms and molecules in the atmosphere, and scattering of waves by rain and fog. Absorption occurs at frequencies specific to particular atoms and molecules. Scattering becomes significant at frequencies where the size of the water droplet is greater than an order of magnitude less than the wavelength of the wave. Consequently attenuation due to rain really becomes significant in the microwave region.

The radio spectrum is normally always drawn logarithmically with the frequency range 10–100 kHz taking as much space as 10–100 GHz. It is, however, important to remember that the communications capacity is proportional to the bandwidth and that the theoretical capacity of the 10–100 GHz band is a million times that of the 10–100 kHz band. The much shorter propagation ranges at 10–100 GHz also mean that those frequencies can be reused easily at relatively short distances, which increases the practical capacity of that band even further.

Figure 7.1 shows the radio spectrum and the position of some of the main communications services. The allocation of frequency bands to particular services depends to a large extent on the range of the service and the volume of communications required. Mobile services and broadcasting have priority for the scarce lower frequencies whereas satellites and high capacity short range point to point links can be accommodated satisfactorily in the microwave region. In future the growth in the use of cable and satellite for broadcasting may enable some of the broadcasting frequencies to be reallocated to mobile communications.

7.3 TECHNOLOGY

7.3.1 Antennas

Antenna design is a function of frequency because the gain pattern of an antenna is determined by its physical shape and the frequency of transmission. Antennas

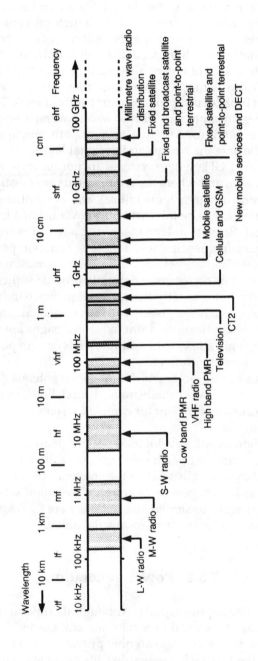

Fig. 7.1 Radio spectrum (showing some of the services described in the text)

are designed for a given frequency and they have a finite bandwidth over which they can meet their design specification. Antenna designs for gains of about 5 dB upwards can be divided into two broad categories: at frequencies below about 1 GHz it is usual for the antenna to be made from a number of separate feed and reflector elements, the most common design for a high gain antenna being the Yagi (like a roof top TV aerial); at frequencies above about 1 GHz, antennas are usually made from a combination of a horn feed and a dish reflector.

At frequencies below 1 GHz antenna technology is not changing very rapidly. The main reason for this is that the most important services at those frequencies are mobile radio and broadcasting which, apart from domestic Television receivers, require mainly low gain antennas with a wide coverage angle and operate in circumstances where fading or loss due to multipath propagation have a much greater effect than the antenna design on the signal levels.

At frequencies above 1 GHz where systems are normally designed to work with straight line transmission and where the requirements are mostly for high gain and low side lobes, antenna design is more critical, and in satellite systems, where the antenna gain pattern needs to be tailored to the coverage of a geographical area, it is particularly critical. Satellite antennas are discussed in more detail in Chapter 13. At these microwave frequencies, it is becoming common practice to use different polarisations to discriminate between different transmissions. Although polarisation discrimination on its own cannot normally provide sufficient discrimination between channels which use exactly the same frequency bands, it can add about 10 dB discrimination, which helps greatly to achieve adequate channel interference. For the less demanding applications, front fed dish designs are normally used, but for more demanding applications with multiple feeds or high polarisation discrimination, offset fed designs are preferable. (see Figure 7.2).

During the last decade, one factor which has helped antenna design greatly is the development of computer-aided mathematical modelling, which is now able to produce results of accuracy sufficient for complex systems to be designed correctly with little experimentation.

The most significant recent development is the phased array antenna which consists of a panel of small flat patch elements (e.g. slots), the individual phase of which can be altered dynamically to steer the antenna's beam. Phased array antennas do not provide as high a gain or as accurate control of sidelobes as the traditional dish reflector designs of similar size and they are therefore used only where there is a need to steer the beam or to have a flat antenna.

7.3.2 Power generation

Power generation is developing rapidly. Although thermionic valves, e.g. devices such as travelling wave tubes and klystrons, are still needed for very high power transmitters, semiconductor devices can now provide powers of up to a few watts at frequencies up to the 1–10 GHz range using silicon, or at even higher frequencies using gallium arsenide which has higher charge mobility than silicon. Where it is necessary to produce higher powers, the outputs of several semiconductor devices

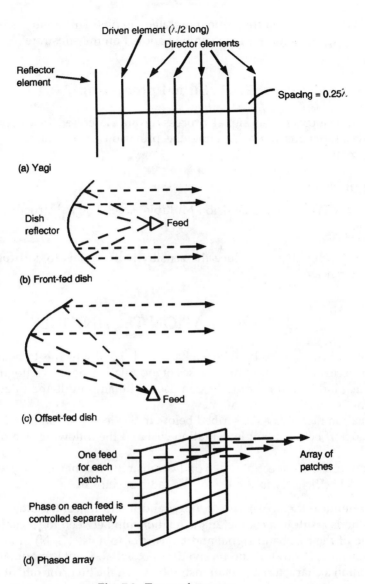

Fig. 7.2 Types of antenna

can be combined, but each stage of combination normally involves a loss of about 1 dB and therefore the benefit of many stages of combination diminishes.

7.3.3 Hybrid integrated circuits

Probably the most important development in mobile communications where very small transmitters and receivers are needed, is the ability to combine both analogue and digital functions on the same integrated circuits. This ability is limited by the dynamic range of signals which can be handled on the substrate.

7.3.4 Signal processing

General developments in signal processing are described in Chapters 6 and 9. These developments are facilitating the provision of the following elements in radio communications:

- digital filters;

- complex TDMA (Time Division Multiple Access) signal structures;

- equalisers;

- coding techniques to reduce the sensitivity of signals to corruption by noise interference or fading.

7.4 MODULATION TECHNIQUES

Modulation is the means by which a baseband signal which is to be transmitted is put on to a carrier. There are two classes of modulating signal: analogue and digital. Analogue modulation is not to be confused with amplitude modulation, which is one form of analogue modulation.

Modulation methods as described below in Sections 7.4.1 and 7.4.2 are designed for a trade off between implementation cost and the following two objectives:

(a) to minimise the spectral usage of the transmitted signal and so give the highest possible efficiency in the use of the radio spectrum;

(b) to minimise the sensitivity to noise and interference. (Analogue modulation methods result in a characteristic relationship between the signal/noise (S/N) ratio of the baseband signal and the carrier to noise (C/N) or carrier to interference (C/I) ratio of the received radio signal. Digital modulation methods result in a characteristic relationship between the bit error rate of the baseband digital signal and the C/N or C/I of the received radio signal.)

7.4.1 Analogue modulation

The simplest form of analogue modulation is amplitude modulation (AM) which involves the baseband signal and the carrier being multiplied together so that the amplitude of the carrier is modulated by the baseband signal. From a frequency point of view, the modulation produces two sidebands, one on each side of the carrier frequency, and in addition this unmodulated carrier frequency is normally also present in the transmitted signal. The bandwidth of the transmitted signal is

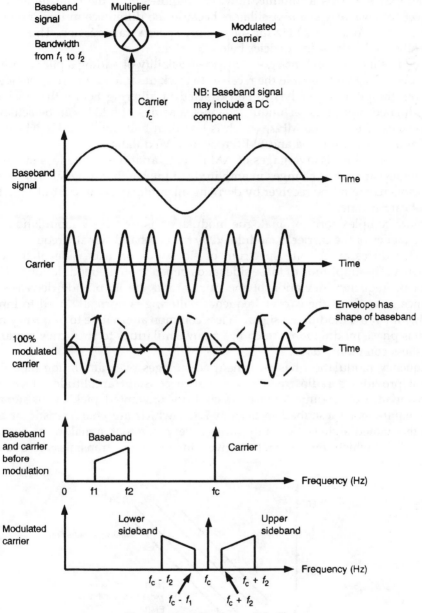

Fig. 7.3 Amplitude modulation

therefore twice the bandwidth of the baseband signal (see Figure 7.3). Amplitude modulated (AM) signals can be demodulated easily by passing the received signal through a diode rectifier and a filter to extract the envelope.

To reduce the bandwidth of the transmitted signal it is possible to transmit only one sideband, in which case the modulation is called Single SideBand Suppressed

Carrier (SSBSC). This technique, however, requires both the transmitter and the receiver to have very stable oscillators because the difference in their frequencies must not exceed about 20 Hz for satisfactory speech transmission. Thus SSBSC is currently used only at frequencies below 30 MHz.

A good deal of research has gone into the feasibility of adding a pilot carrier tone to the SSBSC signal to enable the receiver to track the transmitter frequency. At the receiver, the pilot carrier has to be extracted by filtering before the sideband is demodulated. With this technique, channel spacings of 5 kHz can be achieved at frequencies of up to 200 MHz, which is less than half the 12.5 kHz which is the minimum spacing for AM and FM (Frequency Modulation).

One of the main disadvantages of AM is that variations in the loss of the radio path (fades) appear as changes in amplitude of the baseband signal. This problem can be overcome in the receiver by deriving an automatic gain control signal from a pilot carrier tone.

A more complex form of analogue modulation is frequency modulation where the frequency of the carrier is modulated by the baseband signal. Frequency modulation produces an infinite spectrum of sidebands on each side of the carrier frequency. The amplitude of these sidebands depends on the modulation index and the peak frequency deviation of the carrier, and the amplitude decreases with distance away from the carrier frequency. Filtering is normally used to limit the bandwidth of the modulated signal. A less common alternative to frequency modulation is phase modulation, which has many similarities to frequency modulation but is less commonly used.

Frequency modulation has two main advantages over amplitude modulation. First it provides a radio frequency signal of constant amplitude which means the avoidance of the interference effects from unwanted pick up of signals by other equipment (described in Section 7.4.5) which are characteristic of amplitude modulated signals. Secondly it has a very different signal/noise curve (see Figure 7.4), which means that it gives much better transmission quality at

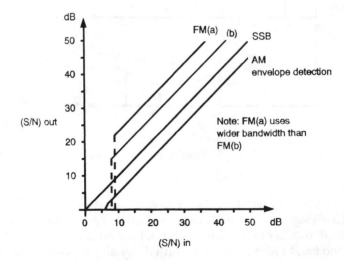

Fig. 7.4 Comparison of amplitude and frequency modulation

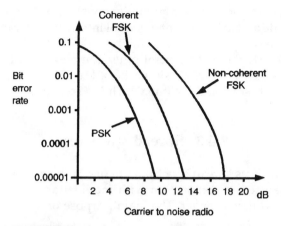

Fig. 7.5 Performance of simple digital modulation schemes

carrier/noise levels greater than about 9dB and discriminates in favour of the wanted signal in the presence of interfering signals. The disadvantage of frequency modulation is that it normally occupies a wider bandwidth than amplitude modulation.

7.4.2 Digital modulation

The two basic types of digital modulation are Frequency Shift Keying (FSK) and Phase Shift Keying (PSK). With frequency or phase shift keying the transmitted signal is switched between two frequencies or phases, depending on whether a 0 or a 1 is to be transmitted. In order to increase the information which can be contained in a given bandwidth of transmitted signal, a more sophisticated alternative to simple binary (two phase) PSK, where the phase of the carrier is changed by 180°, is quaternary PSK where the carrier can have any of four different phases each separated by 90°, and therefore two consecutive bits of baseband signal can be represented by each phase change. The performance of FSK and PSK modulation schemes is shown in Figure 7.5 which relates bit error rate to the received carrier to noise ratio.

The phase changes in PSK described so far introduce an abrupt change (discontinuity) to the transmitted signal which has the effect of increasing the amplitude of the outer parts of the spectrum of the signal, which in turn increases adjacent channel interference. To overcome this effect, the phase changes to the baseband signal can be staggered (one bit instead of two bits per change) so that any single change is never more than 90°, and the baseband digital waveform can be shaped to half cosine pulses before modulation, making the phase modulation continuous rather than discrete so that no abrupt phase changes occur. This technique is called Minimum Shift Keying. If the baseband signal is passed through a Gaussian filter, the modulation is called Gaussian Minimum Shift Keying (GMSK) which is the modulation planned for the new pan-European digital cellular radio system

(GSM— Global System for Mobile communications). The GMSK technique trades reduced bit error rate and increased intersymbol interference for improved adjacent channel interference.

In addition to GMSK there are several other complex digital modulation techniques, many of which have close similarities, but GMSK is probably the best known advanced technique and the one which is likely to have widest use.

7.4.3 Spread spectrum

The techniques known as spread spectrum are not in themselves a form of modulation but they are used in conjunction with modulation schemes and affect the spectrum of the modulated carrier. The main purpose of spreading the spectrum are:

(a) either to desensitise the communications link to fading or interference, or

(b) to reduce the spectral power density (the power transmitted in unit bandwidth), to reduce the interference to other services, or

(c) to make the existence of the transmissions less obvious to an enemy.

In general, the spectral widening produced by the spreading is very much greater than that produced by the modulation. Many of the applications of spread spectrum techniques are in military systems.

There are two main types of spread spectrum technique. The first is frequency hopping, where the frequency of the carrier is changed rapidly over a wide band. If the time spent at any spot frequency is much less than the duration of a fade, and the frequency changes are sufficiently large for there to be no correlation between fades at the different frequencies, then this technique will average out the effect of fades, because the chances are that you will hop out of any frequency which is suffering a severe fade or severe interference. Therefore this technique can reduce the margin which needs to be allowed for fades. It also makes it very difficult for an enemy to deploy spot jamming if he does not know the sequence of frequencies.

The second type of technique is one where the carrier frequency is fixed but the baseband frequency is operated on by a spreading function (e.g. multiplication, addition, or some other function) before the resulting signal is modulated onto the carrier. The bandwidth of the spreading function would normally be at least several times that of the baseband signal and in the digital case it would normally be a random sequence of positive and negative voltages. At the receiver the demodulated signal is operated on by an identical spreading function but using the inverse of the process in order to extract the baseband signal.

Both techniques can be used to enable several communication links to share the same piece of spectrum simultaneously provided that they use uncorrelated frequency hopping or spreading functions. With the second technique the method of spectrum sharing is called Code Division Multiplexing (CDM) and the resultant system is called CDMA (Code Division Multiple Access). One application of CDMA is remote electricity meter reading using the electric main as the transmission medium.

The main reasons for using spread spectrum techniques are identified under (a) to (c) above, and spectrum sharing, which is a form of multiplexing, can best be regarded as a means of extending the capacity of a piece part of the spectrum rather than an effective means of multiplexing in its own right. However, some studies and trials are trying to evaluate how efficient spread spectrum multiplexing is, but there are severe practical problems in mounting a sufficiently intensive trial.

7.4.4 Use of intermediate frequencies

In the explanations given above, we have always referred to the modulation of the carrier as if the final carrier is modulated directly. The explanations have been presented in this way for simplicity. In practice for many systems the carrier is not modulated directly but instead an intermediate frequency (IF) is modulated and the modulated signal is converted up to the frequency of transmission. (The up conversion process is identical to SSB amplitude modulation.) The inverse sequence of operations may take place at the receiver.

One relatively recent development has been the introduction of zero IF or homodyne receivers. With the homodyne technique the local oscillator frequency in the receiver is nominally the same as the RF carrier frequency and the demodulated signal is therefore at baseband. (If the local oscillator is synchronised to the carrier the term synchrodyne is used instead of homodyne but the homodyne approach seems to be the preferred one.) In a superheterodyne receiver with an IF of 13 MHz, say, the local oscillator might be tuned to a frequency of 472 MHz to receive a signal on a carrier at 472 MHz + 13 MHz = 485 MHz. The receiver can equally well receive an 'image' signal at 472 − 13 = 459 MHz and there must be selective tuning in the RF input stage of the receiver to reject any signal at 459 MHz. For a receiver that is required to tune over a wide range of frequencies making the tuning track in sympathy with the local oscillator adds to the cost and complexity of the receiver.

There are several advantages in the homodyne technique:-

(a) There is no image frequency rejection problem because there is no image frequency and this simplifies tuning of the RF stage of the receiver to the point where such tuning may be relatively coarse and simple. The channel selected depends only on the local oscillator frequency and its harmonics if these harmonics are sufficiently strong.

(b) The use of baseband makes digital filtering easier because there is no need for the filter to have to handle the relatively high frequencies of an IF.

(c) The use of digital filters means the filter characteristics are capable of selection under program control to give a bandwidth to match that of the transmitted signal and hence among other things its type of modulation. At the same time the demodulation circuit can be switched to suit that type of modulation.

These techniques together with the use of a frequency synthesizer to generate the local oscillator frequency mean that an almost all digital receiver can be realised. With VLSI as the enabling technology a multi-waveband all purpose receiver on a chip is no longer beyond the bounds of possibility.

7.4.5 Electromagnetic compatibility

We have already referred to the fact that amplitude modulated signals can be demodulated by a simple diode rectifier and filter, but any non-linear device in any type of equipment picking up amplitude modulated radiation has the same effect although the sensitivity of the demodulation depends on the extent of the non-linearity. Consequently AM transmission can cause interference, particularly close to the transmitter, by being demodulated by any non-linear equipment in their vicinity which is not sufficiently well screened. This problem is one of the reasons why FM, the amplitude of which is constant, and so is not demodulated by a simple non-linear device, is preferred to AM.

Unfortunately, many of the problems which spectrum managers thought had been left behind with AM are beginning to recur with those digital systems which use time division multiplexing with bursts of transmission which have an AM-like signal envelope structure. Any non-linear devices, such as some hearing aids and personal stereos, extract the signal envelope which appears as a noise tone with a fundamental at the burst frequency and a high harmonic content.

7.4.6 The future

The trends in modulation systems are fairly straightforward. FM has largely replaced AM, and new digital systems are beginning to replace FM. It is difficult to predict exactly which form of digital modulation will be used most commonly in the future but GMSK is likely to become well established in the mobile field through GSM.

The main advantage of digital techniques are:

(a) reduced sensitivity to co-channel interference so that frequencies can be reused at shorter distances;

(b) compatibility with signals which are already digital for other reasons;

(c) suitability for error detection and correction codes to achieve low error rates.

These advantages are such that digital modulation is expected to give at least a twofold improvement in spectrum usage compared to FM in the pan-European cellular system, even though the digitised speech signal needs about 7 kbit/s compared to the 3.4 kHz for analogue voice. Spread spectrum techniques will become increasingly prevalent in military systems. In civil systems frequency hopping will be used to reduce the margin needed for multipath propagation and spreading functions may be employed to enable a new service to share a band among customers and perhaps to share a band already occupied by an existing service.

7.5 BIBLIOGRAPHY

R. C. Dixon. (1984) *Spread Spectrum Systems*, Wiley.

8 OPEN SYSTEMS INTERCONNECTION

8.1 WHAT IS OSI AND WHAT SHOULD IT DO?

Open Systems Interconnection (OSI) is an architecture or framework for sets of protocols in which all the activities of the communication process are classified in an orderly, consistent manner and as far as possible independently of the information being conveyed. Previously standards have been developed as problems have been recognised and there has been a tendency for standards to be very application specific and perhaps to overlap, each other not always in an entirely consistent way. The number of new data services that have been introduced over the last decade and which are still evolving provides an opportunity for the introduction of protocol standards that fit into the same common framework.

Because telecommunications are international, there has had to be at least a minimal level of standardisation for such services as telephony and telex. Most of this standardisation appears as sets of CCITT Recommendations. CCITT Recommendations are mandatory only in so far as the member countries of CCITT make them mandatory on a national basis. In practice most, if not all, telecommunication administrations adhere to them. CCITT Recommendations prior to 1984 did not, and in the majority of cases still do not, fit into the OSI concept and therefore such services as telephony are incompatible with and are excluded from OSI. OSI is specifically intended for data services and does not apply to existing telephony, but in a multiservice context, e.g. ISDN, OSI principles have to be applied in part to speech. ISDN in providing a telephony interface to the PSTN might be regarded as providing an interface between speech in the OSI context and speech outside that context. A similar situation applies to the older data services, e.g. telex and facsimile, which fall outside the OSI ambit but which can be interfaced to it through standardised gateways.

OSI uses the concept of layering, which assumes an increasing degree of abstraction as one goes from the lowest layer to the highest. At present there is only one OSI model, the seven-layer model, which is described in some detail below. The choice of seven layers was arbitrary, not fundamental, and whether with hindsight one would come to a different decision today is debatable; certainly a seven-layer model can be made to work and the investment in it is a strong disincentive to change.

The disadvantage of OSI is that, because it involves so many interests, the development of OSI standards is a slow process and one involving compromises. These compromises result in options being left open rather than definite decisions being made and some of these options have to be closed later in order to provide successful interworking between networks and between equipment attached to those networks. This has led to a further round of standardisation of what are called 'profiles' or 'functional standards', which are selections of compatible options from the base standards.

The alternative to OSI is a proprietary protocol, such as IBM's Switching Network Architecture (SNA), and this is fine so long as the parties to a transaction all use it, but if they do not then it is necessary to provide protocol transforming gateways between various proprietary networks. Such gateways are expensive, liable to proliferate and not always entirely transparent to the information that they convey.

There is a set of wide area network protocols called TCP/IP (Transmission Control Protocol/Internet Protocol) which were developed by the US Department of Defense and have resulted in a *de facto* standard. TCP/IP provides serious competition to OSI in North America and is receiving acceptance in Europe. A brief comparison between OSI and TCP/IP is made in Section 8.2.5.

Network management is a central concept of OSI but its standardisation has had to be given lower priority than that of the the main communication functions. This means that implementable OSI management is unlikely to be a practical proposition until about 1992 and this incompleteness represents a disadvantage compared with alternative proprietary systems.

The very significant advantage of OSI is that users can procure equipment competitively from different suppliers and be fairly confident that it will interconnect and that they can then connect this equipment to a network and be confident that it will work over that network with some one else's equipment at the other end. The equipment that implements OSI protocols may be located partly in one or more user networks and partly in a PTO network or networks. The corresponding protocols at both ends of a point-to-point communication can be be implemented by physically disparate sets of equipment.

OSI standards are contained in a set of International Standards Organisation (ISO) documents and the reference numbers of some of the principal ones are given in Section 22.3.6. Much of the work on OSI standards was done by CCITT in cooperation with ISO and where this was the case parallel documents in the CCITT X. and F. series are available but only the X. Series ones are listed in Section 22.3.6. These standards only describe *what* the protocols are, they do not describe *how* the OSI protocols are implemented nor how well they perform when they are implemented.

In the next section the seven layer model is described rather briefly; a full description would fill a book on its own, and the final section outlines some of the services that OSI supports.

8.2 OSI PRINCIPLES

The 7-layer model is essentially one of communication between computers (machines) rather than between human operators. This does not mean that humans

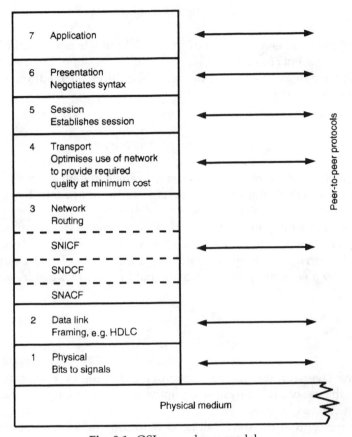

Fig. 8.1 OSI seven-layer model

cannot be direct parties to a transaction but it does mean that a man–machine interface has to exist and has to act as a buffer between the human and a computer based process that is a recognised part of the OSI model.

The division of the model into seven layers implies (a) that each layer performs a specific function, (b) that it does so in isolation from other layers and (c) that communications take place between the same layer numbers at both ends of the connection (this is called peer-to-peer communication). The top, or 7th layer, interfaces to the application processes and represents the source and ultimate destination of an information transfer.

8.2.1 The layers of the OSI model

The following is a description of the physical medium and the seven layers of the model which is mainly in terms of what the layers do. This is followed in Section 8.2.3 by a description of how the layers work. We move upwards from the foot of Figure 8.1.

The physical medium

The physical medium is not a layer of the model but sits underneath Layer-1 and provides a transparent communication path to other systems. Wiring is an example of a physical medium that might be used. The physical medium is sometimes unofficially called 'Layer-0'

Layer-1: The physical layer

The main function of the physical layer is to convert the bit stream received from Layer-2 into a physical signal, e.g. a voltage to be applied to the physical medium. The physical layer also carries out the synchronisation at each end of the link by setting the timing of the transmitted signal and extracting timing from the received signal. The peer-to-peer protocol between the physical layers in each end system can be carried by an out-of-band signalling channel, or by out-of-channel signalling or by a separate pair of wires. Signalling employs a standard line code, e.g. HDB3.

Layer-2: The data link layer

The data link layer controls the operation of a data link, i.e. one or more physical media and the associated physical equipment (e.g. modems). A link does not contain switching.

The main function of the data link layer is to provide a basic framing structure which is used by the network layer. (A frame reference is essential to enable the receiving end of the link to determine which bit is which.) The frame is delineated by a unique flag, and data from the network layer (above) is augmented by bit stuffing to ensure that it does not contain the bit pattern used in the flag. In addition the header added by the data link layer may contain:

- a logical channel number to enable a single physical medium to be time division multiplexed;

- flow control information to control the flow of information to the opposite data link layer to ensure that the storage at each end does not overflow;

- a mechanism for separating control (signalling) data from user data;

- error detection and correction information.

The service provided to the network layer is reliable (i.e. each transmission is acknowledged though not error free) and transparent (i.e. any pattern of data can be sent). The most common form of protocol for the data link layer is High Level Data Link Control (HDLC).

Layer-3: The network layer

The function of the network layer is to establish, maintain and release point to point connections between Network Service Access Points (NSAPs) which are globally unique addresses used by the transport layer. In terms of OSI there is only one global network which embraces all individual networks. Particular networks such as the international public packet switching network are called sub-networks in OSI terminology. The network layer maintains the quality of service requested by the transport layer, notifies the transport layer of errors, controls the sequence of data packets, and carries out flow control.

The network layer itself is subdivided into four sub-layers. From the bottom upwards they are:

- Sub-Network Access Control Function (SNACF), which obtains an end-to-end service across a given sub-network, e.g. one or more interconnected packet switched networks. For example X.25 is run from this layer.

- Sub-Network Dependent Convergence Function (SNDCF) which is concerned with the mapping of a particular type of network service onto a specific sub-network.

- Sub-Network Independent Convergence Function (SNICF), which is concerned with the allocation of a messages to the appropriate type of network service, and

- Routing function, which establishes and clears a virtual call between the NSAPs by selecting the necessary sequence of calls across particular sub-networks and determining the numbers (e.g. X.121 numbers) to be used for the sub-network calls. (Surprisingly perhaps, some of the more basic work on routing is incomplete and this topic is discussed further in Section 8.2.4.)

Each sub-layer adds its own header to execute its peer-to-peer protocol. Links within a given sub-network are connected by a sub-network node e.g. a packet switch (see Figure 8.2). Links between different sub-networks (e.g. a gateway) are connected by a network relay which extends up to the routing function sub-layer. Thus the layers up to the top of the network layer may in practice communicate with layers in intermediate nodes or relays and not directly with the equivalent layers in the distant end system. However, above the network layer all communication is with the distant end user and there are no equivalent layers in an intermediate system.

The boundary between a private network and a public network or between terminal equipment and a public network rarely occurs at the main network-to-transport layer boundary and it is because this interface is usually within the network layer that it has been necessary to define so many sub-layers.

Layer-4: The transport layer

The function of the transport layer is to optimise the use of the network service to provide the required quality of service at minimum cost. The transport layer offers five classes of service:

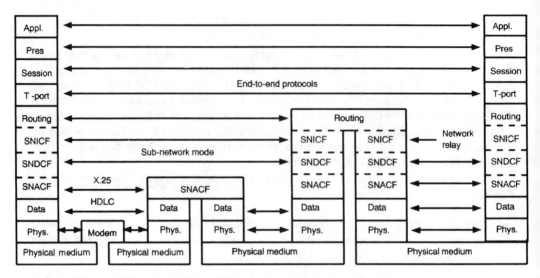

Fig. 8.2 End-to-end connection showing peer protocols

Class 0 (Basic class): No error correction and no multiplexing, class 0 applies when the network connection offers an adequate class of service.

Class 1: Error recovery but no multiplexing. Gives recovery from disconnects which are signalled by the network but does not handle unsignalled disconnects.

Class 2: As class 0 but with multiplexing.

Class 3: Error recovery and multiplexing (Class 1 + Class 2).

Class 4: Signalled and unsignalled error recovery. Multiplexing.

The transport layer is the highest layer which is concerned with the communication medium; the layers above are concerned with the processes that are communicating. The transport layer is potentially capable of relieving the user from having to choose which of a number of potential PTOs should carry a message. The layer can for example be programmed to look up tariff tables to determine which carrier offers the cheapest service for a given route length at a given time of day. To save cost it can multiplex messages (for classes of service of 2 and above) from higher layers onto a common bearer at the network layer.

Layer-5: The session layer

The session layer provides dialogue coordination by establishing, synchronising and releasing the communication session between the communicating processes in the end systems. It determines the mode of operation, e.g. full duplex, half duplex, etc.

For example, the session layer determines whether the transaction is in a conversational mode, i.e. one in which the far end is likely to respond with an answer to an outgoing message or a store and forward mode (e.g. with a message handling system) in which a reply is not part of the protocol.

Layer-6: The presentation layer

The presentation layer is concerned with the method of coding; it negotiates the syntax or language for the communication with the application layer from the list available, or defines the syntax by reference to a common language such as Abstract Syntax Notation (ASN) defined in ISO 8824 and ISO 8825 (X.409).

It should be noted that ASN is both the only syntax or common language for the presentation layer available at present, and a means for defining the transformation from the local syntax or language of the application layer to the concrete syntax (itself ASN) of the presentation layer.

The Presentation Layer might for example negotiate the choice of international alphabet to be used for the content of the message. When, for a particular service, the syntax is fixed, the presentation layer is a null layer.

Layer-7: The application layer

The application layer is the highest layer and the source and ultimate consumer of the OSI services. The application layer allows access from outside OSI to specialised communications services (e.g. the Message Handling System (MHS), see below and in Section 17.5.3, p.305) that are OSI layer-7 standards. The specific software that supports these communications services resides entirely within the application layer. A software interface external to or above the application layer (e.g. an Electronic Data Exchange (EDI) interface to MHS) can take any form; it is not currently subject to OSI standardisation, although in the interests of application portability there is pressure for standardisation in this area and an association called the Application Program Interface Association (APIA) has been formed to further that end.

Depending on the application, the user will pick one of the established communication services or create his own special service. For example, with Electronic Data Interchange (EDI) (see Section 17.5.3, p.308) the current view is that the MHS provides a passable means of transport in the majority of cases but the FTAM might be better in a minority of situations. In the longer term the EDI requirement will result in its own specialised service or services and an MHS service variant (X.435) is under development. The EDI application software, concerned with such things as invoice formats, although standardised by ISO committees, is not part of OSI.

Within the application layer itself there is a need for internal interfaces (see Figure 8.3): on the upper side between a communications service such as MHS or File Transfer Access and Management (FTAM) (see Section 8.3.3) and on the lower side with the presentation layer. On the upper side the interface is provided by Application Service Elements (ASEs) which are mapped onto the presentation layer by three types of operation, each with its own service elements. These operations in the 1988 version of OSI are:

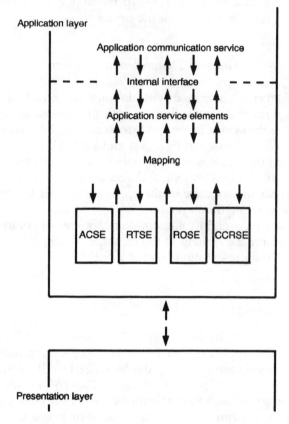

Fig. 8.3 Application layer internal organisation

- *Remote operations.* The Remote Operations Service Element (ROSE) is for asymmetric request–response interactions between end systems,

- *Reliable transfer.* The Reliable Transfer Service Element (RTSE) is for symmetric and asymmetric information transfer in which safe storage at the destination is ensured.

- *Association control.* The Association Control Service Elements (ACSE) provide a means by which all application associations are established, released and managed.

- *Commitment, concurrency and recovery.* The Commitment Concurrency Recovery (CCR) service elements are a means by which activities distributed across several open systems can be completed in the face of failures of individual transactions.

Because the establishment of these services elements lagged behind the standardisation of the lower layers, a number of equivalent proprietary internal application layer interfaces have been developed, but one might expect these to be phased out in time.

8.2.2 Distribution of multiplexing and error control between model layers

As examples of how what may seem to be one function can in fact be distributed throughout the layers of the model we consider multiplexing and error control, but similar considerations apply in a number of other cases.

Multiplexing can be carried out:

- At the transport layer, classes 2, 3 or 4, by mapping different application sessions onto the same virtual circuit. This virtual circuit is between two NSAPs provided by the network layer.

- At the network layer, by multiplexing different virtual circuits identified by different logical channel numbers onto the same data link.

- At the data link layer, by multiplexing different links onto the same bit stream.

- At the physical layer, by multiplexing different bit streams onto the same physical medium by frequency or time division multiplexing.

Error control can be carried out:

- At the application layer by data checking, e.g. a balance for funds transfer.

- At the transport layer with class 4, but note that the flow integrity provided by classes 1 and 3 is also a form of error control.

- At the network layer (but not with X.25).

- At the data link layer, using for example a cyclic redundancy check.

8.2.3 Internal layering concepts

Each layer provides services and protocols. To understand fully how the 7-layer model works, one must understand the internal working of an individual layer (see Figure 8.4). It does not matter which layer we consider, the principles are the same. Consider the Nth layer in each of two end systems which are communicating with each other. The Nth layer in the sending system provides a service to the $(N+1)$th layer. It receives instructions from that layer and executes those instructions by sending peer-to-peer protocol messages to the Nth layer in the receiving end system, and by using the services provided to it by the $(N-1)$th layer.

However, the Nth layer is not able to communicate directly with the Nth layer in the other system because it is not connected directly to it, therefore the peer-to-peer protocol messages take the form of headers which are added to the data which are to be sent. Each packet of data to be sent which is passed down from the $(N+1)$th

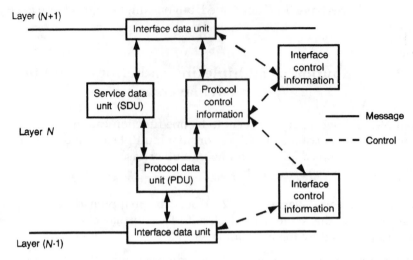

Fig. 8.4 OSI layering concepts

layer is called a Service Data Unit (SDU) and the instructions about the service to be provided are called the Interface Control Information (ICI). The Nth layer uses the ICI and its in-built Protocol Control Information (PCI) to form the header containing the peer-to-peer protocol messages which are added to the SDU.

The header and the SDU together form the Protocol Data Unit (PDU). The PDU is then passed to the $(N-1)$th layer together with instructions on the service to be provided which take the form of new ICI. The SDU of the Nth layer then becomes the SDU of the $(N-1)$th layer.

Thus the data which are passed across the physical medium below the model are a concatenation of the original data from the 7th layer application process and the headers which have been added at each layer. Received packets pass through the Nth layer in the opposite (upward) direction in a similar manner except that information is removed from the packets as they are passed up the hierarchy.

The OSI model and its associated standards define the format of the headers and the logical form of the services and instructions which pass between layers within each end system. The physical form of these services and instructions does not need to be defined because it is not seen by the other system and so the services between the layers can be implemented in different ways by different manufacturers and programmers.

8.2.4 Routing and the connection and connectionless modes

While in general the parameters of a layer are independent of those other layers, in the case of connection and connectionless modes of operation there is some dependency.

- *Connection oriented mode* The path through the switched network is established for the duration of the communication and then released.

- *Connectionless oriented mode.* The destination of the information is contained within a message (e.g. a packet) header and a path through the network is not reserved. In this mode of operation the unit or packet of information is sometimes called a 'datagram'.

The connection oriented mode is normal in Europe, particularly for public services, whereas connectionless oriented mode is common in North America and for LANs generally. Note that PSS (X.25), while it employs packets, is connection oriented because a virtual path is reserved through the network.

The mode of operation chosen for the transport layer is reflected in the three layers above, i.e. session, presentation and application, but the choices of mode for the network and data link layers can be different from that of the transport layer and also can be different from each other. The physical layer is inherently connection oriented.

The function of routing between end systems is carried out at the network layer and the standardisation process for this is still not complete. Obviously there are currently means of routing but there is a need to establish general-purpose routing protocols that will work between end systems through a variety of intermediate networks. Work here is more advanced for the connectionless mode for which there are standards and draft standards but it is not clear how long it will be before corresponding standards for the connection mode become available.

8.2.5 TCP/IP

As mentioned in the introduction, Transmission Control Protocol/Internet Protocol (TCP/IP) provides an alternative set of protocols to those of OSI for data transmission in wide area networks. TCP/IP is not based on a 7-layer model but can be approximated to correspond to layers 7, 5/4, 3 and 2 of the OSI model and does not cover layer 1. TCP provides a service equivalent to connection oriented class 4 transport service (TP4) which runs over a connectionless network using the very simple connectionless, IP, protocol. This form of of connectionless (at the network level) packet (or datagram) operation is well suited for interfacing to IEEE 802.3 (Ethernet) LANs. In some instances TCP/IP may be used in combination with higher layers of the OSI model.

The wide acceptance of TCP/IP and a performance which is acceptable and not significantly inferior to that of TP4 makes it a serious competitor to OSI to the point where a version of TCP/IP might be put forward for consideration by ISO as an addition to the OSI standards.

8.3 OSI IN PRACTICE

A number of other chapters in this book describe network services that use or can use OSI protocols for end-to-end communication and it is not appropriate to repeat

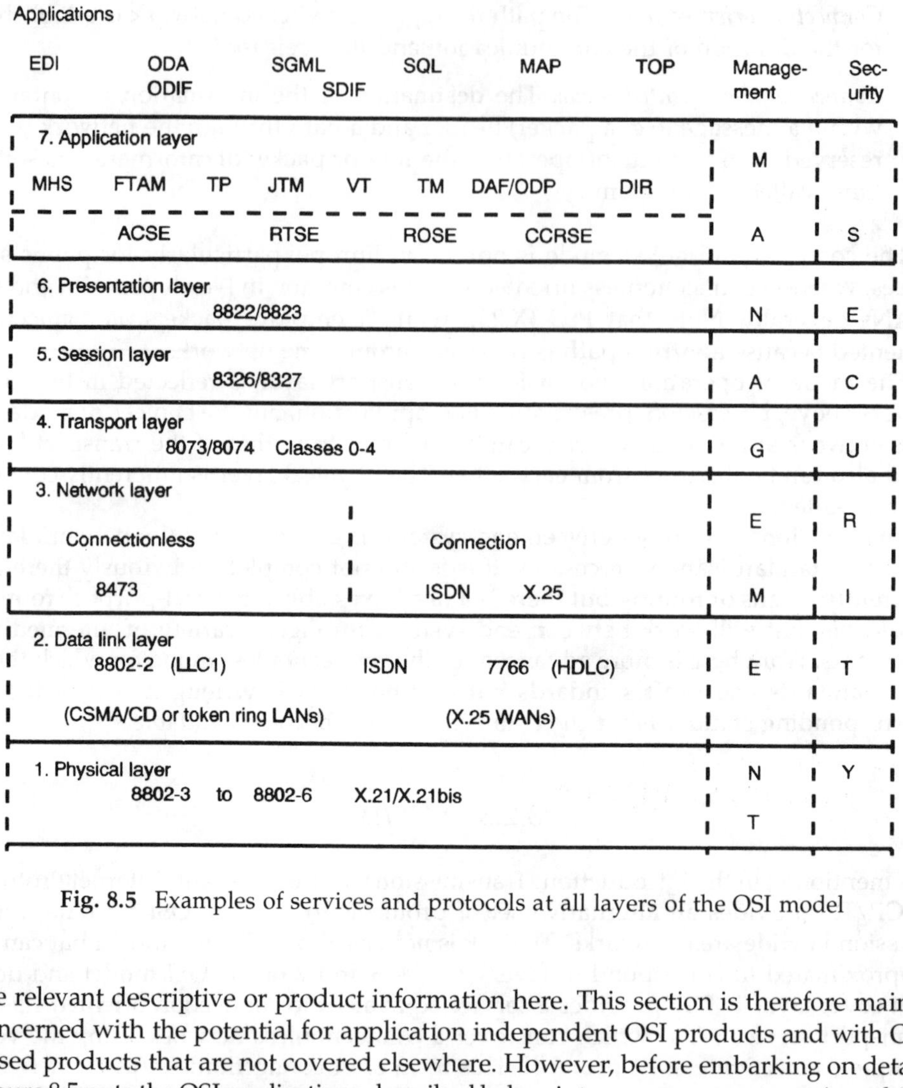

Applications

EDI	ODA ODIF	SGML SDIF	SQL	MAP	TOP	Management	Security

7. Application layer						M	
MHS FTAM TP JTM VT TM DAF/ODP DIR							
ACSE RTSE ROSE CCRSE						A	S
6. Presentation layer 8822/8823						N	E
5. Session layer 8326/8327						A	C
4. Transport layer 8073/8074 Classes 0-4						G	U
3. Network layer — Connectionless | Connection — 8473 | ISDN X.25						E M	R I
2. Data link layer — 8802-2 (LLC1) ISDN 7766 (HDLC) — (CSMA/CD or token ring LANs) (X.25 WANs)						E	T
1. Physical layer 8802-3 to 8802-6 X.21/X.21bis						N T	Y

Fig. 8.5 Examples of services and protocols at all layers of the OSI model

the relevant descriptive or product information here. This section is therefore mainly concerned with the potential for application independent OSI products and with OSI based products that are not covered elsewhere. However, before embarking on detail, Figure 8.5 puts the OSI applications described below into a common perspective which should prove useful when reading the following sections. There is no significance in the vertical relationship between the items in the individual layers of Figure 8.5, but a significant relationship is intended in the example of a profile given in Figure 8.6. In both figures references are made to the ISO document number (see Section 22.3.6) or other specification descriptor if the latter is more widely known.

8.3.1 Application independent products

Some suppliers produce products that are unbundled from an application and provide the necessary support for specific upper layers or a set of specific upper layers. Such products could include:

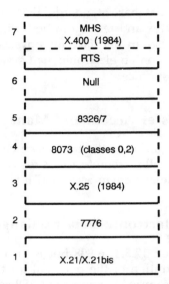

Fig. 8.6 An example of a profile for a message handling system using a wide area network

- application layer support services and interfaces;
- presentation layer;
- session layer;
- transport layer for specified classes of service.

The demands placed on these layers do vary depending on the application, for example MHS (X.400 (1984)) does not need a presentation layer, and hence a general-purpose product may be somewhat redundant. It is clearly most attractive when end systems support a number of applications calling upon a variety of layer 4–7 facilities.

The various OSI applications are dealt with comprehensively in later chapters; what follows is a brief outline of the major applications and the relevant references to other parts of this work.

8.3.2 Message Handling System (MHS) X.400

MHS is the first OSI application to have received widespread acceptance; it receives treatment in Section 17.5.3, p.305, with particular reference to of the Interpersonal Message Service (IPMS). MHS is fast becoming the main means of inter-communication between the various E- Mail services which are described in Section 17.5.3, p.305.

IPMS is only one of a number of services MHS is capable of supporting; EDI is discussed in Section 17.5.3, p.308, and other possibilities are the various document interchange and relational database standards. IPMS provides interfaces to non-OSI services; specifically telex and teletext in both directions and fax in the outgoing direction. There is also an interface for physical delivery; this would for example enable persons not on an electronic mail system to have their messages delivered to the nearest post office.

8.3.3   File Transfer Access and Management (FTAM)

This service is designed to facilitate the transfer of computer files between non-colocated computers and to allow remote access to, and the management of, computer data bases. Further details are in Section 17.6.1.

8.3.4   Electronic Data Exchange (EDI)

EDI is considered in Section 17.5.3, p.308 but is also discussed in Section 8.2.1 (layer-7) above as an illustration of the potential for the layer-7 communication services, MHS and FTAM, to provide support to EDI as an end application.

8.3.5   Transaction processing

In Transaction Processing (TP) one end of the connection is usually a human being operating a keyboard and the other end is a machine, and hence it finds application in enquiry and reservation systems. Section 17.5.3, p.309 contains further information.

8.3.6 Job Transfer and Manipulation (JTM)

Remote data entry includes the input of pure data and commands to carry out specific tasks such as batch processing. At the procedural level the ISO OSI Job Transfer and Manipulation (JTM) standardisation activity is relevant and should reach full completion by the end of 1991. Most of such activities are likely to be in-house and therefore represent traffic on private networks, and the need for any sort of public service appears distant. Compared with other OSI applications JTM has received very little product support but products are available. It is not treated further here.

8.3.7   Support to documentation interchange and data base access

Standardisation of Office Documentation Architecture (ODA) (ISO 8613) and the Standardised Graphical Mark-up Language (SGML) (ISO 8879) provides a common means of organising and expressing the contents of documents, but there is in addition the need for a 'service' to enable such documents to be conveyed between users. The Office Documentation Interchange Format (ODIF) provides a set of communication protocols for ODA.

Standardisation of (relational) data base access is likely to be based on the Standard Query Language (SQL) which is the subject of standardisation by the American National Standards Institute (ANSI). For SGML and SQL the transport protocol is expected to be the Standardised Documentation Interchange Format SDIF (ISO 9069).

The service used for conveyance for all three standard services might be MHS or FTAM (or a variant of one of them) with the emphasis on MHS for short transactions and FTAM for long ones. ODA/ODIF and SGML/SDIF products are available. Some further information can be found in Section 17.5.3, p.307.

8.3.8 Virtual Terminals(VT)

The Virtual Terminal Protocols (VTP) that provide the means of data entry or access from a remote terminal to an application in a host computer are also the subject of OSI standardisation. (In principle VTP can also be used between two terminals or two applications.) VTs are treated further in Sections 9.2 and Chapter 17.6.2.

8.3.9 Manufacturing and design protocols

The need to communicate manufacturing and design information has led to the development of two OSI based standards. Manufacturing Automation Protocol (MAP) covers the conveyance of manufacturing information either within or between organisations. Technical Office Protocol (TOP) is more concerned with design information. Further details are given in Section 17.6.3.

8.3.10 Terminal Management (TM)

Terminal Management (TM) is concerned with the control of many simultaneous dialogues between terminals or between terminals and network servers. The dialogue itself may be in terms of, for example, the VT standard (Section 8.3.8 above) or in a non-OSI format such as videotex or teletex. The managed equipment can of course be located anywhere and the managing processes are partly distributed but need one centralised coordinating process called a Terminal Management Control Process (TMCP) which must be centralised.

It will be the end of 1993 before the current standardisation activity is complete and therefore unlikely that products will be available much before that date. This topic is not treated further here.

8.3.11 Distributed Application Framework (DAF) and Open Distributed Processing (ODP)

In the early OSI work the emphasis was on communication between a pair of end systems, and FTAM and VT for example are based on this premiss. To cover the case where there are more than two end systems in simultaneous communication, work has started on a Framework for the support of Distributed Applications (DAF). This work is under the auspices of CCITT with ISO cooperation and output is expected in 1992. The resultant recommendations will be published in the X.900 series and where appropriate also as ISO standards.

The scope of ISO's work on Open Distributed Processing (ODP) is somewhat broader and more ambitious than DAF and to a longer time scale; substantive output is anticipated in 1993/94. This work could be of major importance, comparable with OSI itself. The basic concept is one of a multiple-client–multiple-server model in which clients, users of local applications, can request services from remote servers, for example relational data bases. Clearly the standardisation of the protocols to enable this to be done is the key to future success and the Remote Procedure Call (RPC) is likely to become the most common mechanism for interworking. Many of the OSI services discussed here will become candidates for inclusion in ODP interface standards. There is a UK DTI initiative, ODSA (Open Distributed Systems Architecture), which is contributing to the ISO standardisation activity. There are, however, several proprietary architectures competing with ODP and it remains to be seen how widely ODP will be accepted.

A major commercial driving force for DAF and ODP arises from the widespread adoption of PCs; clearly the capabilities of a PC can be much enhanced if it has access to information outside of its directly accessible memory.

8.3.12   Directory service

Directory service is an important OSI application particularly in the immediate context of MHS; it is treated further in Chapter 21.

8.3.13   Network management

OSI network management has lagged behind most of the other OSI work but a clear picture of how it works is beginning to emerge and the subject is treated in Chapter 26.

8.3.14   Security

The OSI security architecture is described in principle in ISO 7498/2 which was published in 1988 but very little progress has been made on any detailed implementation until comparatively recently and this work initially at least is concentrating on layers 3 and 4. In the meantime various security layer-7 protocols that are specific to an OSI application have been or are being developed. The topic is treated further under network management in Section 26.2.

8.3.15 OSI legislation

Governments have put money into developing and promoting OSI and some governments make it mandatory for bodies in the public sector to specify IT equipment that conforms to OSI standards. In the UK and the USA the government procurement specifications are both called 'GOSIP' (Government OSI Procurement), while the EC use the acronym 'EPHOS' (European Handbook for Open Systems). There is an EC directive (88/295/EEC) which demands standards conformance for a wide range of equipment procurements but each EC member country will have to implement this directive by passing its own legislation.

In the USA GOSIP is mandatory for central government agencies but each state can legislate as it wishes. However the need to communicate with central government puts considerable pressure on States to conform to the GOSIP specifications.

In 1992 it is too early to say how effective these policies are and it will take several more years for a clear position to be established.

8.4 THE FUTURE FOR OSI AND OPEN SYSTEMS ENVIRONMENT—THE NEXT STEP

OSI products have taken a long time to reach the market place, but there is every reason to believe that the pace of their acceptance is accelerating. The number of applications and the number of bearer or telecommunication services supported under the OSI heading is continually increasing and the base standards continue to be refined and extended. An example of this trend is the ESPRIT Y-NET project which is a four year project aimed to provide OSI services to ESPRIT participants and other European Community research and development workers. X.400, FTAM and directory services are included.

One may expect to find OSI software products not only running on computer mainframes but also on PCs and work stations. Certainly when planning for data services on private networks or for customer premises equipment one needs to consider what OSI can offer, particularly in terms of an ability to inter-work with other business organisations both in the UK and world-wide.

The Open Systems Environment (OSE) concept is gaining momentum in Europe and the USA. OSE extends open systems beyond the OSI reference model to provide greater portability and interoperability of applications software. This is an extension away from the world of telecommunications towards that of IT but it will undoubtedly have repercussions on OSI because of the ever-increasing breadth of the information that telecommunication networks have to carry and the new interfaces that arise from these needs.

8.5 BIBLIOGRAPHY

The list of references to the relevant standards is given in Section 22.3.6; the works cited below are of a general nature.

Availability of standards

The British Standards Publication '*BITS*' contains from time to time lists of standards documents of ISO, European and British origin that are available from BSI. OMNICOM International Ltd also provide a documentation service.

Product information

The DTI Central Computer and Telecommunications Agency (CCTA) produce comprehensive OSI product reports from time to time, early ones were published by HMSO but more recently they have been published in loose-leaf form by Blenheim Online Publications.

OSI technology updates

Technology Appraisals Ltd publish *OSN: The Open System Newsletter* every month which contains tutorial material, standard statuses, and new product information.

9 TERMINAL EQUIPMENT

9.1 INTRODUCTION

This chapter includes more than terminals at the network periphery (i.e. the man–machine interface such as the telephone) and encompasses terminals associated with the transmission network that may or may not be owned by the user depending on their position in the network. The choice as to what is covered is strongly influenced by the impact of VLSI as a technology which has led to the treatment of the following topics:

1. Terminals that form the user man–machine interface if only because the investment in them is becoming an increasing proportion of the total network cost and because of the many new facilities such terminals will offer.

2. Digital filters because of the impact they are having on the design of terminal equipment, particularly in the mobile radio field.

3. Adaptive equalisers because they are particularly important where echo cancellation is needed.

9.2 THE USER INTERFACE

There is a very wide range of terminals that compose the man–machine interface ranging from the simple telephone through 'feature phones' and work stations to more special-purpose terminals such as radio phones, facsimile machines, telephone answering machines and printers. The emphasis here is on terminals operated by humans in the office or home.

Twelve years ago the majority of network terminals in the UK were simple telephone handsets selling probably at less than £30 at today's values. That situation is rapidly changing and terminals are increasingly becoming complex and multipurpose. While the telephone terminal is still to a large extent divorced from terminals concerned with data that situation is likely to change too and the ICL One Per Desk concept (in which the telephone is combined with a display, keyboard and microprocessor) started the trend here. Equipment manufactures have not been slow to realise that with the decreasing cost of switching and transmission, the investment in the network is increasingly in the local area and in terminals. The acceptance by the customer that a sophisticated terminal gives value for money means that the investment in work stations could in the longer term be greater than in the rest of the network as a whole. Furthermore the average life of a terminal is probable less than that of other network components, giving considerable scope for a replacement market.

The main components of future terminals are:

- the telephone transducers;

- the display;

- the keyboard;

- the processor support;

- the housing, power supplies and general support.

Little need be said about the telephone transducers except that the technology has improved markedly over the last fifteen years and the carbon microphone is certainly obsolescent if not obsolete. Loudspeaking features are of course becoming increasingly common. The introduction of high quality speech services will mean improved transducer performance and the ability to change the handset's overall performance when switching between current PSTN quality and high quality. The realisation of suitable designs is not likely to be inhibited by the need for advanced technology.

The display is an area where significant technical advances may be confidently anticipated arising from the replacement of the Cathode Ray Tube (CRT) by a flat panel display, probably based on liquid crystal technology. Liquid crystal displays have already received acceptance in portable computers but have not yet matched the resolving power of CRTs. What is required in the future is an A4 display with the order of 10^6 pixels and so far only in the laboratory have semectic liquid crystal displays achieved an equivalent pixel density. Most existing liquid crystal displays use nematic liquid crystals which are limited in pixel count by the need to refresh pixels using multiplexing techniques. Semectic liquid crystals do not need refreshing and therefore show considerable promise for future high pixel count displays. (The situation is analogous to that of dynamic and static RAM except that dynamic RAM does not run up against the refresh timing problem that limits the performance of nematic liquid crystals.) The next step would be to make coloured liquid crystal displays and the techniques for doing this are again available in the laboratory. The disadvantage of flicker and the purported health hazards associated with CRTs would then be obviated and one would have a smaller, lighter, potentially much more portable terminal—perhaps one that could be taken home in a briefcase.

Liquid crystal displays are inherently formed of pixels and these can be addressed and changed individually, whereas for a CRT a scanning process is involved, which means that to change one pixel the whole display has to be rewritten. The discussion of picture coding in Section 6.3.8 shows that the ability to change pixels individually is a distinct advantage and suggests that liquid crystal displays are well suited to video telephones.

A few years ago it was confidently stated by certain telecommunication marketing men that a typewriter (QWERTY) keyboard was quite unacceptable for a product intended for telecommunication use. History is proving them wrong and the use of a QWERTY keyboard plus functional keys and multipurpose key overlays is now widely accepted. However it would be premature to expect any startling technological developments in this field. The inhibiting effect that a limited number of telephone keypad buttons has on the provision of additional features is perhaps not generally realised and no longer applies with the more user friendly interface of a keyboard and display. The implications of this will become apparent in later chapters.

A wide variety of Optical Character Recognition (OCR) systems are available for document reading. In order to recognise printed characters it is necessary to have hardware in the form of a document scanner which turns the printed information into an electrical signal and software that processes that signal and recognise the characters. There is a wide choice of both hardware and software with hardware prices ranging from about £600 to over £10 000 and software from about £250 up to several thousand pounds. Some software is constrained in that it can recognise only a limited set of fonts and point sizes. All systems are prone to errors which can be significant for long documents. For facsimile (fax) reception it would be particularly advantageous to convert incoming documents into word processor compatible form using an OCR; however the quality of the received characters after transmission over the PSTN often leaves much to be desired and the resultant error rate may be unacceptable.

Speech recognition machines have been on the horizon for the last 30 years, and there are at last beginning to be products available that are commercially viable. For simple control functions (e.g. to replace dialling), a limited vocabulary of a few tens of words is all that is necessary and for a relatively low price there are machines on the market that are either speaker independent or have to be trained to recognise an individual or limited set of speakers. One product, a dictation system, claims a vocabulary of 30 000 words, which is for most purposes a substitute for a keyboard. It is however currently many orders of magnitude more expensive than a keyboard. Speech recognition involves complex algorithms and the resultant machines rely heavily on VLSI, memory and microprocessor technology. Some systems echo back words as they are recognised via a speech synthesiser to the speaker and this enables overall system performance to be improved because an error correction routine can be entered when a word is recognised incorrectly.

It is a *sine qua non* that, other than for the most basic unit, there will be at least one microprocessor per terminal. More important than the hardware configuration, however, is the software that runs on the terminal and the provision of a so-called user friendly interface. The fashionable concept here is called Windows, Icons, Mouse and Pointers (WIMP). Windows allow information to be overlaid on the screen, much as you can place several pieces of paper in front of yourself on a desk; Icons are ideographs or simple compact pictorial representations of features available to the user; a Mouse is a means of selecting and moving a Pointer to the object you want implemented. (A mouse is an easier to use alternative to the four shift keys and the entry key on the keyboard.) For example, for telephone service you might move the mouse so that the pointer points to the telephone icon and press a button. A considerable obstacle to progress in this area is the widespread adoption of what are generally known as IBM PCs, which include those made by IBM and copies made by a large number of other manufacturers. The MS-DOS operating system of the IBM PC does not lend itself particularly well to the WIMP, concept and special programs such as Windows which provide the WIMP features took some time to develop. In the meantime the Apple Macintosh range of PCs which are derived from the original work undertaken by the Xerox Corporation on the problem of a good user interface had established a significant share of the PC market. The ideal terminal should be sufficiently self-explanatory that it does not need an instruction book; the reality of the existing situation is that some PC systems not only need an instruction book, they need another book to explain the instruction book!

The need for standardisation of the user–terminal interface led to ISO developing Virtual Terminal Protocols (VTP) (see Section 8.3.8) which were intended to include graphics and raster/photographic images as well as text either separately or in a combined class. Only the basic class which handles text has been standardised and it is now proposed to integrate X-Windows, which is the subject of ISO standardisation in its own right, into the VTP standard.

The housing, power supplies etc. are only important in so far as they represent a minimum irreducible cost below which, whatever the advances in display and computer technology, the cost of the terminal cannot fall.

Functionally a terminal should permit easy access to all office and telecommunication facilities and allow information to be moved freely between them. We shall see in Section 17.5.3, p.307, how various standards that are currently being established should provide the framework that can enable this to happen.

9.3 DIGITAL FILTERS

9.3.1 Introduction

Filters are essential in the selection of one signal from others in a medium where the available spectrum is shared on a frequency division basis and filtering techniques are being revolutionised by digital filtering.

Digital filtering is not so much a technology as a technique; however, it is a technique tied to VLSI technology. Filtering of signals is an extremely common requirement in telecommunications; for example, in order to conserve bandwidth in radio systems or FDM line transmission systems, the speech signal is confined to cover a range from 300 Hz to 3.4 kHz. Because the microphone in the telephone handset responds to frequencies outside that range, a band-pass filter is needed so that the speech signal can confined before being transmitted over a limited bandwidth channel.

9.3.2 Competing technologies

Until fairly recently filters have been constructed in analogue form using inductors, capacitors and resistors. Such filters are usually less than ideal mainly because it is not possible to make inductors of zero resistance. One way of avoiding the use of inductors is to employ operational amplifiers as components in what are known as active filters. Because VLSI is more suited to digital than to analogue circuits and because the telecommunications environment is becoming increasingly digital, digital filtering is now in general a more attractive proposition than analogue filtering.

9.3.3 Digital filtering principles

The use of digital techniques implies that the signal is sampled and quantised and that the samples can be treated as binary numbers. These binary numbers can be handled in exactly the same way that a general-purpose digital computer handles numbers. For filtering it is necessary to be able to carry out the following basic operations on the sampled signal:

- multiply/divide;
- add/subtract;
- delay.

With these three operations it is possible to emulate any filter that can be constructed from inductors, capacitors and resistors without the imperfections normally to be associated with those components. However precision does depend on the sampling rate and the number of quantisation levels and the filtering operation must be performed within a time limit; hence precision is a function of the available processing power. Add and multiply can be performed in arithmetic units and delay simply by storage and retrieval from memory; thus filtering can be performed in the central processing unit of a digital computer and it is perfectly feasible to carry out filtering operations using an off-the-shelf microprocessor. However because speed is usually of the essence, most digital filters are realised using special purpose VLSI in configurations tailored to the requirements of one particular filter design.

9.3.4 Transversal filters

One of the commonest forms of digital filter is the transversal filter, one form of which is as shown in Figure 9.1. It contains (conceptually at least) digital delay lines along which successive speech samples travel under the control of a clock signal. At each section a weighted tap takes the output of its section to be summed at a common point. The two sections (a) and (b) are connected in cascade to form two polynomials, these being the numerator and denominator respectively of the filter transfer function, which can be represented by

$$G(z) = \frac{a_0 + a_1 z^{-1} + \ldots + a_n z^{-n}}{1 + b_1 z^{-1} + \ldots + b_m z^{-m}} = \frac{X(z)}{Y(z)}$$

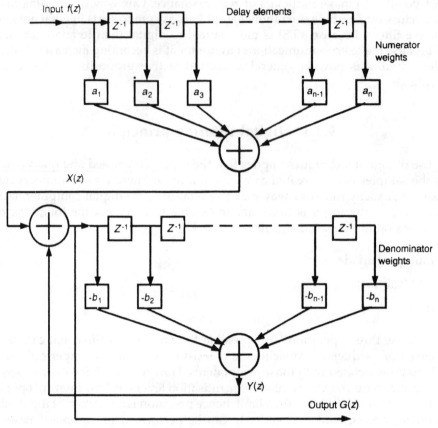

Fig. 9.1 Transversal filter

say, where z^n represents a delay of n sampling time units.

The assignment of values to the coefficients $a_0, a_1, \ldots a_n$ and $b_1, \ldots, b_m$ determine the transfer characteristic. The numerator determines the zeros, or anti-resonances, in the transfer function and the denominator determines the poles , or resonances, in the transfer function. A particularly common variant has the denominator of the transfer function effectively set to one and is known as a finite (duration) impulse response filter or an FIR filter in contra distinction to an IIR (infinite impulse response) filter. The FIR filter is unconditionally stable whereas the IIR is not. The term 'infinite' is used (perhaps 'potentially infinite' would have been better) because if the denominator is zero then $G(z)$ becomes infinite, the filter becomes unstable and its response lasts for ever.

9.3.5 Program control of digital filter characteristics

One immediate advantage of the digital approach is that the characteristic of the filter can be changed under program control simply by changing the values of the weights representing the polynomial coefficients. This can, if the processing power is available, allow one transversal filter to emulate a bank of filters on a time sharing basis.

Another important advantage of this programmability is that it is possible to make the filter an adaptive one; that is the characteristics of the filter can be changed to meet the particular requirements of an environment that varies in some way. For example, digital pulses when transmitted over a transmission system become distorted to an extent that depends on the number and quality of the links involved. This distortion is due the frequency and phase response of the links and can be compensated for by an 'equaliser' which in effect is a filter having a characteristic that is the inverse of that of the transmission medium. A fixed equaliser is unsatisfactory because of the variable nature of the distortion, but by sending at the start of transmission a known test pattern, the filter at the receiving end of the link can be made to compensate for the distortion of the transmission medium on a per-connection basis. We go on to consider equalisers more generally in Section 9.4 below. Note, however, that for an equaliser, because the form that the adaption of the filter will take is not completely known in advance, the unconditionally stable FIR filter is normally used in preference to the IIR filter.

9.3.6 Conclusions

Because of their flexibility and because they can be realised very compactly using VLSI, digital filters have already established a strong position and show promise of extending this position much further. Particularly in the field of mobile communications, system designs are being attempted which would be unthinkable without digital filter technology.

9.4 ADAPTIVE EQUALISERS

9.4.1 Introduction

There are a number of problems in telecommunications and elsewhere where it is desirable either to correct a signal or to remove an unwanted signal in the presence of a wanted one. Four specific examples where adaptive equalisation/cancellation techniques can be effectively used are in:

1. correction by equalisation of signals distorted unpredictably during transmission;

2. noise cancellation in aircraft cockpits;

3. echo cancellation on long distance communication links;

4. cancellation of unwanted signals arising from multipath propagation.

The second of these is not relevant to the current context but illustrates the widespread applicability of the basic technique. The essence of equalisation is to cancel the frequency distortion produced in another part of the system by frequency distortion of equal but opposite sign, and an example of this in the form of the equalisation in a transmission system is mentioned at the end of Section 9.3 above. Cancellation involves the injection of a signal in antiphase with the unwanted one so as to cancel it. The adaptive element comes in when that unwanted signal is not totally predictable and therefore different parameter values are required for the equaliser varying with time and/or circumstance. An essential component of all current equaliser designs is the adaptive digital filter as described in Section 9.3.5.

9.4.2 Unwanted signal cancellation

Echo cancellers

By way of an example it will be assumed that the echo is a delayed and distorted version on the return path of the original signal that was sent out on the forward path. In a long-distance speech connection a major source of echo is the hybrid that connects the two-wire transmission points at the ends of the circuit to the four-wire transmission path which is the major and middle part (see Figure 9.2). Echo cancellation can take place at the near end or, with the cooperation of the called party's administration, at the far end. Clearly there is a fair degree of autocorrelation between the echo and the original which is useful in determining the round trip delay (usually negligible at the far end but very significant at the near end). Once this delay is established the cancellation signal can be formed in terms of a delayed, attenuated and shaped (in terms of frequency spectrum) version of the original, so as to cancel the echo. Thus by in effect simulating the performance of

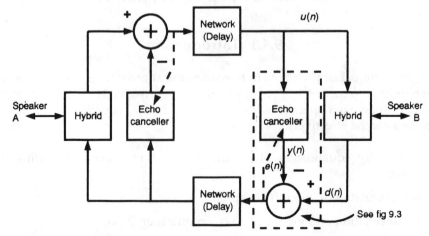

Fig. 9.2 Echo cancellers in an end-to-end speech connection (with far end cancellation)

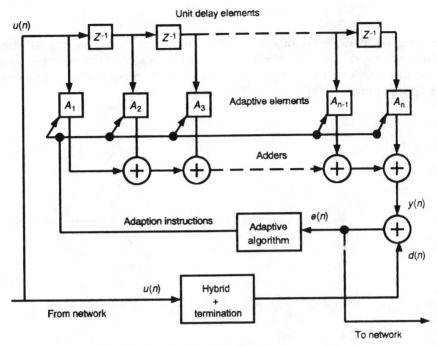

Fig. 9.3 An adaptive filter in an echo canceller

the transmission path to the right of the hybrid (speaker A talking to speaker B), a signal can be produced which is the inverse of the echo and can be used to cancel it. Figure 9.2 shows an arrangement for far end cancellation. The box marked 'echo canceller' contains the adaptive digital filter, which can take the form shown in Figure 9.3. At the establishment of a new connection the properties of the echo path are not known so there is a period during which the adaption process has to take place. The adaptive equaliser needs to converge in a matter of seconds to the best approximation it can give to complete cancellation. Normally the adaptive algorithm involves minimising the RMS error signal.

Multipath propagation

In the GSM pan-European mobile radio system the digital signal is subject to interference arising from multipath propagation and adaptive equalisers are proposed as a means of cancelling this interference. In this case because one end of the link is a vehicle that can be travelling at high speed, extremely rapid convergence times, possibly as short as a few tens of milliseconds, are necessary and a special framing sequence of bits is sent at the start of every block of speech information to provide a known signal which enables rapid convergence to be achieved.

Implementation

Estimating the parameters of the equaliser and performing the digital filtering operations is clearly calculation intensive, so that it should be no surprise by now to learn that a cost effective implementation depends on digital signal processing using special purpose VLSI.

10 BUILDING WIRING

10.1 INTRODUCTION

It is easy to underestimate the importance of building wiring in terms of the investment that it entails, the effect it can have on the performance of the equipment that it connects and the management that it needs. Because wiring installation is labour intensive and wiring components have a large material content the installed cost of wiring may be comparable with the cost of the equipment that it connects. The cost of installing large building wiring systems can run into millions of pounds but in this chapter the emphasis is on the technical and management issues that enable a wiring investment to be used most effectively.

In this context the term 'wiring' means the totality of an installation which includes various types of cable, e.g. twisted pair, coaxial and optical fibre, and components such as terminal blocks, connectors and patch panels that provide the flexibility points in a building. In an installation of any appreciable size, the importance of flexibility for rearrangement must be strongly emphasised. Until fairly recently most building wiring has been for analogue telephony only and a large body of practice has been built up in that context, but this will not be considered in any detail. The increasing introduction of digital services means that new wiring technologies need to be considered and fresh problems arise because of potentially undesirable interactions between services when they share wiring. These undesirable interactions are expressed in terms of crosstalk attenuation and much of the technical content of what follows is concerned with that topic, which assumes a much greater importance than hitherto. This chapter treats the topic from the point of view of the designer of a building wiring scheme and concentrates on novel technical issues but starts by providing some background on structured wiring, regulatory requirements and standards.

The emphasis is almost entirely on wiring carrying information at base band. There are a minority of situations where information may be distributed at carrier frequencies, using for example a CATV distribution system, but that falls outside the scope of this chapter.

10.2 STRUCTURED WIRING, REGULATIONS AND STANDARDS

Most building wiring has interfaces with one or more public telecommunication networks and the demarcation point between the public network and the private building wiring has to be rigorously specified for legal reasons. In the UK this point is called the Network Termination Point (NTP). In the majority of buildings with a PBX, the NTP is located on the Test Jack Frame (TJF) of the PBX, but for recently installed large installations it may be located elsewhere. In any event the TJF is most likely to be the hub of the building wiring and provides a flexibility point or main node at which changes to extension wiring can be made. Cables between buildings, or in multi story buildings cables between floors, are called backbone cables (risers) which are joined on each floor to horizontals. There are flexibility points where the backbone joins the horizontal where rearrangement of connections for a particular floor can be made independently (see Figure 10.1). These points may be called floor distribution points or local nodes. There may also be flexibility points in the backbone itself and these are called intermediate nodes. Rearrangement is usually by means of patch panels which are matrices of plugs or switches or thier equivalent. Typically horizontals consist of four twisted pairs and the backbone cable contains a greater number of pairs, sufficient to meet current demand with an allowance for growth. When the distance between buildings becomes large, i.e. 1 to 3 km, a higher tier in the wiring structure (not shown in the figure) is likely to be required.

Most structured wiring schemes are based on a star topology but can accommodate ring and bus topologies folded within the basic star.

A number of suppliers, for example Alcatel, AT&T, BT, Bull, DEC, Ericsson, IBM, Northern Telecom, Pirelli, Siemens and Unisys, market structured wiring systems in which all the piece parts are supplied and when assembled give an overall performance for the system that is closely specified. The alternative is for the instal-

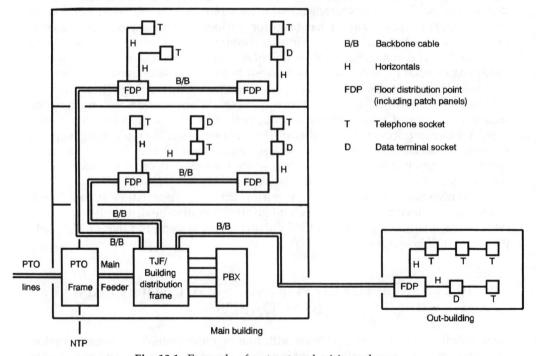

Fig. 10.1 Example of a structured wiring scheme

ler to buy piece parts from component suppliers and put together an *ad hoc* arrangement to suit a particular site. The structured approach is usually considered to increase in attractiveness as the building and the number of services carried gets larger.

In the UK wiring has to comply with either or both of two OFTEL General Approvals, NS/G/23/L/100005 for analogue circuit wiring and NS/G/1235/M/100009 for digital circuit wiring. Installation is covered by a British Standard code of practice, BS 6701, and safety, which is mandatory, by BS 6301. OFTEL also publish a *Wiring Code* (OFTEL, 1991). Part 1 deals with domestic and other simple installations, Part 2 with business and complex installations and Part 3 with shared wiring which is also covered by a third General Approval NS/G/1235/N/100018. The Wiring Code contains among other things guidelines for the maximum length of wiring that is applicable in various circumstances.

In the USA there are two Standards EIA/TIA 568 and 570 for commercial and residential/light commercial wiring respectively. EIA/TIA 568 'Commercial Building Wiring Standards' applies to wiring schemes connecting a mixture of analogue and digital services, and although it is intended specifically for the USA it contains material of wide interest which is relevant in any context.

In Europe the CEN/CENELEC TC115 committee is producing a document called *The Design and Configuration of Customer Premises Cabling*, and Parts 1 and 2 (EN 50098–1 and EN 50098–2) which deal with ISDN basic rate and primary rate respectively is at an advanced state of preparation. TC115 is charged with preparing harmonised standards for customer premises cabling and took over the work of a previous *ad hoc* committee, TC108.

Internationally, ISO/IEC/JTC1 Committee SC25/WG3 is developing wiring standards. Their work has taken EIA/TIA 568 as a basis and is using a top-down approach as the basis for the requirements. BSI Committee IST/6/10/1 provides the UK support for this work. Of particular note is the division of the spectrum into four bands as follows:

- *Class 1* Includes the speech band and long cable lengths. Class 1 wiring arrangements may allow working up to frequencies of about 100 kHz.

- *Class 2* Includes medium bit rate and medium distance applications such as basic rate ISDN extended S-bus.

- *Class 3* Includes high bit rate short distance applications such as 10Base-T and token ring.

- *Class 4* Includes very high bit rate, short distance transmission such as FDDI.

These divisions are not precise, still the subject of discussion and overlap somewhat; one might hope for some refinement of the standard in this area. It should not be assumed that two signal types, each of which is suitable for use in isolation with a given class of cable, can be shared within a cable sheath of that same category.

ISO SC25's approach is one of treating performance requirements for the totality of the wiring (i.e. end-to-end for the whole system) rather than considering the performance of individual components, such as cable, in isolation. While this

Table 10.1 Comparison of cable classifications

ANSI/EIA/TIA Category	NEMA Title	UL Level	Application
1	—	I	Analogue voice and data up to about 20 kbits/s (BT CW 1308B)
2	—	II	ISDN and data up to a rate of about 4 Mbit/s (Similar to IBM Type 3, BT CW 1700)
3	Standard premises telecom. cable	III	Data rates up to about 16 Mbit/s
4	Low-loss premises telecom. cable	IV	16 Mbit/s token ring. Data rates up to about 20 Mbit/s
5	Low loss extended frequency telecom. cable	V	Twisted pair FDDI. Data rates above 20 Mbit/s (4 pairs)

enables end-to-end performance to be guaranteed it confines applicability to structured wiring schemes. For this reason the categories 1 to 4 above do not match categories for cable in isolation that are used by various trade associations in the USA. These classifications are summarised in Table 10.1. A category corresponds to a set of cable specifications rather than a single specification.

10.3 GENERAL PROPERTIES OF WIRING

10.3.1 Types of cable and their characteristics

Cables are assemblages of conductors, screens, strength members and insulating material that carry one or more circuits within a sheath. A circuit consists of two or more conductors and provides the connection between nodes and terminal points in the wiring layout. In selecting a cable for a particular application the following parameters may need to be considered:

1. *Loop resistance* The resistance in ohms of a specified length of cable; for example $250\,\Omega/\text{km}$.

2. *Attenuation* The loss in dB/km at a frequency or frequencies in the pass band of the signal using the cable. (At frequencies above about 5 kHz loss increases approximately as the square root of the frequency.)

3. *Characteristic impedance* For telephony this is normally taken to be $600\,\Omega$ but for digital services using higher frequencies the characteristic impedance is a

function of the cable dimensions and it is necessary to ensure that the cable is properly terminated either separately or by matching the terminal apparatus impedance. Typical high frequency characteristic impedances are in the range 50–80 Ω for coax and 100 to 150 Ω for twisted pairs.

4. *Crosstalk* This may be expressed directly as a loss in decibels between circuits for a specified length of a cable at a specified frequency or it may be expressed indirectly (for twisted pair cable) as a capacitance out-of-balance for a specified length. A worse case figure may be given or a median figure with some indication of spread.

5. *Delay* This is expressed in suitable units such μs/km. It is perhaps worthy of note that because delay is proportional to the square root of the dielectric constant of the cable insulating material, PVC insulated cable with a dielectric constant of about 6 has 1.6 times more delay than cable with a dielectric constant of 2.3 (e.g. polythene).

Cable is categorised as follows:

- *Flat copper cable* When this is approved for use in building wiring it is likely to provide the final under carpet connection between a wall socket and a piece of terminal apparatus. Because of high levels of crosstalk and radiation it is only suitable for low frequencies and relatively short runs of not more than about 3 metres.

- *Unscreened Twisted Pair (UTP)* This is by far the most commonly used type of cable being ubiquitous for telephony applications and widely used for LANs etc. It is even under consideration for a version of FDDI working at 100 MHz. There are many varieties; the number of pairs within a sheath and high frequency performance are important options for selection.

- *Screened Twisted Pair (STP)* Screening is intended to reduce crosstalk and/or external radiation (EMC). Screening may be round pairs, quads, etc., or there may be a single outer screen or both inner and outer screens. Individual screening reduces intra- and inter-service crosstalk while screening of sets of pairs can be made to reduce inter-service crosstalk.

- *Transverse screened* Transverse screens partition the cable into sections. For example with a four wire circuit all the transmit pairs may be on one side of the screen and all the receive pairs on the other. There is often an outer screen as well to which the transverse screen may be connected. This type of cable is widely used in the public network but its use in building wiring is speculative.

- *Coaxial* For coaxial cable the screen acts as the earth return providing an unbalanced two wire circuit. Coaxial cable is most efficacious at frequencies well in excess of 1 MHz and its use in building wiring for other than multi megabit and radio frequency applications is unlikely.

- *Optical fibre* Optical techniques allow the multiplexing of a large number of circuits onto the optical medium, it takes up little space, is not sensitive to external radiation and does not radiate. An opaque outer coating can be provided to prevent crosstalk between fibres in a common sheath. Its cost relative to alternative solutions prevents it from being more widely adopted.

Additionally to the basic types of cable given above there are composite cables containing mixtures of the basic types and specially protected or strengthened versions for purposes such as outdoor use.

10.3.2 Other components

Components associated with wiring schemes whether those schemes are structured or not can include the following:

- *Connectors* In the UK plugs for telephones and telex terminals conform to BS 6317. In North America RJ45 (ISO 8877) connectors are widely used for analogue and digital circuits in structured wiring schemes. The OFTEL wiring Code, Part 2 contains information on the types of connector recommended by CCITT for use with various X., V., and I. series interchange circuits and by the UK PTOs for G.703 interfaces.

- *Terminating blocks, boxes and distribution frames* The PTOs and customer's wiring is connected to terminating blocks in boxes or on frames. There are various proprietary products covering a wide range of sizes, e.g. boxes for from 2 pairs up to 100 pairs and frames for up to 1600 pairs or more. With some designs there are plugs to provide connection and disconnection but others involve the use of tools to achieve this end.
  The frames include the test jack frame (TJF) and for most installations in the UK the network terminating and test apparatus (NTTA) (which is at the NTP) is colocated with the TJF. The NTTA provides a means by which PTOs can test their part of the network in isolation from private apparatus. When the NTTA is not colocated with the TJF, testing the PTO's wiring in isolation from the private wiring is more difficult and currently requires the cooperation of the customer, but developments in NTTA design should result in a feature that enables PTOs to test their part of the network in isolation by remote control.

- *Patching apparatus* Currently most patching is manual, using patch panels and patching leads or manually operated switches. The PTOs have begun to introduce cross-connects with centralised control for the corresponding function in their own networks and it is possible that larger private network owners will follow suit.

- *Line terminations, transformers and baluns* Most line terminations are resistors with a value that is specified in the relevant (e.g. CCITT) specification. Baluns provide balanced to unbalanced transformation and vice versa. There may be a move away from the use of transformers as baluns towards active semiconductor circuits that perform an equivalent function.

- *Signal regenerators, line drivers, electrooptic and optoelectronic converters* If the attenuation introduced by building wiring is excessive or the signal received at the private network interface is weak, there may be a need for signal regeneration involving active components including line drivers. If building wiring involves optical fibre then conversion apparatus is necessary. No very major developments in this area are anticipated but costs, particularly for the optical–electronic interfaces, may be expected to continue to fall.

- *Overvoltage protection* There is a well established position on overvoltage protection components for analogue voiceband circuits where gas discharge tubes are the normal means of primary protection and CCITT Recommendations K.11 and K.12 are relevant. OFTEL (1991), Part 2, contains details of the maximum impairments that the overvoltage protection circuit should be allowed to introduce in its quiescent state. For digital circuits there is no well established position on overvoltage protection and the components used for analogue circuits may not always be suitable, particularly at high data rates where their capacitance may prove to be excessive.

10.4 PERFORMANCE REQUIREMENTS

Performance is considered under the following headings:

- end-to-end transmission and noise performance of a cable circuit within a building in isolation from other circuits;
- crosstalk between circuits within a cable sheath;
- the performance of wiring components other than cable;
- EMC—building wiring as a source and sink of interference;
- earthing.

10.4.1 Transmission performance

The major considerations are in-band attenuation in all cases and delay in some cases. In any transaction involving an external connection, and in particular an international call, the building wiring plays only a small part of a situation in which an apportionment of end-to-end impairments is made by the CCITT. Part of this total is given to the national networks and this is then divided between public and private networks and the private networks include private wiring. As well as apportioning impairments the CCITT defines a number of interchange circuits and in some cases makes recommendations for the length of wiring that is attached to them. National and international standards bodies may take the CCITT recommendations further in terms of detail.

Wiring requirements depend on the mixture of signals that the wiring carries and while it is sometimes convenient to express such requirements in terms of services, e.g. analogue voice band and packet switching, a more fundamental approach which is preferred whenever applicable is to consider the lower layers of the OSI model where there may be commonality between the interchange circuits for a number of data services.

The OSI lower layer approach leads to categorisation in terms of the interface types that are listed below. In the context of this list, the distinction between balanced and unbalanced circuits is important and is even more important, as we shall see later, in the context of crosstalk behaviour. In a balanced circuit the mean potential of the conductor pair is zero with respect to earth; whereas with an unbalanced circuit there is a net potential between the conductor pair and earth.

Interfaces to serial transmission media

Analogue voice band services These include speech, fax (other than Group IV), and voice frequency data transmissions. Unscreened twisted-pair wiring is the norm for distribution and in the UK cable to BT specifications CW 1308B or CW 1700 is widely used. The North American specification EIA/TIA 568 is similar to CW 1700 in its twisted-pair requirement but has slightly different specification for a cable used horizontally as opposed to in the backbone. For a conductor diameter of 0.5 mm, a 1 km loop has a resistance of about 200 Ω and an attenuation in the voice band of about 2 dB.

The connection between a telephone and the wall or floor socket that contains the ringing components is a three wire unbalanced circuit and because it is unbalanced is likely to be subject to length restrictions on other grounds than attenuation (see below under crosstalk). Note too that during 10 ips dialling, going on-hook or off-hook, or while a ringing signal is present, because the two-wire circuit is unbalanced these signals can be a significant source of impulsive noise that can interfere with other, particularly data, circuits.

Telex The considerations that apply to analogue voice band wiring, excluding those of ringing, apply to telex.

ISDN The relevant wiring is usually attached to the interface at the S or T reference point. The annex to CCITT Recommendation I.430 gives limits on attenuation and delay for various wiring topologies which can be point to multipoint as well as point to point. These are general requirements and in order to determine a distance limit for a specific installations, the wiring performance must be specified. The European TC 115 standard referred to in Section 10.2 above is applicable when twisted pairs are used and contains limitations on length for a variety of wiring topologies.

The majority of wiring contained within the digital line section and connected to what is commonly called the 'U' reference point is likely to be in the public network but if it is private wiring then CCITT Recommendations G.960–961 may be the basis for a design. G.961 currently contains six different options for line codes.

V.10 Interchange Circuits CCITT Recommendation V.10 Appendix II contains recommendations on the length of wiring that may be attached to a V.10 interchange circuit. This length depends on data rate for data rates above 1000 bits/s and should not exceed 1 km below 1000 bits/s. The V.10 interchange circuit is unbalanced and twisted pairs are the norm. The same considerations apply to X.26.

V.11 Interchange Circuits CCITT Recommendation V.11 Appendix I contains recommendations for the length of wiring that may be attached to a V.11 interchange circuit. For data rates in excess of 10 kbit/s the length depends on data rate and on whether the interchange circuit is terminated or not. Below 10 kbit/s the length is limited to 1 km. The V.11 circuit is balanced and normally uses twisted pairs. The same considerations apply to X.27.

BS 7248 (ISO 8492), *Multipoint interconnection of data communication equipment by twisted pair cable*, may apply when V.11 is used in a multipoint configuration.

V.28 Interchange Circuits CCITT Recommendation V.28 contains no length restrictions but gives a load capacitance limit of 2500 pF which includes the capacitance of interconnecting cable. A limit of 15 m is however considered to be advisable. V.28 circuits are unbalanced and might be expected to use twisted pairs.

V.36 Interchange Circuits. CCITT Recommendation V.36 contains no length restrictions but a limit of 60 m is considered to be advisable. V.36 is not recommended for new designs and the alternative Recommendations are V.37 for data rates of up to 72 kbit/s and V.37 for data rates above 72 kbit/s. For V.36 and V.37, interchange circuits according to V.10 and V.11 are recommended, see above.

G.703 Physical and Electrical Characteristics of Hierarchical Digital Interfaces Limitations on wiring length are implicit in the attenuation limits which are 3 dB maximum at 64 kbit/s (codirectional) with twisted pairs of 120 characteristic impedance and 6 db maximum at 2048 or 8448 kbit/s with 75 Ω coaxial cable.

IEEE 802.3 (CMSA/CD) Ethernet LAN The system is designed to operate up to distances of 2.5 km but repeaters are required for distances in excess of 500 m, or even less, depending on the number of stations attached and the transmission medium. The attenuation between repeaters must not exceed 6.5 dB at 5 MHz and 8 dB at 10 MHz. There are many options for the transmission medium as follows:

- *10base5* This means 10 Mbit/s with up to five 100 m segments. The transmission medium is 50 Ω coaxial cable. Up to 100 stations per segment are allowed with a minimum interstation gap of 2.5 m.

- *10base2* This is sometimes known as 'cheapernet' or 'thinwire' and up to two segments using thin coaxial cable (RG58) of 50 Ω characteristic impedance are allowed. A segment is limited to 185 m. 30 stations are allowed with an interstation gap of 0.5 m minimum

- *1base5* This is known as 'StarLan' and uses twisted pair at 1 Mbit/s with up to 5 segments. Repeaters for segments are known as StarLan hubs and stations are individually attached to these hubs to give a tree structure rather than the bus structure of the other configurations. Individual stations can be attached up to 250 m from the hub. While there can be up to five hubs the hub tier depth is limited to two, thus no station can more than 500 m from the base hub.

- *10baseT* The 'T' stands for twisted pair and the arrangement works at 10 Mbit/s. The distance between repeaters is not more than 100 m. Up to 5 repeaters are allowed and there are rules for configuring systems containing a mixture of twisted pair and the other types of transmission media described above and below.

- *10broad36* This allows a 3.6 km segment carrying data at 10 Mbit/s modulated onto a carrier at a frequency in the CATV band. This is perhaps only of interest for a large site or campus where a CATV system is present for other reasons.

- *Fibre-optic Ethernet (10baseF)* The original Ethernet concept restricted the use of fibre optics to links connecting remote segments with no provision for the attachment of stations, but that situation is changing with the IEEE 10baseF standard. The 10baseF standard is making slow progress but the signs are that there will be three variants, 10baseFP for end systems or repeater connections, 10baseFB for backbone systems and 10baseFL for medium attachment. There are however a number of proprietary solutions pre-dating any standard using optical fibres and optical couplers.

IEEE 802.4 (Token Bus) LAN The system is designed to operate over distances of several hundred metres depending very much on the actual network topology, the number of intermediate stations (which introduce about 0.5 dB insertion loss), the tap losses (typically 28 dB end-to-end) and on whether repeaters are employed. The baseband transmission medium is 75 Ω coaxial cable with short stubs consisting of 37 to 51 Ω drop cable. There are also two broadband options.

IEEE 802.5 (Token Ring) LAN The system is designed to operate over distances of several hundred metres, perhaps up to 2 km. Attached to the ring are a number of wiring concentrators (called Multi-station Access Units (MAU)) which can be many hundreds of metres apart as determined by net transmission loss (repeatered or unrepeatered). Normally stations are attached to the MAUs by 150 Ω screened twisted pair when the number of stations is limited to 8 per MAU (about 72 for the system) all within a distance of about 100 m from an MAU. The overall cable attenuation end-to-end should not exceed 26 dB net at 4 MHz within a total allowable attenuation of 29 dB.

IEEE 802.7, BS:6531, Part 2 (Slotted Ring) LAN Operation is over distances of up to 10 km using two twisted pairs plus a signalling line. Screening of the twisted pairs is optional. The characteristic impedance can lie between 90 and 150 Ω but each physical segment must consist of cable having the same nominal impedance. Repeaters are used for long distances and to connect physical segments using different cable types.

Fibre Distributed Data Interface (FDDI) There is an ANSI standard X3.166-1989. Optical transmission is by multimode fibre with a preferred core diameter of 62.5 μm and a cladding diameter of 125 μm. The wavelength is in the 1270–1380 nm range and the optical fibre bandwidth is at least 500 MHz/km. The maximum loss of the cable is 2.5 dB/km and the maximum transmitted power is 7 dBm. The line data rate is 125 Mbaud using a 4B/5B binary coding scheme in which a 5th bit is added (for control purposes) to the 4 user bits. A link length of 2 km can readily be achieved even in the presence of up to six stations in the bypass mode (i.e. non-operational) when they each introduce a switching loss of up to 2.5 dB.

Twisted-Pair Distributed Data Interface (TPDDI) (also known as Cable Distributed Data Interface, CDDI) Work on a standard is proceeding in the ANSI X3T9.5 sub-committee. The use of twisted-pair cable applies only to horizontals for distances of up to about 100 m; the backbone cabling is always intended to be optical fibre (FDDI). In early 1992 the preferred options for line codes were (1) a two-level non-return to zero (NRZI) code, (2) a three-level (MLT-3) code and (3) a 32-level (CAP-32) code. At that time the MLT-3 code appeared the more likely choice. There is a considerable EMC problem at a 100 Mbaud rate, hence the need to lower the line transmission rate by multilevel coding. For the same reason scrambling is likely to be used to flatten the line spectrum during repetitive bit sequences such as those observed during idling.

Interfaces to parallel transmission media

SCSI Data rates are up to 5 Mbyte/s or 40 Mbyte/s (double SCSI II) for a distance of up to 6 m and both way transmission. A 50 pin connector is required.

Centronix Transmission is generally one way for data rates of up to 200 kbyte/s and a distance of up to 15 m. A 50 pin connector is employed.

HIPPI (HIgh Performance Parallel Interface) Transmission is one way and there is either a 32 bit data bus for a transfer rate of 800 Mbit/s or a 64 bit data bus for 1600 Mbit/s transfer rate and a distance of up to 25 m. The 32 bit version uses 50 twisted pairs. HIPPI (sometimes called HPPI) is being standardised by ANSI Task Group X3T9.3. HIPPI is intended to provide interconnections between super computers and to provide an interface from super computers to the SDH at 1.2 Gbit/s.

10.4.2 Crosstalk

Crosstalk, or unwanted coupling, between circuits sharing the same cable sheath has been a subject of interest to PTT transmission engineers for a long time and a considerable body of knowledge (dating back to the beginning of the century) is available from their work. However at a detailed level this work applies to much longer lengths of cable than are to be found in most buildings. With the introduction of a variety of data services into private networks there is a requirement, based largely on cost considerations, to share analogue speech with data and data of one type with data of another type within the one cable sheath. While the types of cable mentioned in Section 10.3.1 are designed to allow the one type of signal to share a sheath it does not automatically follow that they are suitable for a mixture of different types of signal.

Crosstalk is the result of near-field coupling between transmission media and applies to metallic and optical bearers. Far-field coupling, which may be correlated with near-field coupling, is of concern in the context of EMC. The major topic in this section is crosstalk between copper bearers, most specifically twisted pairs, but we shall also touch on optical fibre and in Section 10.5 on EMC.

A circuit that gives rise to crosstalk is called a disturbing circuit and a circuit receiving crosstalk is called a disturbed circuit. Because in general there will be more than two active pairs in a cable sheath there will be a multiplicity of disturbing circuits for every disturbed circuit. The factors involved in crosstalk assessment are:

- the crosstalk properties of the cable;
- the spectra of the signals involved in the crosstalk;
- the noise requirements or margins for the disturbed circuit.

Crosstalk attenuation between a pair of circuits, the disturbing circuit and the disturbed circuit is expressed in terms of near end, far end or effective level far end as follows (where the near end is the sending end of the disturbing circuit):

- *Near End CrossTalk (NEXT) attenuation* This is the ratio of the voltage at the near end of the disturbed circuit to the voltage at the near end of the disturbing circuit (normally expressed in dB).

- *Far End CrossTalk (FEXT) attenuation* This is the ratio of the voltage at the far end of the disturbed circuit to the voltage at the near end of the disturbing circuit.

- *Equivalent Level Far End CrossTalk (ELFEXT) attenuation* This is the ratio of the voltage at the far end of the disturbed circuit to the voltage at the far end of the disturbing circuit.

The choice between ELFEXT and FEXT as the appropriate measure of far end crosstalk depends on the application. ELFEXT is the more appropriate measure if signal to noise ratio is the main concern and FEXT is the better choice if for example the absolute voltage at the far end is the determining factor for bit error rate. ELFEXT gives a lower crosstalk attenuation value than FEXT for a correctly terminated cable.

The choice between NEXT and FEXT/ELFEXT may also be relevant and depends on which end of the disturbed line is the receiving end and how that end relates to the sending end of the disturbing line. A major objective in limiting crosstalk in building wiring is to prevent that crosstalk from getting into the public network, and when that is the case the NTP can be regarded as being at the receiving end of the disturbed line. The disturbing circuit can be an incoming line from some other public network service or an outgoing line from the local end. The latter, because it has not been attenuated by transmission through the network should have the higher level, and in this case FEXT/ELFEXT would seem to be a more appropriate measure than NEXT. However, the more general case is one where transmit circuits interfere with receive circuits at the terminal equipment (rather than the NTP) in which case NEXT attenuation is the critical parameter. For short lines (and in this context building wiring is 'short') terminated in their high-frequency characteristic impedance, NEXT tends to be worse than FEXT and it is perhaps more convenient to measure NEXT than FEXT. For these reasons NEXT is being specified in emerging cable standards. In any event NEXT and FEXT are highly correlated in most circumstances.

In a cableform containing a large number of bearers the crosstalk between any two bearers is far from constant and depends on their 'lay' within the cableform. Two bearers that run close together for most of the cable length are bound to have a much lower crosstalk attenuation than two that are at opposite ends of a diameter. Much of the art and science of cable design is in the manufacturing process which, for example by varying the twist angle between adjacent pairs, reduces the spread in crosstalk attenuation. There is a major measure of agreement in the literature that crosstalk attenuation figures correspond to a truncated normal (Gaussian) distribution. For example the mean crosstalk attenuation might be 90 dB with a standard deviation of 12 dB truncated at 3.5 standard deviations giving a minimum crosstalk attenuation of 48 dB. While a statistical approach is suitable for cable-forms with a relatively large number of bearers, some caution is required when the number of bearers is small, say less than about 10. However, there is evidence that while the mean crosstalk attenuation is less for a small pair cable, the standard deviation decreases too, in which case the application of statistical data based on a large pair cable to a small pair cable of the same basic construction may well be accurate enough for most purposes.

For design purposes a worst case figure rather than the median is needed and the worst case figure can either be derived in terms of a number of standard deviations (or the equivalent percentile) from a mean or for smaller cableforms might be determined from measurements of a representative sample of cables.

Twisted pairs

For both balanced and unbalanced circuits, twisting a pair of wires together is a very cost effective means of reducing the far-field from the point of view of radiation. From the point of view of crosstalk attenuation the question of the balance/unbalanced state of the disturbed and disturbing transmit and receive circuits is of major importance and we must consider three cases:

1. both circuits balanced;

2. one circuit balanced and the other not;

3. both circuits unbalanced.

In going from (1) to (2), as a rough rule of thumb, the worst case crosstalk attenuation reduces by 20 dB and the mean by about 25 dB. In going from 2 to 3 the crosstalk attenuation may decrease by as much as a further 40 dB. If by this rule the absolute attenuation were to be less than about 10 dB the rule would cease to be accurate enough. However, unbalanced circuits are unlikely to be operated at frequencies high enough for the attenuation to be as low as 10 dB.

Irrespective of whether the transmit and receive circuits are balanced or not, there are two components of crosstalk, a transverse term between the two conductors of a pair and a longitudinal component common to both conductors of the pair. In the balanced case (3), if the equipment terminating the wiring has a signal balance about earth of much better than 20 dB—and this is usually the situation in practice—then the longitudinal component of crosstalk has negligible effect and only the transverse component is significant. However, the longitudinal component is a source of far-field radiation.

Each twisted pair has self-capacitance and self-inductance and mutual capacitance and inductance with other pairs in the cableform. The self and mutual capacitance and inductance are determined by the geometry of the cable and they are all interrelated. The self components are determined by conductor spacing, conductor diameter and the dielectric and magnetic constants of the surrounding media. The mutual components are proportional to the self components but are modified by the twist and lay of individual pairs in relation to each other. It is possible to show that when the disturbed and disturbing pair are terminated in their high frequency characteristic impedance, i.e. a value in the 100–150 Ω range, the effect of capacitive and inductive coupling are about equal. For circuits working in the analogue voice band and terminated in 600 Ω capacitive coupling is dominant. This means that at audio frequencies a suitable measure of crosstalk is out-of-balance capacitance. Circuits, such as ISDN circuits, working at a relatively high data rate are terminated in a much lower impedance, and capacitance alone is not a suitable measure of crosstalk. Cable carrying signals with spectra extending to frequencies above, say, about 20 kHz are best characterised directly in terms of crosstalk as measured at some specific frequency and cable length.

Generally crosstalk increases with the length of the cable, but NEXT behaviour in particular is dominated by cable attenuation but also depends on whether the cable is long enough for phase shifts along it to be significant. For short lengths and low frequencies, voltage or current addition of crosstalk terms applies and for long lengths and high frequencies crosstalk terms can be regarded as uncorrelated and hence are added on a power sum basis. Figure 10.2 gives examples of NEXT behaviour under both such extreme conditions. Loop attenuation of the cable at the reference length and for the frequency under consideration is the parameter in Figure 10.2. The results are normalised to the reference cable length and a length of the order of 200–300 m corresponds to the attenuation values chosen for illustrative purposes. ELFEXT attenuation decreases at about 20 dB/decade in the short length, low frequency case and at about 10 dB decade in the long length, high frequency case with values (e.g. 15/decade) in the intermediate length/frequency region between these extremes.

Reducing the impedance in which a cable is terminated reduces crosstalk, for example in reducing the terminating impedance from 600 Ω to 50 Ω FEXT might be reduced by some 10 dB up to a frequency of about 10 kHz. Above 30 kHz the effect of the terminating impedance is not very significant.

The effects of screening

The effects of screening twisted pairs are twofold; screening can be used as a means of ensuring physical segregation of unlike circuits in the cableform and it much

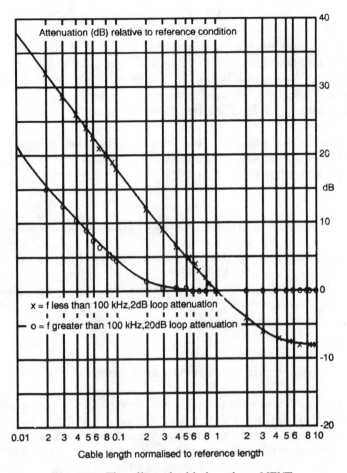

Fig. 10.2 The effect of cable length on NEXT

reduces the mutual capacitance between any circuits that are screened from each other. For individually screened pairs the mutual inductance is largely unaffected up to frequencies above about 100 kHz where eddy-current losses in the screen become significant. Hence, if mutual inductance is a major coupling mechanism, there may be little or no advantage to be gained in screening individual pairs at least up to frequencies of the 100 kHz order. However, above 100 kHz in the 1 to 10 MHz region screening is usually very effective. The effects of physical segregation are not generally quantifiable, being very dependent on actual cable geometries and because of the variability of lay are of a statistical nature.

For audio frequency circuits terminated in 600 Ω mutual capacitance is dominant and screening is therefore effective in reducing in-band crosstalk between like circuits. It is also effective when the disturbed circuit is terminated in a lower impedance than 600 Ω but coupling in the opposite direction from a lower impedance circuit into a 600 Ω circuit is less affected by screening.

10.4.3 Coaxial cable

It may come as something of a surprise to find that if coaxial cables are laid side-by-side crosstalk at lower frequencies (e.g. below 100 kHz) can be worse than for twisted pairs. Again this is largely because there is mutual inductance between them and because they are unbalanced. At frequencies in excess of about 5 MHz coaxial cables should give crosstalk performance superior to that of twisted pairs.

In terminating coaxial cable the outer braid is sometimes stripped back to form a pigtail and this can be a source of appreciable crosstalk and radiation at frequencies above about 10 MHz (Paul, 1980).

10.4.4 Optical fibre

Light travelling along an optical fibre is not confined entirely to the fibre core and multimode fibres in particular, because the core to cladding ratio is relatively low, are prone to crosstalk. However, the provision of an outer opaque cladding on the fibre is a simple means of overcoming this problem and allows optical fibres to be closely packed together in a tube without a need to control fibre lay.

10.4.5 Signal spectra and noise margins

Crosstalk introduced into a disturbed circuit is regarded as noise on that circuit and hence crosstalk limits are determined by the noise limits acceptable for the disturbed circuit. Both in-band and out-band noise may need to be considered and here the situation for analogue and digital services is likely to be somewhat different. We start by considering crosstalk margins when an analogue voice band circuit is the disturbed circuit, and while this is a very specific example it provides a good illustration of the principles involved.

Figure 10.3 shows the noise out-of-band noise mask of BS 6305 together with a figure for in-band noise obtained from BS 6450. This mask applies to equipment noise at the NTP for terminal equipment attached via a PBX to PTO lines in the UK. Figure 10.4 shows a worst-case frequency/crosstalk attenuation characteristic for the wiring used to connect the equipment to the PTO network. Figure 10.5 is the result of combining the the first two figures by adding the crosstalk attenuation to the noise mask. The resultant upper line of Figure 10.5 is the threshold for the disturbing circuit above which noise introduced into the disturbed circuit could constitute a problem. Figure 10.5 contains points calculated for the short term (125 µs) spectrum of an ISDN signal at the S/T reference point. It will be seen from this illustration that, because the points on the ISDN spectrum at about 50 kHz and 100 kHz lie above the threshold, crosstalk between ISDN and a voice circuit does constitute an out-of-band noise problem. There is no in-band problem, however. Strictly, a long term spectrum for the ISDN signal should have been used because BS 6305 requires that the noise be averaged over a 10 s period, but it is believed that a long term spectrum would give a similar result.

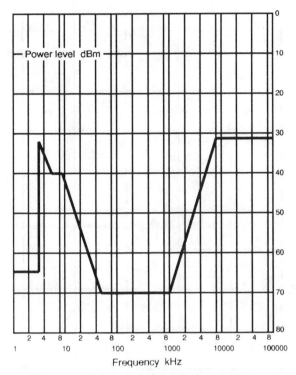

Fig. 10.3 Noise mask for voiceband circuits

In setting the limit in Figure 10.5 noise introduced by the equipment which would add on an RMS basis to the crosstalk noise has been ignored. This can be justified on the ground that a combination of worst-case equipment noise with worst-case wiring noise has a very low probability.

The above example treats the problem as being in the frequency domain by working in terms of the spectrum of the noise rather than the waveform as would be the case if one were to work in the time domain. When the disturbing signal is a continuous data stream, e.g. 64 kbit/s digitally encoded information, the disturbing signal can be expressed in terms of a long-term spectrum averaged over a period of, say, minutes, or it can be treated as a short-term spectrum averaged over, say, a few frame periods, i.e. a period of the one millisecond order. Ideally the short-term spectrum should be taken and the worst-case interfering pattern should be found. In most cases a bad pattern ,if not the worst pattern, is a string of all 1s, all 0s or an alternating pattern such as 1,0,1,0, etc., depending on what line code is used. From such patterns the waveform of the disturbing 'noise' pulse can be calculated. When the disturbed circuit is a digital circuit it is the duration and amplitude of these noise pulses in relation to the duration of the sampling window and threshold level of the disturbed circuit's receiver that is critical.

Speech circuits acting as disturbers produce noise due to calling, dialling, ringing, ring trip and clearing that is impulsive in nature and relatively infrequent in time.

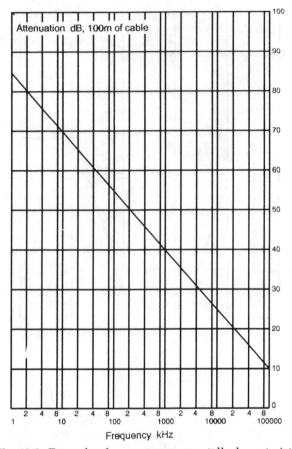

Fig. 10.4 Example of a worst-case crosstalk characteristic

In this case a time domain approach is the only viable one, the disturbance can be expressed in terms of the number of impulses in a specified time above a specific threshold. If the disturbed circuit is a digital circuit than this specified threshold needs to relate to the performance of the disturbed circuit's receiver.

BS 6450 Part 2 sets limits on impulsive noise that PBX speech circuits are allowed to produce in the speech band which is expressed in terms of the number of impulses above a level of − 40 dBm in a a period of 15 min. How representative this is of impulsive noise from speech circuits in a practical situation, outside the speech band, involving attachment to the PSTN and to terminal equipment over long lengths of cable is debatable. There is experimental evidence that voltages of the order of 0.5–1 V can be introduced into a disturbed circuit consisting of a balanced twisted pair, albeit at a very low rate, a rate that depends on call attempts in disturbing circuits.

For digital disturbed circuits errors are due to in-band noise in those circuits and there is no evidence of a need to establish out-band criteria in this case. When the

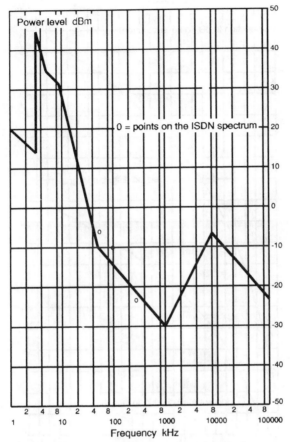

Fig. 10.5 Threshold diagram for voice band circuits

disturbing circuit is another digital circuit then its in-band spectrum is what matters, but for analogue speech circuit disturbance most of the offending signal's spectrum is above the speech band.

The seriousness of bit errors in digital circuits depends upon error detection and correction procedures at layers in the OSI model above layer 1 and it may be uneconomic to establish a crosstalk specification on the basis of layer 1 performance in isolation.

Multiple disturbers

Figure 10.5 deals with the case of a single disturber but noise from multiple disturbers can make the situation worse. Two cases may need to be considered:

1. where the noise from the disturbers is uncorrelated and can be summed on a power or root mean square basis;

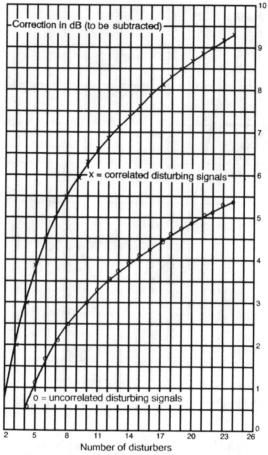

Number of disturbers

Fig. 10.6 The effect of multiple disturbers

2. where the noise from the disturbers is correlated and must be summed directly.

In the latter case it is assumed that the noise can be added with random sign because the crosstalk coupling in the cable can be of either sign and there is no good reason to suppose that the sign is systematic.

If there are n disturbers in a cable of N pairs and the the worst case crosstalk attenuation for the pth percentile (e.g. 95%) is C dB then the crosstalk attenuation C_n for n disturbers for the pth percentile is given by:

$$\text{for } n < n_t, \ C_n = C$$

$$\text{for } n \geqslant n_t, \ n < N, \ C_n = C - A \left(\log(n) + B \right)$$

where n_t is a threshold value and A and B are constants. For uncorrelated disturbers n_t is likely to be about 3 and for correlated disturbers $n_t = 1$. Figure 10.6 shows a typical example of the above formula for each case.

For impulsive noise having (as is normally the case) a low duty cycle (i.e. a duration very short compared with the average interval between pulses) the probability of getting coincident pulses is generally sufficiently low not to cause a significant error rate, but the number of potentially interfering pulses increases linearly with the number of disturbers. However, it has been demonstrated (Conte, 1986) that if the crosstalk attenuation is such as to make the error probability associated with the worst case disturber very low, then the effect of the other disturbers is negligibly small and no correction need be made for them. Of course if the error probability with a single disturber is high then the effect of multiple disturbers causing significant error rates must be summed. Not enough information is currently available on this topic to allow a quantitative approach to be adopted.

10.4.6 Examples of crosstalk margin determination

Rather different procedures apply depending on whether:

1. the disturbing circuit is digital and the disturbed circuit is analogue;

2. the disturbing circuit is analogue and the disturbed circuit is digital; or

3. both circuits are digital.

Each of these cases is illustrated by an example below.

Digital disturber—analogue disturbed as illustrated by the ISDN S/T bus interfering with analogue voiceband circuits

Figures similar to 10.3, 10.4, 10.5 and 10.6 can provide the basis for the method needed to determine whether a cable of specified crosstalk performance can safely be used in a sharing mode. However, Figure 10.4 relates to a particular cable length and a correction for other cable lengths is required. Figure 10.2 shows how the NEXT attenuation margin varies with cable length.

The S/T bus example has already been used as an illustration in Section 10.4.5 and is a case where the spectrum of the noise in the analogue voice-band circuit is the determining factor. We shall assume that the reference length is 100 m and wish to determine the maximum length of cable that is allowable assuming that there are 5 ISDN circuits acting as disturbers.

- From Figure 10.5, the point in the ISDN spectrum that is most removed from (above) the noise mask is at 100 kHz and therefore a reduction in crosstalk of the difference, i.e. about 4 dB, is required for a single disturber.

- Assuming five uncorrelated disturbers, from Figure 10.6 a further reduction in crosstalk of 1 dB is necessary, giving 5 dB in all.

- Taking the $f < 100$ kHz criterion, a loop attenuation of 2 dB, and assuming that Figure 10.2 applies to a reference length of 100 m, it will be seen that the length of the cable would have to be reduced from 100 m to about 50 m to achieve a 5 dB reduction in crosstalk.

The alternative would be to use a different type of twisted pair cable with a higher crosstalk attenuation. The next grade of cable is likely to have about 10 dB more NEXT attenuation and in the example given above would be satisfactory without any conditions being attached.

Analogue disturber—digital disturbed as illustrated by impulsive interference between an analogue circuit and the ISDN S/T bus

This is the converse of the above example and we need to consider the waveform of impulsive interference from analogue voiceband circuits adding to ISDN signals on a bus attached to the S/T reference point. The effect of this addition depends on the sign (or phase) and magnitude of the impulsive interference in relation to the sign of the ISDN signal and the receiver margin. Unfortunately there is a lack of both amplitude and spectral information from which to quantify this interference, although it is known that when NEXT attenuation is near to the limit assumed for the example above, analogue circuits can cause data errors particularly during the time that a call is being established on them. Thus data errors could occur and if they do, then the acceptability of this depends upon the traffic levels on the disturber(s) and on error correction procedures for the ISDN circuit.

Both circuits digital as illustrated by V.11 interfering with ISDN S/T bus and vice versa

The procedure for calculating the crosstalk margin is as follows:

- Determine the interchange circuit parameters, particularly the peak voltage, rise time, and pulse repetition frequency for the disturbing circuit and the noise voltage margin on the disturbed circuit.

- Determine the repetitive waveform on the disturber which produces the worst interfering pattern. In this case this is assumed to be a pattern of alternative 1s and 0s. Find the Fourier series expansion terms of this trapezoidal waveform and the relative phase shift of the Fourier terms for NEXT, regarding the cable as a transmission line.

- Determine the attenuation of the Fourier terms from the NEXT attenuation characteristic and then carry out a vector addition of them. Determine the peak voltage from this addition.

- Determine the margin between the peak voltage and the threshold voltage of the disturbed circuit.

Table 10.2 V.11 and ISDN S/T bus. Threshold margins for maximum cable length—single disturber.

Data rate (kbit/s)	Line termination (Ω)	Max. V.11 length (m)	Threshold margin (dB)
100	110	1000	5.6
100	unterminated	100	20
1000	110	100	8

In this example the answer depends on the data rate at which the V.11 circuit operates, the maximum allowable cable length at that data rate and on whether the cable is terminated or not.

Calculated margins for some conditions are given in Table 10.2 and these show that the V.11/ISDN combination. There is a margin of 8 dB for a 100 m length of cable at a data rate of 1 Mbit/s but V.11 is limited to 100 m at 1 Mbit/s in the CCITT V.11 recommendation and hence a longer length than 100 m cannot be considered. At the ISDN data rate of 96 kbit/s and for a length of 300 m, the converse arrangement with ISDN as the disturbing circuit and V.11 as the disturbed circuit has an ample margin of about 39 dB. (A cable having similar crosstalk characteristics to that of Figure 10.2 is assumed but the results are based on a reference length of 300 m rather than 100 m.)

10.4.7 Other wiring components

At analogue voice band frequencies and low data rates the potential transmission impairments introduced by other components, such as connectors, patch panels, switches and overvoltage protection devices can be expressed in terms of lumped series and shunt impedances, of which shunt capacitance is often the most important. Limits can be set on these lumped impedances that will make the effect of the mismatch that they introduce insignificant. At the highest data rates components need to treated as being distributed rather than lumped and should be matched to the characteristic impedance of the circuit in which they are connected. Any mismatch can be determined by measuring the reflection produced when a component is inserted between two sections of cable having the required characteristic impedance. At intermediate data rates there is a grey area where either a lumped or distributed approach may be valid. Certainly if a component is satisfactory at the highest data rates it is most unlikely that it will cause serious insertion loss problems at a lower data rate.

The work of ISO/IEC/JTC SC25/WG3 outlined in Section 10.2 is based on the division of the spectrum into four bands and requires that wiring and wiring components be specified as being suitable for one or more of these bands. The reflection technique mentioned above is one of the measurement methods described for use in proving the suitability of wiring components.

Most of the wiring components listed in Section 10.3.2 are unlikely to be major sources of crosstalk, a possible exception being patch panels. In some designs of patch panel 'there is a matrix of crossovers consisting of tracks on opposite sides of a printed circuit board with a means for bridging these tracks where a connection is to be made. The capacitances associated with these crossovers (and crossovers in general) are a source of loss and crosstalk and when signals have strong high frequency components in their spectrum the effect of these capacitances needs to be considered.

At data rates of about 10 Mbit/s and higher non-concentric connectors (e.g. edge connectors) can be a significant source of crosstalk and an allowance may need to be made for them. Unlike cable, such connectors can be treated as lumped, rather than distributed, components. A relatively small lumped capacitance produces a significant impairment because the harmonics of the disturbing waveform at the near end add in the correct phase to produce a sharp spike whereas on the cable the harmonics are phase-shifted by different amounts and the wave shape is less well preserved.

One of the advantages claimed for structured wiring schemes is that one does not run the risk of incompatibility between impedances that is present when assembling *ad hoc* collections of components.

10.5 ELECTROMAGNETIC COMPATIBILITY

Electromagnetic compatibility (EMC) defines the interaction between a system (in this particular instance the wiring) and the electromagnetic environment in which that system operates. There are two concerns: emission and immunity. Existing wiring practice would not be acceptable unless it ensured a good measure of immunity from, for example, rf fields, and the major concern in this context is emission.

From 1992 EMC is regulated by an EC Directive which in the UK is likely to result in an amendment to the *Wireless Telegraphy Act*. A radiation limit will be defined at a distance of 10 m in an open area test site as defined in CISPR 16. The limit is defined over the frequency range from 30 to 1000 MHz for two classes of equipment; Class A, for a commercial/light industrial environment and Class B for a domestic environment. Structured building wiring is likely to have to have a performance corresponding to Class A which is the less onerous requirement. At the time of writing the requirements for wiring are beginning to emerge and are given in Table 10.3, where it will be seen that they are expressed in terms of a longitudinal current. It is presumed that the limits apply to lengths of cable of the order of one wavelength at the relevant frequency. The actual implementation of these requirements has been postponed to 1996. It is too early to say how onerous

Table 10.3 EMC—Longitudinal current limits

Frenquency band (MHz)	Maximum current in dB relative to 1 μA
0.15–0.5	51
0.5–30	45
30–1000	35

the requirements are but it would be most surprising if they were to have a serious impact on building wiring design at date rate below about 1 Mbit/s.

The major source of radiation from unscreened wiring is the longitudinal induced common mode signal which, as we have seen above, can arise from crosstalk. Thus there is likely to be correlation between radiated energy and crosstalk because crosstalk always manifests itself as a relatively strong unbalanced component in the disturbed circuit. However, cables are not the only potential source of radiation and attention may need to be given to screening patch panels and other situations in which the spacing between conductors of a pair can be much greater than it is in a cable.

As long as there are no requirements below 30 MHz it is most unlikely that any of the existing PTT network services as delivered to the user will constitute a problem, but for some high data rate LANS (e.g. FDDI at 100 MHz on twisted pair) a clear position will have to be established. If EMC is a problem, then screening will act to reduce it by at least an order of magnitude, possibly two orders of magnitude, but it cannot be assumed that screening will completely eliminate radiation and particular care needs to be taken when designing the earthing arrangements for screens (see Section 10.6).

10.6 EARTHING

The two main topics of concern in the design of earthing arrangements are protective earths for safety (including equipment protection) and functional earths for prevention of coupling between different earthing paths. The latter may also have a bearing on EMC. Safety is the subject of national regulations and is not an appropriate topic for description here, it is however paramount in the design of all earthing systems. Two functional earthing techniques are discussed here:

- earthing of screens;
- common bonding networks.

The former is not new but is included because there appears to be a general lack of appreciation of the principles behind current practice.

10.6.1 Earthing of screens

When cable connects equipment over appreciable distances and is screened either with an 'inactive' screen, e.g. a screen around a balanced pair, or an 'active' screen that forms one side of the signal circuit, as with a coaxial cable, then the question of how such screens should be earthed has to be resolved. There is a problem when the equipments at the ends of the cable have circuits in common in addition to the one being connected by the cable because earth currents from elsewhere can cause interference if they flow through the screen. In practice an additional circuit in common is often the mains power supply and it is necessary to try and prevent

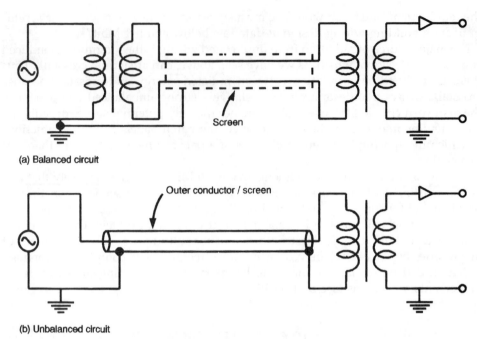

(a) Balanced circuit

(b) Unbalanced circuit

Fig. 10.7 Screen connections—balanced and unbalanced circuits

earth currents associated with the mains flowing through cable screens. Safety issues aside, this is essential only if the spectrum of the mains signal overlaps that of the wanted signal. Because of the many non-linear devices that load mains supplies the mains has harmonics extending far beyond its fundamental frequency of 50 or 60 Hz. As long as the length of the screened cable is much less than a quarter wavelength at the highest frequency of interest, earthing the screen at one end only should not limit its effectiveness and it is normal practice to earth the screen at the transmitting end on unidirectional circuits. When the screen is round bidirectional circuits the screen is normally connected to earth at the major node. For example in a screened connection from a public exchange to a private network the earth connection is at the public exchange. In a connection from a PBX to a terminal equipment the earth would be at the PBX. While an inactive screen is left floating at the receive end (Figure 10.7a), an active screen must be connected in order to complete the signal path and it is necessary to provide a transformer (or equivalent device) to provide what is in effect a balanced to unbalanced transformation, (see Figure 10.7b).

From the point of view of EMC it is preferable to earth screens or to connect them to the signal common return at both ends and this is normally the practice at frequencies above 30 MHz.

10.6.2 Common bonding networks

It is common practice in buildings containing large amounts of equipment for each work area to have its equipment earthed locally to an earth point (or earth points) which is joined to the building earth by a separate connection giving what is in effect a star earth configuration. Recent work in ETSI/EE2 has led to a new approach using mesh connection resulting in what is called a Common Bonding Network (CBN). The objective is to make the earth impedance at every point in the equipment area as low as possible by providing numerous earth paths bonded together (whether protective or functional) at every opportunity. Reducing the earth impedance is advantageous from the point of view of both reducing coupling between apparatus and improving the EMC situation.

There are likely to be regulatory and managerial objections to the adoption of this approach but it would seem from the purely technical point of view to be basically sound. It remains to be seen whether the introduction of CBNs will change current practice for the earthing of screens within buildings; earthing of screens between buildings is unlikely to be affected.

10.7 MANAGEMENT OF WIRING

The management of large wiring installations is a major undertaking calling for computer based tools for record keeping (e.g. site layout plans, equipment identities and equipment connection rules), the labelling of all terminations and strict adherence to laid down procedures when any changes are made. The automation of wiring changes under computer control is not very practicable at the moment but is a goal which if achieved should lead to much more accurate records. In older installations it is not unusual to find that there is more unused cable lying under the floors or in the ducts than is actually in use.

Precautions are necessary to prevent the connection of terminals to the wrong service because this can result in network harm and equipment damage, particularly if a data terminal is connected to an analogue speech circuit. Making the relevant plugs physically incompatible or electrically foolproof is the solution at the apparatus socket but at floor distribution points greater reliance may have to be placed on the skill of maintenance personnel and good record keeping.

When communications is a key factor in a large business, redundant wiring schemes that continue to provide a degree of service when part of the building has been destroyed by a catastrophe, such as fire, are called for. Reliability can be enhanced by multiple PTO cable entry points to the building, several PBXs that share traffic and interwork, and the partitioning of wiring on any one floor between these PBXs.

10.8 CONCLUSIONS

There is no doubt that building wiring is a major growth area both in terms of business turnover and in technical activity. The increased importance and complexity of the average business's telecommunications equipment has led to a marked increase in standardisation activities and in the demand for new and flexible wiring schemes.

A key technical issue is crosstalk in shared cabling using twisted pairs and as we have seen this a very complicated subject and one for which much of the hidden complexity has only begun to be touched upon here. The difficulty arises from the inherently complex and distributed nature of the equivalent circuit of a multicore cable. Very few people could have forecast twenty years ago that twisted-pair cable would be made to operate satisfactorily over such a wide frequency range in a building environment. The pressure to achieve this ubiquity is largely an economic one.

There is no such thing as a perfectly screened cable although there are situations, e.g. optical fibre, where the screening can be made perfect enough from the practical point of view. It does not pay to take anything for granted as far as novel cabling arrangements are concerned.

It remains to be seen whether EMC is a real problem as far as wiring is concerned and it is to be hoped that the optimistic view expressed here proves to be correct; certainly common bonding networks should help.

The management of complex wiring installations is a topic for which there is much scope for further development which could lead to a greater degree of automation in the handling of wiring system rearrangements.

10.9 BIBLIOGRAPHY

Bell Telephone Laboratories (1970/71) *Transmission Systems for Communication*, 4th edn, Chapter 11.

C. R. Paul (1980) Effect of pigtails on crosstalk to braided-shield cables, *IEEE Transactions on Electromagnetic Compatibility*, **EMC-22** (3), 161–172.

R. A Conte (1986) A crosstalk model for balanced digital transmission on multipair cable, *AT&T Technical Journal*, **65** (3), 41–59.

OFTEL *The Wiring Code Parts 1 & 2* (1991), *Part 3* (1992) Export House, 50 Ludgate Hill, London EC4M 7JJ.

11 PROCESSORS AND SOFTWARE

11.1 INTRODUCTION

Processors and the software that runs on them are increasingly a part of telecommunications products and here we concentrate on processors as used for the control of switching equipment in exchanges and software from the point of view of its design performance and flexibility. Because VLSI design is important and shares common ground with software design it too is briefly treated under the software heading. However, we start by describing the various types of processor that can be used in telecommunication systems.

11.2 PROCESSOR TERMINOLOGY

Processors or central processors can range in processing power from super computers and array processors though mainframes and minicomputers to microprocessors. However, in comparing, say, a mainframe with a microprocessor it is not only the processing power that is at issue. A mainframe has associated with it the ability to interwork with peripherals and memory as a system whereas a microprocessor can be regarded as a component around which a system can be built.

Super computers are very powerful mainframes that find application in tackling highly calculation-intensive numerical problems, such as weather forecasting, and as far as is known have not been used for telecommunications.

Nowadays array processors are arrays of microprocessors configured to perform calculation-intensive tasks sometimes of a highly specialised nature. The two most common architectures for array processor are single instruction multiple data (SIMD) and multiple instruction multiple data (MIMD). With SIMD the same instruction is applied to multiple sets of data simultaneously, each set of data residing in the memory of a different processor in the array. MIMD machines assign tasks between processors in the array and task-passing mechanisms are important. The transputer is an example of a microprocessor that can be configured as an array processor and which has its input/output circuits designed to make task passing efficient. SIMD machine are usually application specific whereas MIMD machines can be made more general purpose. Although not cheap, array processors are generally much less expensive than super computers and are used for signal processing applications, particularly in a military context.

Mainframes provide the central processing function for data processing machines. They form the heart of most general-purpose computing systems.

Minicomputers are traditionally less powerful versions of mainframes with the emphasis perhaps more on scientific and engineering applications rather than data processing. Processors providing the centralised control of telephone exchanges are often minicomputers or their specialised equivalent.

Microprocessors (microcomputers) are a VLSI component that provides the central processing function on a single chip. Their importance in the context of telecommunications and generally cannot be over emphasised.

Figure 11.1 shows trends in computer processing power ranging from super computers to microprocessors and there is no sign yet of any fall-off in increasing performance with time. Of particular interest is the fact that microprocessor performance has overtaken that of minicomputers and mainframes. This does not signal the demise of minicomputers or mainframes because as mentioned above a

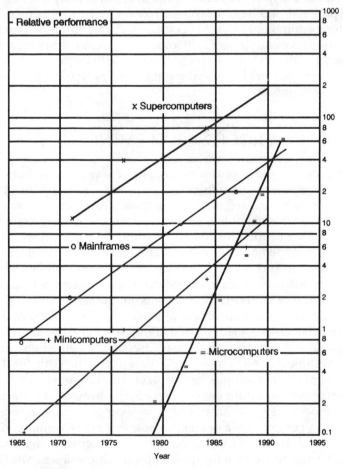

Fig. 11.1 Trends in computer performance. (After Hennessy and Jouppi, 1981)

computer system consists of a lot more than the central processor but it does mean that microprocessor solutions to major computing tasks must be regarded very seriously.

Particularly in the context of microprocessors the term RISC is frequently encountered and stands for Reduced Instruction Set Computer. In the early days of microcomputers a large number of instructions in the instruction set was considered to be a good selling point and some machines had hundreds of instructions, some of them very complex ones. It was then realised that in most programs the machine spent 90% of its time executing perhaps a dozen instructions, that complex instructions could be realised as a sequence of simple instructions and that the machine would run a lot faster with a small instruction set, hence the RISC approach. The limited instruction set of a RISC machine can make assembly language programs more tedious to write but this only affects a small minority of programmers.

There is a wide variety of microprocessors with specialised internal architectures, some aimed at specific applications such as digital signal processing (e.g. fast Fourier transforms) and others aimed at particular types of advance programming techniques (e.g. data flow architectures).

11.3 PROCESSOR ARCHITECTURE AND PERFORMANCE

11.3.1 Trends towards decentralisation

In the sphere of business computing there has been a marked trend over the last ten years or so from centralised mainframes to personal computers (PCs), almost to the point where there is one PC per desk. The PC has not rendered the mainframe obsolete and perhaps never will; each will continue to find its own particular niche. There is a similar, but perhaps less obvious, trend in the field of telecommunications.

11.3.2 Requirements for telecommunication switching control systems

The use of computers in telecommunications for real-time processing started with the Bell ESS No. 1 telephone switching system in 1964. ESS No. 1 did not use a commercial computer for the switching control function because the requirement was, and still is, to have a very high availability compared with what is normally acceptable for business processing. The usual means of achieving high availability is to have a redundant control configuration containing a multiplicity of cooperating processing elements arranged so that the control subsystem as a whole continues to function in the presence of at least one failure. There is often a need to provide special assembly language instructions to control the switching hardware efficiently. These requirements have led to the widespread development of special-purpose processors for telecommunication switching systems.

The main function of the control subsystem can be summarised as being 'call processing', i.e. call set-up, call progression and call release. Performance is measured in terms of the number of call attempts that can be handled in the busy hour, expressed as Busy Hour Call Attempts (BHCA). As well as providing higher reliability, multiple processing elements can cooperate to share the processing load and hence more calls can be handled. When a processor fails the traffic handling capacity will be reduced but the system will continue to function; the effect is sometimes known as 'graceful degradation'.

11.3.3 Control configurations

Centralised control systems are realised as co-located multicomputer or multiprocessor configurations. A multicomputer (see Figure 11.2a) is several computers each of which is self-contained, whereas a multiprocessor (see Figure 11.2b) is a number of central processing units (CPUs) sharing common memory. Multicomputers intercommunicate via data links while multiprocessors communicate via their shared memory. This interprocessor communication eventually leads to a bottle-neck or 'contention' which results in diminishing returns as more processors are added to the configuration. A practical limit is probably about five for multicomputers and ten to twenty for multiprocessors. When a control configuration reaches a limiting size there are two ways in which more power can be provided:

1. by increasing the power of individual processing elements;
2. by delegating the more routine and peripheral tasks to slave processors, which may be remote from the site containing the central control hardware.

In practice both approaches are used. The delegation of routine tasks to slave processors means that less computing power is required at the centre and hence the need for more processors is reduced and the contention problem is also re-

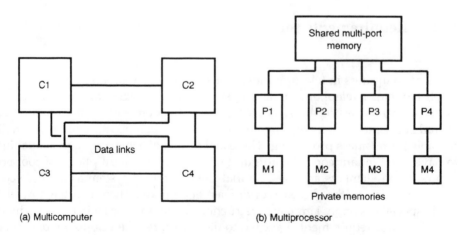

(a) Multicomputer (b) Multiprocessor

Fig. 11.2 Multicomputer and multiprocessor communication

duced. The situation is further eased if these central processors are more powerful because interprocessor communication at a given traffic level is then reduced. The fact that the delegated tasks are mostly concerned with the controlled hardware means that specialisation can be confined to the slave processors and the central control hardware no longer needs a special-purpose design and general-purpose commercial computer hardware is therefore suitable. Commercial designs of advanced performance are normally available on a shorter time scale than are special-purpose ones. Commercial designs are already being used for management and maintenance functions in public networks and it is probably only a matter of time before they take on the real-time call control function. They already do that for PBXs.

The slave processors are generally off-the-shelf commercial microprocessors mounted on boards containing perhaps special-purpose VLSI logic to interface with the hardware. Typically a slave processor may handle a few tens to a few hundreds of lines; its failure is less catastrophic and less likely because it has fewer components than the failure of a large single processor handling many thousands of lines. If necessary slave processors can be used in a dual configuration to increase availability.

11.3.4 Completely distributed control

There is really no limit to which decentralisation can go and private systems using LANs for transmission have been proposed in which there is no central control at all; each terminal has its own complete set of system software and can communicate with its peers on an equal basis. Some terminals can be assigned special functions, e.g. acting as a gateway to the public network or providing the operator facilities (see Figure 11.3). The price that has to be paid for this is the extra memory needed to replicate most of the software at every terminal. While memory is rapidly getting cheaper the facilities that are expected are becoming more and more complex. For other than the very smallest systems, a compromise, in which frequently used memory is distributed towards the periphery of the network and rarely used memory is concentrated at the centre, looks to be a cost effective solution and one that should endure for some time to come. There are certainly a number of network functions such as call charging, billing, number translation and directory where too much distribution could prove counter productive.

11.3.5 Feature enhancement and the role of software

As Section 11.4 on software complexity explains, software does not always provide the flexible solution that one might have been led to expect. Once switching systems are installed, whether they be PBXs or public exchanges, it is sometimes easier to enhance them by peripheral devices or 'black boxes' rather than to change the centralised call processing software. It is understandable that such features as short code dialling should be contained within the handset on electromechanical exchanges where replacement by stored programme control systems is taking many

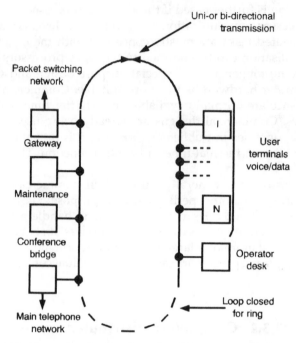

Fig. 11.3 Example of a distributed architecture applied to a LAN

years, but it is less understandable that black boxes should be added to stored program controlled exchanges to provide such features as conferencing or voice message recording. That this type of approach is economic is in large measure due to the availability of cheap microprocessors which are ideally suited to peripheral functions of fair complexity. When a large number of microprocessors would be required, as for the short code dialling case, a VLSI development may well be justifiable and a solution using an off-the-shelf microprocessor may not be adopted at all or, if it is, perhaps only as an interim measure while the market is being tested and the VLSI circuit is being developed.

11.3.6 Benefits of and trends in increased processing power

The processing power of VLSI circuitry can be confidently predicted to go on increasing for a very considerable time to come but, for telecommunication applications, processing power is not a major limitation as it is say in weather forecasting or in some types of signal processing. Increase in processing power can, however, be beneficial in two ways; firstly to reduce the cost of the call processing subsystem by reducing the number of processors needed to achieve a given throughput at a given availability. Secondly, as indicated in the discussion on software in Section 11.3.2, it may be used in part to make the programmer's task easier and hence development costs lower by providing a higher level of abstraction in, for example, the call processing algorithms.

There are in telecommunications a number of peripheral computation-intensive tasks, e.g. adaptive echo cancellers or speech codecs (see Sections 9.4 and 6.3), that are required in sufficiently large numbers as to justify a VLSI development. On the other hand, there are a number of potential centralised functions such as directories where, as will be discussed later, large amounts of power may be required. Whether the function is general purpose or specialised, peripheral or central, the solution lies in VLSI technology and it is therefore relevant to consider a forecast based on extrapolating past behaviour for increase in power and decrease in cost per unit power. Perhaps the most relevant measure as discussed previously in Section 1.2.4 is gate.hertz per unit area of chip. The gates are a measure of processing complexity, the hertz, usually expressed in terms of the clock rate, are a measure of speed and the unit area might be assumed to equate roughly to a constant cost. Gate.hertz/mm^2 are therefore a measure of processing power and the reciprocal is a measure of reducing cost per computed function. Figure 1.3 shows past history based on this measure and a projection to the year 2000. Improvements are clearly quite dramatic and make feasible systems that are of enormous complexity compared with what was possible even five years ago.

11.4 SOFTWARE

11.4.1 Introduction

Software, although not a component part in the same physical sense as a transistor, is nevertheless a vital element in most systems. Software has limitations and it is mainly in its design process that problems arise. We need to understand why this is so and to have a broad understanding of the main constituents of software programs.

The number of telecommunications products that do not contain software of some form or the other is diminishing fast. Even something as simple as a telephone handset will have a stored program if it contains automatic features such as short code dialling or last number repeat. Software design has long been a profession but only comparatively recently has it been widely recognised that such design needs to be controlled using the types of discipline familiar to the engineer; hence the term 'software engineering'. The following is not intended to be in anyway a treatise on how to design software but rather to explain why software design is difficult and how the problems of complexity might be solved.

11.4.2 Software languages

The press is full of talk of fourth generation software languages (which should not be confused with computer hardware which has reached the fifth generation). What does this mean? Previous language generations were:

1. *Binary code*, in which the program instructions are expressed as a pattern of 1s and 0s or bits, or machine code in which the binary numbers are replaced in a slightly more digestible form, e.g. as hexadecimal numbers (i.e. numbers to base 16).

2. *Assembly language*, in which the programmer writes instructions in alpha-numeric form which the assembler (a piece of software) translates into binary code. These instructions are particular to the processor being programmed and generally bear a direct relationship to machine code instructions.

3. *High level languages* such as Algol, Fortran or Cobol, which are intended to be processor independent but may be application specific (e.g. Cobol for commercial and Fortran for scientific computing). Statements in high level languages are transformed in general into many machine code instructions by a piece of software which usually takes the form of a compiler.

Fourth generation languages are high level languages too but differ in the way that they express the problem. Whereas a third generation language such as Fortran specifies the problem and the way that problem is to be solved as a sequence of serial instructions to be run on the target machine; fourth generation languages tend to be based on a particular type of calculus and merely state the problem much as it might be written in a mathematical text book and leave the question of how to turn that problem into a set of machine instructions to the compiler. (For this reason some fourth generation languages may be referred to as 'declarative languages'.) Typical fourth generation languages are Lisp, Prolog and Hope. Again each of these languages is application specific but in a way that depends fairly directly on its mathematical formulation.

In going from assembler language to a third generation language and then to the fourth generation, i.e. in going from the particular to the general, there is an increasing loss of efficiency in the way the problem is transferred to the machine; this affects adversely both storage and run time. The advantage of the third generation language over the first two, in addition to a measure of machine independence, is that it enables the problem to be expressed in more user friendly terms and hence it should be possible to write programs faster and more accurately. The fourth generation language goes further; because there is no inherent serialism in the application programmers input it should enable the problem to be solved faster using a number of processors working in parallel. Furthermore, because of its sound mathematical basis it lends itself more readily to having software algorithms checked for accuracy or more strictly subjected to 'a proof of correctness'. One of the obstacles to the efficient realisation of fourth generation languages is that there is gap in our ability to translate problems from those languages into instructions to be executed on processors and inexperience in the design of multiprocessor architectures to match specific languages. Research on both these topics is being very actively pursued, not least in Japan, where fifth generation computers in the Icot (the institute for new generation computer technology) project are aiming to combine the latest advances in highly parallel computer architectures with the use of Prolog as the fourth generation language.

The higher demands placed on storage by high-level languages offset to some extent the remarkable reductions in the cost of storage that have and will continue to take place.

11.4.3 The problem of complexity

Why does software take so long to develop, why is it so prone to errors, why in spite of the improvements in technique does the situation not improve very much and why is the flexibility claimed for it in its early days not realised? In a word the answer is complexity. The improvements in technique are not used so much to reduce errors but to write larger and more complicated programs. The art of the possible holds sway over the science of the prudent. In order to write large software programs the problem has to be broken up into manageable entities usually known as 'modules' each of which is given to a software designer to implement. Unless the interfaces to these modules are accurately and completely specified before they are written, there is a major problem and resultant chaos. Thus a very considerable amount of preliminary descriptive specification work needs to be done before attempting to write code. There is ample evidence that a pound spent at this stage of the design is worth many hundreds or even thousands of pounds which are the costs if problems have to be put right or changes or additions made at a late stage in the development or worse still in the field. Nevertheless pressure to meet time scales often results in short cuts that are regretted later. The communication problem can be alleviated if small teams of very able programmers are employed in preference to larger teams of less able ones. Some of the most successful software packages have in fact been the product of one person's labour.

The longer-term answer is covered by the broad heading of software engineering, where such techniques as structured design, frequent design audits and proofs of correctness are taken as a matter of course. There is a need to provide sophisticated tools or design environments in which software engineering can take place and a number of such developments are under way with various acronyms such as IPSE (Integrated Program Support Environment).

11.4.4 Operating systems

A means of shielding the application programmer from machine-specific knowledge and the more routine aspects of complexity is the provision of an operating system. Operating systems perform such functions as input/ouput, store management, multi-user operation and scheduling. All but the most simple software packages tend these days to run under an operating system and operating systems vary in complexity enormously from DOS say on a PC via UNIX on a PC or mini to an IBM or ICL operating system on a mainframe. Again there is a price to be paid in terms of storage and run-time efficiency in going from the particular to the general. There is pressure to standardise on operating systems and to make them machine independent and the front runner in this context is currently UNIX, although UNIX is becoming subject to criticism on the grounds that in the standardisation process it has become too complicated, inefficient and large.

Because.of their ubiquity, operating systems are a favourite target for attacking a system using a software virus. The virus may exist in any piece of purchased software (but is most likely in pirated software) and once that piece of software has been loaded into the computer, the virus perpetrator is able to use his knowledge of the operating system to create the maximum havoc. To date telecommunication control software, largely because of its highly specialised nature, has not been reported as suffering viral attack. But in the future with telecommunication systems using UNIX, say, they may become more vulnerable. Software viruses can only enter via a program and so programs being conveyed from one point to another by a telecommunications network are vulnerable to the clever criminal whilst in transit.

11.4.5 Expert systems and Intelligent Knowledge Based Systems (IKBS)

Expert systems and more recently IKBS have shown considerably more promise than performance. Expert systems represent a particular class of IKBS but without attempting to draw a firm distinction between the two it should suffice to say that both are aimed at providing specialised knowledge in an intelligent fashion by making use of the latest software and knowledge engineering concepts. Expert systems for example are available to help in medical diagnosis and fault finding in telephone exchanges. The specialised knowledge of the medical consultant or the exchange designer is encapsulated in software in a way that makes it easy for those less expert to retrieve it readily. It has not proved to be a simple task in general to formulate the rules and provide satisfactory man–machine interfaces. As a tool expert systems seem to be of more use as a kind of reference book than they are as teaching aids.

IKBS aims are to develop a common framework and man–machine interface in which expert systems and other systems such as image understanding/recognition systems can be developed and implemented using common software and artificial intelligence concepts. It is early days to predict what the future of IKBS is going to be.

11.4.6 Neural nets

Pattern recognition is relevant in the context of optical character readers (OCR) (see Section 9.2) and it seems appropriate to say a few words on the topic of neural networks (or nets) which could be important here (although neural nets are not the basis of currently available OCR equipment). Human and animal brains are much better at pattern recognition than are computers and it is a fairly obvious strategy to try to model a computer on the human brain. The workings of the brain are extremely complex and biologists are still at a fairly low level of understanding.

Nevertheless there have been numerous attempts to build electronic and optical devices that correspond to basic brain building blocks which consist of brain cells joined to each other by connections called 'synapses'. Synapses are a very complex pattern of weighted interconnections so that a nerve cell can be represented as a threshold device connected to many other nerve cells by simulated synapses. Pattern recognition devices of this type are 'taught' by being presented with a set of reference patterns, during which time the weights are adjusted to achieve maximum separation between what are defined as unlike patterns and minimum separation between what are regarded by humans as the same pattern (e.g. the same alphanumeric character in different type fonts).

Figure 11.4 shows how neural nets are generally organised into layers with pattern receptors as the first layer and decisions coming from the last layer. Figure 11.5 shows a hypothetical case where the system is required to recognise a damaged character 'R', which it does by giving that result a higher probability than the nearest neighbours 'P' and 'B'.

There is some controversy as to whether the building blocks should be close analogies of brain cells or should be chosen to match what current technology can provide efficiently. In either case the resultant machine is distinguished from a conventional computer in that its 'program' resides in its connectivity rather than in a separate entity called a memory. In other words memory and processing are fully distributed. This means that unlike digital computers, which can fail completely if one component fails, neural nets degrade gradually as components fail.

The use of neural nets for expert systems is possible; instead of rules having to be written as algorithms as for a conventional computer, the machine is taught by

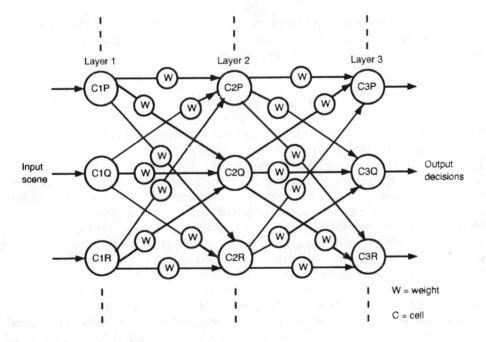

Fig. 11.4 Three level nerve cell—synapse analogue

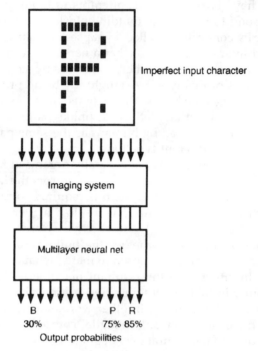

Fig. 11.5 Conceptual diagram of a character recogniser based on a neural network

example as very often human beings are by watching over someone else's shoulder. Which of these methods will prove the more satisfactory in the long term remains to be seen; there should be scope for both.

11.4.7 VLSI design

VLSI circuit design is included here because it involves orders of complexity comparable with those that apply to software and thus gives rise to similar problems. The situation is not as bad because layout rules, for example, must be fully specified and it is a mechanical process, much easier than proof of correctness for software, to ensure that they are obeyed. The solution is to provide VLSI development systems that are fully software based using software packages that will interface to each other as the design proceeds through its various stages. Starting with a green fields situation this would be relatively easy to achieve, but in practice with various evolutionary proprietary systems dedicated to a particular phase of the design cycle and running on some specific make of processor it is not so easy. The effects of making mistakes are serious because the time to reprocess (i.e. refabricate) a prototype circuit can be many weeks and if faults are found after the product has been sold in quantity the cost of replacement can be very high indeed. The term 'silicon compiler' is sometimes used in connection with the automated fabrication of VLSI circuits.

This raises the issue of testability, which obviously increases in difficulty with complexity; one answer to this problem, particularly applicable as discussed above to WSI, is to provide a measure of testing built into the chip itself.

11.4.8 Summary

This chapter has laid considerable emphasis on the question of complexity for both software and VLSI design. We have seen that the tools for dealing with complexity have difficulty in keeping up with the possibilities that advanced technology offers. With product life decreasing to the point where a telecommunication system's life expectancy has fallen from say 20 years to less than 10, the cost and time scale of development becomes a much greater proportion of the product life cycle. When the complexity lies in software the reusability of that software in another hardware environment becomes an important consideration. We have seen that software is in fact becoming very much more divorced from the machine on which it runs, making its reuse potential much greater. Of course new software algorithms will always be required and the software closely associated with the switching hardware will need to change as the hardware changes, but there will be enduring software associated, for example, with common channel signalling, which may well be applicable to several generations of hardware.

The increasing intelligence potentially available from software, whether realised in conventional computer form or as pattern recognition devices such as neural nets, will surely impact the telecommunication network as a whole so as to give it more intelligence too.

11.5 BIBLIOGRAPHY

J. L. Hennessy and N. L. Jouppi (1991) Computer technology and architecture: an evolving interaction, *IEEE Computer*, September, 18–29.
C. B. Jones (1980) *Software Development—a Rigorous Approach*, Prentice-Hall.

Part 2

NETWORKS AND SERVICES

OVERVIEW

After a century of fairly slow growth in choice for networks and services, in the 1980s we experienced and in the 1990s we expect to experience a major expansion in both fields. In Part 2 we examine what is new in the context of what went before.

It is expedient in a number of cases to treat the network and service aspects under the same heading and we do so in Chapters 12 on mobile systems, 13 on satellites, 14 on cable TV, 15 on ISDN, 18 on intelligent networks and 20 on UPT. Chapters 16 and 17 on data and multiservice divide roughly into network and service aspects respectively, but 17 contains some service aspects of networks discussed in other chapters where these form part of the overall picture. In Chapters 19 on VPN and centrex and 21 on directories the emphasis is on services.

Among the networks treated in Chapter 16 are passive optical networks, LANs MANs and WANs. Services treated in Chapter 17 include voice, video telephony, messaging, bulk data, document and videotex.

12 MOBILE RADIO SYSTEMS

12.1 INTRODUCTION

This chapter covers the more advanced forms of mobile radio systems ranging from radio paging systems and private mobile radio, through analogue cellular to the new pan European digital cellular system formerly called Groupe Speciale Mobile (GSM), after the name of the CEPT committee where the standardisation work started (GSM is now called Global System for Mobile communications), Personal Communications Networks (PCN), and the research programme on the Universal Mobile Telecommunications System (UMTS).

It is not possible to cover such a wide range of systems in great depth within a single chapter but the aim is to give sufficient information for the reader to understand the main features and basic operation of the systems. The reader is assumed to have read the chapter on radio technology and so be familiar with basic radio concepts and terminology.

12.2 PAGING SYSTEMS

A paging system is a one-way radio system for transmitting data. Paging systems were developed initially for short range (1 km) applications, e.g. hospitals, but the technology has developed such that wide area and nationwide systems are now available.

From the customer's point of view, pagers can fall into three categories: those which emit a single tone, those which emit a range of tones with each tone signifying a particular pre-defined message, e.g. ring the office, and those which display an alphanumeric message.

There are many different standards for paging signals but basically they fall into two types.

Tone paging

This is not to be confused with the reference above to tones emitted for the user. In tone paging, the transmitter sends out signals in the form of a sequence of M

tones, from a defined set of N tones, each tone with a duration of T seconds, and these tones are modulated onto a carrier. The transmitter is thus able to select one out of a total of NM pagers, and a calling rate of $1/M(T + T_1)$ calls per second is possible where T_1 is the guard time to mark the end of a paging signal. There are several different standards, but values of $M = 5$, $N = 10$ and $T = 30$–100 ms are common giving calling rates of 1 to 5 calls per second.

Digital paging

In digital paging, the transmitter sends out a stream of bits which contain the address of the pager and can also contain a message. Again there are several different standards and the following section gives a brief description of the POCSAG standard which was developed for the British Post Office and has been accepted as a CCIR Radio Paging Code No. 1.

12.2.1 POCSAG or CCIR Radio Paging Code No. 1

In POCSAG the transmitter sends out bursts of batches of code words (see Figure 12.1). Each burst begins with a 1.125 s preamble and is followed by a number of batches each of 1.0625 s duration. The bit rate is 512 bit/s and FSK modulation is used. Each batch consists of 17 code words, the first of which is a synchronisation code word. The remaining 16 code words of 32 bits each contain an initial bit to

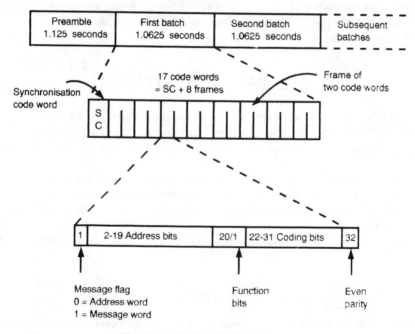

Fig. 12.1 POCSAG signal format

say if the code word is an address word or message word, 20 address or information bits, 10 coding bits for error detection and correction (using a BCH code), and a final parity bit. The 16 code words form 8 frames (2 code words each) and an individual pager will be addressed only in a specified frame so it needs to activate all its circuitry only for the duration of that frame thus increasing battery life. Message words always follow the address word for the pager in question, and there is no limit to message length. The message can use the full range of ASCII or International Alphabet No. 5 characters. POCSAG can handle two million pagers and provide a calling rate of 15 calls/s if no messages are sent. An uprated version of POCSAG operating at 1024 bits/s has recently been defined.

Signals to a pager will be subject to loss owing to screening and multipath propagation and, if all the transmitters use the same frequency, to interference effects between transmitters. As a pager moves around, the strongest signal may be received from one transmitter at one minute and another the next. To ensure that this effect does not disturb the reception of the signal or affect the order of the bits received, the transmitters must be approximately synchronised or quasi-synchronous and the total spread in the time of arrival of signals must not exceed 25% of the duration of a bit. This requirement constrains the maximum differential time delay to about 240 µs for POCSAG, which in turn means that transmitters must be no more than 40 miles apart. Transmitter powers are typically 100 W and frequencies at 138 MHz and 153 MHz are used.

To make more effective use of the paging channels. the UK can be divided into zones with transmission bursts in adjacent zones being sequential while different transmissions in well-spaced zones take place simultaneously.

Several countries within Europe, including the UK, France and Germany, are implementing a pan-European analogue tone paging system called Euromessage which operates at 466 MHz with a high speed version of POCSAG (CCIR No. 1).

12.2.2 ERMES

For the longer term the Paging System (PS) technical committee of ETSI has nearly completed the definition of a new digital paging system called European Radio MEssage System (ERMES) which will offer tone and alphanumeric services and is scheduled to commence operation in late 1992 on eight channels in the band 169.6–169.8 MHz with a further eight channels to be added in about 1995.

The effective radiated power (ERP) of the transmitters will be about 100 W, and channel spacing will be 25 kHz and the co-channel protection ratio needed will be about 15–20 dB. Transmissions will be at 6.25 kbit/s (or 3.125 kbaud/s) using a 4-level pulse amplitude modulation/frequency modulation scheme. Area coverage and the sharing of the available frequencies by different operators can be achieved by either frequency division or time division and neighbouring transmitters within an area will operate quasi-synchronously. It will be possible to have flexible time division arrangements to cater for variations in traffic density.

The ERMES system is defined in ETS 300 133, which has the following parts:

Part 1: General Aspects

Part 2: Service Aspects

Part 3: Network Aspects

Part 4: Air Interface Specification

Part 5: Receiver Conformance Specification

Part 6: Base Station Specification

Part 7: Operation and Maintenance Aspects

12.3 PRIVATE MOBILE RADIO

The earlier types of private mobile radio system were ones where radio channels were allocated to individual users who operated their own base stations. During the mid-1980s trunked systems were introduced because they provided much more efficient use of the radio spectrum.

12.3.1 Exclusive channel private mobile radio

The older types of PMR systems provide two way simplex communications in the vhf and uhf frequency bands. The early systems used AM but most systems now use FM. The channel spacing is normally 12.5 kHz.

12.3.2 Trunked private mobile radio

UK system

In the UK a new system of trunked Private Mobile Radio (PMR) has been introduced in a part (sub-band 2) of the band (174–225 MHz) used formerly for Band III 405-line television. This service is called trunked because private users such as taxi firms share the same base station and radio channels instead of having their own base stations and frequency allocations. Trunking therefore increases the efficiency of use of the spectrum. In addition to trunking the new PMR system provides automatic queuing of calls, which allows the system to be operated closer to maximum capacity and gives more graceful degradation in the quality of service in times of severe congestion.

The trunked system is run by licensed operators who interconnect suitably located base stations by a network of leased lines and switches which may be enhanced PBXs. Because a mobile may travel to any area the network maintains a record of the location of the mobiles so that it can route calls to the most appropriate base station. The dispatchers on the customer's premises can be connected to the network either by leased line or by radio, in which case they appear themselves as a mobile. Figure 12.2 shows a trunked PMR system. Individual base stations can provide coverage of up to about 25 miles radius.

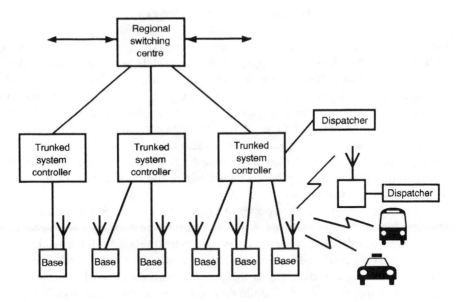

Notes:
(1) Trunked system controller normally serves up to 20 bases
(2) Dispatcher may be connected to TSC by radio or private circuit

Fig. 12.2 Trunked PMR network

The UK trunked PMR service operates in the band 192–207 MHz and uses frequency modulation with a channel spacing of 12.5 kHz. Speech communications are half duplex (only one way at a time), with an average call duration of about 30 s which is much less than for cellular. Calls cannot exceed a duration that is determined by a time-out set by the service provider. A dedicated channel is used for system control. The air interface is defined in MPT 1343, and the signalling protocol in MPT 1327. (The MPT standards are obtainable from Radiocommunications Agency, Department of Trade and Industry, Waterloo Bridge House, Waterloo Road, London SE1 8UA. MPT stands for Ministry of Posts and Telecommunications—a name which dates from the early 1970s.)

Trunked PMR schemes are also being used for private applications by the electricity, gas and coal industries in the bands 139.5–140.5 MHz and 148–149 MHz.

The performance of the new trunked PMR system is superior to that of the older systems and the volume of traffic is expected to grow rapidly.

New European system

The RES6 committee of ETSI is studying the development of a digital trunked PMR system and plans to publish draft standards at the end of 1992.

12.3.3 Mobile data service

Nearly all private mobile radio systems at present provide voice services, but interest in data services is beginning to develop. Data services could be designed to communicate and display short messages of text which would be coded in character form and they could be used widely to replace many of the current applications for private voice services such as giving instructions to delivery vans or taxis. From the point of view of the user they would have the advantage of providing a message which could be stored and so overcome the problem of a voice message being forgotten. However because a driver would need to stop to enter a message, they would tend to be preferred where the majority of traffic went from the base station to the mobile.

The efficiency of use of the spectrum would be very high for data because the messages would normally be much shorter than voice messages and because short delays could be tolerated allowing messages to be queued with the result that very high usage levels could be achieved. Consequently many more subscribers could be served from a given bandwidth than would be the case with voice.

In the UK, three licences for mobile data services at uhf and one in Band III have recently been issued.

12.4   CORDLESS TELEPHONES AND TELEPOINT

12.4.1 Analogue cordless telephones CT1

Analogue cordless telephones operate as extensions to the Public Switched Telephone Network using frequency modulation. The following details refer to the UK system; different national standards are used in other countries.

UK handsets operate on any of eight channels in two widely separated bands. The base units transmit in the band 1.642–1.782 MHz (20 kHz spacing) while the portable units use the band 47.456–47.543 MHz (12.5 kHz spacing). The maximum handset power is 10 mW and the maximum range is around 200 m. Sets are built to operate on only one or two of the eight available channels; when a set can operate on two channels, the user can select the channel with a switch. This facility is valuable if another cordless telephone in the vicinity is also assigned to one of the channels. Standards MPT 1322 and MPT 1371 apply.

An improved version of the analogue cordless telephone was introduced after the basic version had been available for several years and is known colloquially as CT1.5. The improvements include the use of more codes to identify handset-base station pairs for greater security, and much more frequent handshaking during a call to protect against the effects of interference from another handset-base station pair which, in the earlier versions, could capture the call and prevent the user from clearing the PSTN call. A longer range version is now under consideration.

12.4.2 Digital cordless telephones CT2

The digital cordless telephone known as CT2 is a new technology with at least three applications—the conventional cordless telephone, the Telepoint service, and the cordless PABX.

The Telepoint service is a new service introduced in 1989 in the UK, whereby the owner of a CT2 handset and a subscriber to the service can make calls away from home via a public base station. Initially different CT2 standards were developed by Ferranti and Shaye but subsequently agreement was reached on a Common Air Interface (MPT 1375) which all Telepoint services had to provide from 31 December 1990. The use of the Common Air Interface was intended to facilitate roaming between systems and so improve service coverage for the user. Although the UK Government issued four licences for the operation of Telepoint services, the three licensees who began operations have all withdrawn their services because the number of subscribers was very low. The remaining licensee, Hutchison, has recently commenced operations. Despite these problems, there is considerable interest in Telepoint in some other European countries, for example France where the service is known as Pointel.

In the UK, CT2 has been allocated a 4 MHz band from 864 to 868 MHz which contains 40 channels of 100 kHz each. CT2 is the first operational radio system to use time division duplexing where both the handset and the base station transmit and receive on the same RF channel using time interleaved bursts (ping-pong) in alternative directions, with in this case with a burst frequency of 500 Hz. Both the handset and base station have a peak power of 10 mW and a mean power of 5 mW.

The system also has a novel self-trunking algorithm. Each mobile terminal and base station is capable of operating on all 40 channels, and when a call is initiated from the mobile terminal, that terminal scans the band and selects the channel with the lowest signal power and attempts to establish a call. Consequently in a dense traffic situation, a caller who is closer to a base station may cause a call from a more distant caller to be terminated prematurely. CT2 has a maximum range of up to a few hundred metres but in a congested area the effective maximum range may be considerably lower owing to adjacent channel interference, and the theoretical maximum of 40 channels operating simultaneously may not be achieved unless all users are at a similar range.

The transmission rate of the system is 72 kbit/s and the modulation is Gaussian shaped FSK. The burst lengths (except Multiplex 3) are 66 or 68 bits and the bursts in each direction are interleaved with a guard time equivalent to 12 or 8 bits respectively.

Three different burst structures are defined (see Figure 12.3). Multiplex 3, which is a 10 ms burst of synchronisation and signalling (D channel) followed by a 4 ms receive slot, is used only from the handset to the base station to initiate a call. Multiplex 2 is a standard 1 ms burst of synchronisation and signalling which the base station uses to initiate a call and which both base station and handset use for signalling before switching to Multiplex 1. Because the handset is not synchronised to the base station when the handset attempts to initiate a call, Multiplex 3 contains sub-muxs which are sent four times to ensure that at least one sub-mux is received satisfactorily during a period when the base station is not transmitting. The base

(a) Multiplex 3 (handset to base only)

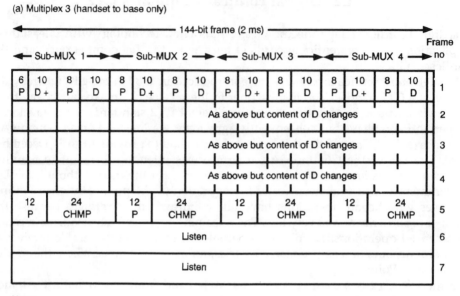

Notes:
(1) The 20 bits of the D-channel are repeated in each sub-MUX of each frame.
(2) D-channel always begins with a sync word in slots marked +.
(3) P = preamble in syn channel.
(4) CHMP = channel marker for portable.

(b) Multiplex 2

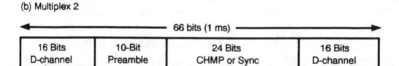

(c) Multiplexes 1.2 and 1.4

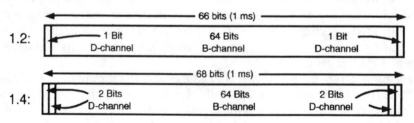

Fig. 12.3 CT2 frame structure

station then re-initiates the call using Multiplex 2. Both Multiplexes 2 and 3 are used only for call set up. Multiplex 1 is used to carry the B (32 kbit/s) and D (2 kbit/s or 1 kbit/s) channels over an established link. Speech is carried as 32 kbit/s ADPCM. The signalling in the D channel follows the ISDN packet form fairly closely. The system includes complex handshaking and authorisation procedures for security.

Although CT2 is capable of incoming call handling, and incoming call handling is used for cordless telephones and PBXs, Telepoint was only permitted initially to handle outgoing calls, but in 1990 the UK Government agreed to allow it to handle incoming calls to visitors who log-on to the nearest base station, and in the Duopoly Review the Government indicated its intention to permit Neighbourhood Telepoint services with incoming call capability. Manufacturers are expected to include message pagers in the more expensive handsets to enable users to receive 'call me back' requests.

The CT2 standard has been adopted by ETSI as an interim European standard I-ETS 300 131.

12.4.3 Digital European Cordless Telephone (DECT)

The RES3 Technical Subcommittee of the European Technical Standards Institute (ETSI) has prepared standards for a European digital cordless telephone system known as DECT for operation in a 20 MHz band from 1.880 to 1.900 GHz. DECT is defined in ETS 300 175, which has the following parts:

Part 1: Overview

Part 2: Physical layer

Part 3: Medium Access Control layer

Part 4: Data Link Control layer

Part 5: Network layer

Part 6: Identities and Addressing

Part 7: Security features

Part 8: Speech Coding and transmission

Part 9: Public Access Profile

The approval test specification is given in ETS 300 176.

The basic principles of DECT are similar to CT2 except that the signal structure is narrow band TDMA (as opposed to one channel per carrier) with 12 channels sharing a single carrier with a transmitted (or net) bit rate of 1152 kbit/s. Channel spacing is 1.728 MHz and several carriers may be used simultaneously at different frequencies in the band, so that the overall scheme is TDMA–FDM. Each of the 12 TDMA channels normally carries 32 kbit/s ADPCM speech but there can be more channels carrying GSM half-rate speech at 7 kbit/s. The use of narrow band TDMA means that there can be flexibility in bit rate and bit rates for data of up to 384 kbit/s can be accommodated. Like CT2, DECT uses time division duplexing with both directions of transmission being time interleaved on the same frequency. DECT also uses a self-trunking technique (or dynamic channel allocation) where each base station and handset has access to the whole band and is able to select any unused TDMA channel at any frequency, although this sharing of carriers through TDMA makes the allocation algorithms more complex.

The transmitters have a peak power of 250 mW (or 10 mW mean per channel). The maximum range is about 400 m which is determined by the absence of equalisation to control multipath propagation and by the guard times between the transmit and receive parts of the time division duplexed frame structure. The modulation method is GMSK.

DECT does not include equalisation, frequency hopping or bit interleaving which are used in GSM . Consequently the transmission delay is limited to about 15 ms, which is sufficiently low to avoid making the use of echo cancellers essential. In addition to the use of a low bit rate coding for speech, DECT is designed to be as compatible as possible with ISDN signalling and data transmission.

12.4.4   CT3

CT3 is a proprietary digital system developed by Ericssons. CT3 was developed slightly earlier than DECT and was intended to form the basis for the DECT standards although it uses a different frequency band. Its development has undoubtedly influenced the work on the DECT standards, but a proposal for ETSI to adopt the CT3 specifications as interim standards was rejected, and Ericssons are now planning to realign their products to comply with the DECT standards.

12.5   CELLULAR RADIO

12.5.1   Analogue cellular radio

Cellular radio is a two-way mobile telephone system which normally provides contiguous coverage through the use of overlapping radio cells. The available frequency band is divided up between a number of cells which form a repeat pattern, and the band is reused many times as the pattern is repeated across the country. The size of the cells can be varied according to the traffic density with smaller cells and a lower power transmission used in urban areas giving a smaller repeat pattern and greater capacity, and larger cells used in rural areas. Thus in urban areas with dense traffic the cell size is determined by the traffic density, whereas in rural areas it is determined by the propagation limit, which in turn is set by the power in the link budget.

Figure 12.4 shows a typical cellular network. Communication to the mobiles in each cell takes place through a base station. Base stations can be located centrally in the cells, in which case there is one cell per base station, or they can be located at the corner of a cell in which case one base station can serve several cells. The basic cell structure and layout is normally hexagonal and some cells may be divided into three or six sectors. A seven cell repeat pattern is common, although either a four cell pattern with threefold sectorisation or a three cell pattern with six-fold sectorisation may be used in areas of dense traffic. Functionally a sector is

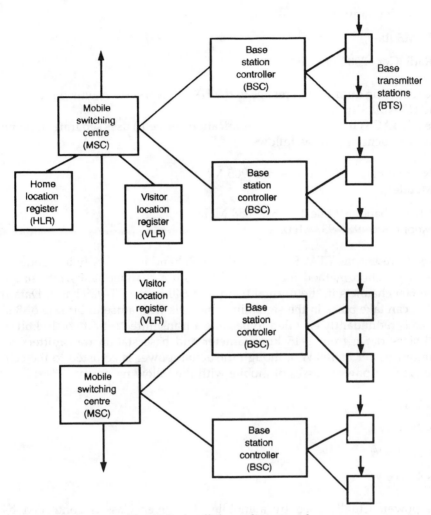

Fig. 12.4 A cellular radio network

very similar to a cell except that the sectors that make up a cell are all fed from the same base station site.

The base stations are connected by fixed links to switching centres for mobiles called (somewhat ambiguously) mobile switching centres (MSCs) which maintain records of the location of the mobiles in vehicle location registers, route calls between mobiles and the PSTN, carry out call accounting, and manage the hand off procedure to maintain a call which is in progress when a mobile moves from one cell to another.

There are several different types of analogue cellular systems:

- TACS UK, Ireland;
- NMT450 and NMT900 Scandinavia, Benelux, Spain, Austria;

- C450 Germany;

- RTMS Italy;

- RadioCom 2000 France.

The following information relates to the UK's TACS system, which is a derivative of the US AMPS system.

The UK TACS (Total Access Communications System) uses separate transmit and receive frequency bands as follows:

- Normal Band: Mobile Tx—890–905 MHz
  Mobile Rx—935–950 MHz

- ETACS Band: Mobile Tx—872–888 MHz
  Mobile Rx—917–933 MHz

where at present the ETACS (Extended TACS) band is used only in London.

The modulation method for voice is FM with a channel bandwidth of 25 kHz giving 600 channels in the normal band and 640 in the ETACS band. Data transmission can take place in the signalling channels with a useful bit rate of 8 kbit/s but this is redundantly encoded to become a transmitted bit rate of 16 kbit/s.

Cell sizes range from 1–15 km diameter and base station transmitters have a maximum power of 100 W, although the actual power is adjusted to the cell size. There are four power classes of mobile with the following powers:

Class 1: 10 W

Class 2: 4 W

Class 3: 1.6 W

Class 4: 0.6 W

The power actually used by a mobile of a given class is controlled by the base station and so the mobile will frequently operate at a power lower than the maximum for its class. All mobiles can tune to any channel except that some earlier models are limited to the normal band and cannot work in the ETACS band.

Within each network 21 radio channels are assigned for system management and signalling and all such channel assignments are programmed in to the mobiles. These channels carry three different logical control channels which are normally multiplexed onto the same radio channel. They are:

- dedicated control channel for the mobile to monitor the base station signal strength and for other functions;

- paging channel to alert mobiles to an incoming call;

- access channel for mobiles to acknowledge paging messages and to request outgoing calls.

When the mobile is idling (not making a call), it monitors the paging channel to see if it needs to receive an incoming call. If the received signal strength drops below an acceptable level, it scans for a stronger signal and re-registers via the access channel.

Mobiles are capable of operating on only one channel at a time. Calls are established via the control channel but on completion of the establishment protocol, the mobile switches to the voice channel assigned by the MSC. During a call, the MSC monitors the signal strength received at the base station in use and at adjacent base stations, and when it becomes necessary to hand-off the call to an adjacent base station, the MSC allocates a new channel and instructs the mobile which frequency to re-tune to. During hand off there is a 400 ms break in transmission. Some signalling can take place on the voice channel during the course of a call.

All established voice channels carry one of three Supervisory Audio Tones (SATs) of around 6 kHz (above the voice band). A single SAT is transmitted by base stations over a cluster of cells which form a repeat pattern. If there is a deep fade in a transmission between a base and a mobile such that the mobile receives a stronger signal on the same frequency from a distant base station which is communicating with another mobile but with a different SAT, the mobile mutes the voice communications to prevent the other conversation being overheard.

12.5.2 Pan-European digital cellular radio (GSM)

The Pan-European digital cellular radio system is known as GSM after the body called Groupe Speciale Mobile, which was initially under CEPT but is now under ETSI and which was given the task of specifying a European cellular radio system for operation from 1991. (The system has been renamed Global System for Mobile communications.) A central requirement for GSM was that it should provide pan European roaming so that subscribers could use their equipment anywhere in Europe without having to make special arrangements with the local operator. Another requirement was that the efficiency of use of the spectrum should be considerably better than that for analogue cellular radio.

The ETSI standards for GSM are listed in Table 12.1.

Radio aspects

GSM uses the following frequency bands:

Mobile Tx: 905–915 MHz

Mobile Rx: 950–960 MHz

which are adjacent to the UK's bands for analogue cellular radio and which will facilitate the transfer of spectrum from the analogue service to GSM in the future when subscribers migrate in that direction. Unlike CT2 and DECT, GSM does not use time division duplexing.

GSM mobiles are divided into the following power classes:

Table 12.1 ETSI GSM standards

I–ETS 300 020 Pt1	Mobile station type approvals procedure principles and mobile station conformity specifications (NET 10 Pt1)
I–ETS 300 020 Pt2	Mobile station conformance test system, system simulator specification (NET 10 Pt2)
I–ETS 300 021	Mobile station–base station system interface data link layer
I–ETS 300 022	Mobile radio interface layer 3
I–ETS 300 023	Point-to-point short-message service support on mobile radio interface
I–ETS 300 024	Short-message service cell broadcast support on mobile radio interface
I–ETS 300 025	Rate adaptation on the mobile station–base station system interface
I–ETS 300 026	Radio link protocol for data and telematic services on the mobile station–base station system interface and the base station–mobile services switching centre interface
I–ETS 300 027	Mobile radio layer 3 supplementary services specification, formats and coding
I–ETS 300 028	Mobile radio interface layer 3 call offering supplementary services specification
I–ETS 300 029	Mobile radio interface layer 3 call restriction supplementary services specification
I–ETS 300 030	Multiplexing and multiple access on the radio path
I–ETS 300 031	Channel coding
I–ETS 300 032	Modulation
I–ETS 300 033	Radio transmission and reception
I–ETS 300 034	Radio subsystem link control
I–ETS 300 035	Radio subsystem synchronisation
I–ETS 300 036	Full rate speech coding
I–ETS 300 037	Substitution and muting of lost frames for full rate speech traffic channels
I–ETS 300 038	Comfort noise aspects for full rate speech coding
I–ETS 300 039	Discontinuous transmission for full rate speech traffic channels
I–ETS 300 040	Voice activity detection
I–ETS 300 041	General on-terminal adaptation functions for mobile stations
I–ETS 300 042	Terminal adaptation functions for services using asynchronous bearer capabilities
I–ETS 300 043	Terminal adaptation functions for services using synchronous bearer capabilities
I–ETS 300 044	Mobile application part
I–ETS 300 045	Subscriber identity module—Mobile equipment interface
I–ETS 300 068	Man–machine interface of the mobile station
I–ETS 300 069	Technical realisation of the short message service cell broadcast
I–ETS 300 070	Technical realisation of facsimile Group 3 transparent
I–ETS 300 071	Technical realisation of facsimile Group 3 non-transparent

Class 1: 20 W peak (for vehicles)

Class 2: 8 W peak

Class 3: 5 W peak (for hand portables)

Class 4: 2 W peak

Class 5: 0.8 W peak.

The system is designed for a maximum range of 35 km and like analogue cellular radio the transmitter power is controlled by the base station within the limits of the mobile's class.

The choice of radio transmission method for GSM gave rise to considerable debate, there being three main options:

- FDMA/SCPC (Frequency Division Multiple Access/Single Channel Per Carrier) where each channel uses a separate carrier;

- wideband TDMA where a single carrier is modulated to cover the whole band;

- narrowband TDMA (or FDM-TDMA) where a number of carriers operate at different frequencies and each is shared by several channels using TDMA.

The FDMA/SCPC approach has several disadvantages. First it requires high performance filtering to achieve a narrow bandwidth and a high performance diplexer to protect its receiver from its own transmitter. Neither are capable of being realised in VLSI and would add cost and physical volume to the handset. Second the bandwidth is within the coherence bandwidth of Rayleigh fading so that a fade would affect the whole signal. (A fade occurs with multipath propagation when signals travelling over different paths cancel each other.) Third it offered little scope for flexibility in bit rate for the signals carried.

At the other extreme, because the bandwidth exceeds the coherence bandwidth of the fade, wideband TDMA offers the opportunity to use equalisation to overcome the Rayleigh fading. It is suitable for realisation in VLSI and offers bit rate flexibility. However, it has the disadvantages of requiring a high peak power and large amounts of processing which is demanding in terms of battery life in hand-held portables.

The system chosen is a narrowband TDMA structure with 8 mobiles sharing a 200 kHz radio channel; 200 kHz is sufficient to exceed the coherence bandwidth of the fading so that an equaliser can be used to good effect. The mobile's transmitter and receiver are allocated both different carrier frequencies and time slots, therefore the transmitter and receiver do not operate simultaneously and diplexers are unnecessary. Base stations, which cannot avoid simultaneous transmission and reception, do need diplexers. (Diplexers are very high performance filters that protect a receiver input from being swamped by the transmitter output. They are often bulky, lossy and expensive.)

The further the mobile is from the base station, the longer signals take to arrive at it, and so TDMA requires accurate synchronisation of the transmissions from the mobiles to ensure that received signals do not overlap at the base station. The times for transmission are calculated at the base station and synchronisation information is sent to the mobile. Without such a procedure it would be necessary to allow relatively long guard times which would increase with the size of the cell.

Multipath propagation results in a dispersion of the signal arrival time which is much greater than the duration of one bit. An equaliser is used to model the path and correct for the dispersion. Because with a fast moving vehicle the geometry of the path can change rapidly, a training or initialising sequence is included in every time slot in order to set up the equaliser's parameters.

To reduce further the effects of multipath propagation and to spread the effect of interference (which may be more severe at some frequencies than others), frequency hopping may be used with the frequency changes being made between bursts.

Speech is coded using a regular pulse excitation long term prediction linear predictive coder into a data stream of 13 kbit/s, although there is an option in future to halve the coding rate to double the spectrum efficiency and system capacity, (see Section 6.3.7). Figure 12.5 (which should be read from the bottom up) shows the GSM frame structure. The coded speech is divided into 20 ms blocks of 260 bits each and the first 182 bits of each block are extended to 189 by parity and spare bits and are then encoded for error detection and correction using a convolutional code which adds a further 189 bits, and then recombined with the remaining 78 bits to give a total error-protected block length of 456 bits. The bits that are error protected are those for which accuracy is the more critical to the voice coder.

The 456 bit blocks are then divided into 8 sub-blocks of 57 bits which are diagonally interleaved with sub blocks from the adjacent blocks. The purpose of this block interleaving is to spread and so dilute by dispersion in time the effect of any deep fade and so give the coding scheme more opportunity to work successfully.

Pairs of the sub blocks of 57 bits each are combined with a 26 bit training sequence for the equaliser and 2 control bits and 3 header and 3 tail bits into a time slot of 148 bits. Time slots from each of the eight mobiles are combined sequentially with guard times equivalent to 8.25 bits into a TDMA frame with 4.6 ms duration. 24 of these TDMA frames are combined with 2 supervisory and control frames which have a channel-associated structure and contain the signalling information for the 8 mobiles. The overall bit rate is 270 kbit/s.

In addition to the control frames referred to above, the system includes:

- dedicated control channel for registration, location updating, authentication and call set-up;

- broadcast control channel (base to mobile only) to give base station identity and information;

- random access channel (mobile to base only) to request service;

- access grant channel (base to mobile only) to assign a dedicated control channel for signalling;

- paging channel (base to mobile only) to indicate an incoming call;

- fast signalling channels which replace the traffic channel during hand-over.

From the point of view of performance there are two significant features.

1. The full rate speech coder introduces a noticeable quantisation distortion. Quantisation distortion is normally measured in quantisation distortion units

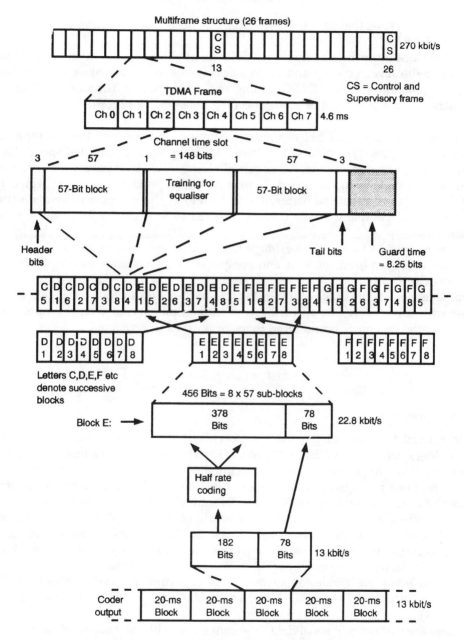

Fig. 12.5 GSM frame structure (mobile to base)

where 1 QDU represents the distortion introduced by one standard PCM 64 kbit/s codec. QDUs are considered by some experts as not being an entirely appropriate method of measuring the distortion introduced by the GSM coder and CCITT may need to revise their approach to transmission planning for distortion in order to provide a sensible accommodation for GSM and other similar systems with low bit rate coding. Nevertheless the GSM coder has been given a rating equivalent to about 7 QDUs.

2. The GSM introduces a significant delay of about 90 ms of which up to 30 ms is due to the speech coding and up to 40 ms is due to the sub-block interleaving, and the remainder to the TDMA framing and other sources. The size of this delay is such as to make the use of echo cancelling essential for speech.

In terms of technology, GSM is a most demanding system with the full range of digital techniques, viz: equalisation, frequency hopping, sophisticated speech coding, error correction coding, echo cancellation, block interleaving and advanced modulation, being provided to maximise the performance. The degree of processing is such that the battery current drain of the integrated circuits in the mobile is comparable with the current required to provide the RF power for the transmitter.

From the point of view of investment GSM is also most interesting in that by far the main cost has been software and computer aided hardware development, for the system control and the design of the integrated circuits respectively. The cost profile represents the shape of things to come by showing a very high development cost with a relatively low recurring cost for equipment and with the difference between the two more pronounced than for most earlier systems.

Fixed network aspects

Network structure

The general structure of the fixed network is the same as in Figure 12.4. The signalling system is based on CCITT No. 7. Each cell is served by a Base Transceiver Station (BTS) which contains little or no intelligence. A group of BTS is connected to a Base Station Controller (BSC) whose function is to manage the radio frequencies and channels used by the BTS. The BTS are connected to a Mobile Switching Centre (MSC), or more accurately a switching centre for mobiles. The MSCs are switches with extra intelligence or data base facilities having typically five times the processing power that would be found in an ordinary switch. The MSCs together form an intelligent network.

Certain MSCs are connected to the PSTN and contain gateway functions for handling calls to and from the PSTN. These MSCs are known as Gateway MSCs (GMSCs).

There are two main types of data base concerned with the routing of calls. There is a Home Location Register (HLR) that holds data on all the mobiles that are customers of the network, including information on the services to which the mobile may have access, and the current location of the mobile. Different parts of the HLR may be held at different points in the network, but each mobile has only one file in the HLR. The file length is typically 200 bytes.

The second type of register is the Visitor Location Register (VLR). There are VLRs associated with each MSC and it is common practice for them to be integrated into the switch. Whenever a mobile is active (logged on and able to make or receive a call) most of the data about the mobile that is held in the HLR is downloaded (copied) into the VLR of the MSC in whose area the mobile is.

The GSM networks use CCITT Signalling System No. 7. A special part of that system called the Mobile Applications Part (MAP) is used for the various messages and interactions that take place between the MSCs, HLRs and VLRs. Many of these interactions take place before a traffic route is allocated and may follow different routes from the traffic.

Authentication system

The GSM system contains a highly secure authentication system. Each customer has a smart card that contains a secret number known only to the HLR that holds his file. The HLR generates challenge response pairs, which are numbers that are related by a mathematical algorithm that works in only one direction and is controlled by the key or secret number. The algorithm is also programmed into the mobile equipment. An Authentication Centre may be associated with the HLR.

The HLR passes challenge response pairs to the VLR and the VLR uses these pairs to check that the mobile is allowed to make or receive a call; normally this check is carried out at call set-up time.

During authentication, the VLR sends a challenge number to the mobile. The mobile processes the number using the algorithm and its secret number or key, and returns the response. This response number is checked against the response number given by the HLR to ensure that the mobile has the correct secret number or key and is authorised to use the system.

The secret number is never transmitted and never leaves the HLR. Because the mathematical algorithm works only one way (one-way trap door function) it is impossible to determine the key from the challenge–response pair.

Smart card system

A feature of GSM that gives great flexibility is that the subscriber's details are held on a smart card and are not stored permanently in the mobile equipment. This means that the subscriber can readily change or borrow equipment. However, some networks run an equipment identity register, associated with the HLR, to check automatically that equipment is not stolen.

Numbering

Each subscriber has a directory number that is allocated nationally. In the UK, the directory number has the form:

{network identity}{HLR location}{customer no}

and it is expected that the network identity will be a three-figure code (National Numbering Group) initially, although arrangements could change when the UK moves to 10 digit numbering. National dialling would use the prefix '0', and incoming international calls '+44'.

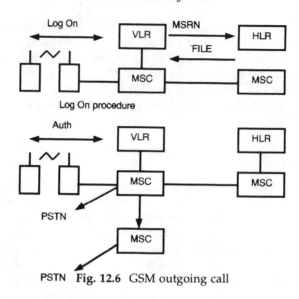

Fig. 12.6 GSM outgoing call

Roaming

The call routing and roaming arrangements are best explained by describing the procedure for establishing outgoing and incoming calls.

Outgoing calls When an outgoing call is made, the mobile concerned will already be logged on to the system. Thus the VLR of the MSC in whose area the mobile is will contain a file on the mobile. This will be the case irrespective of whether the mobile is in the area of its home MSC, or of a distant MSC on its own network or on any other network on which it is roaming (see Figure 12.6).

When the mobile requests a call set-up, the MSC conducts the authentication process described above, and then when this is complete, commences to establish the call. Depending on the number dialled, the call may be to

- a local fixed network;

- a trunk destination;

- an international destination;

- another mobile.

The route chosen will be determined by routing tables in the processor of the MSC that are prepared as a result of commercial arrangements concluded by the GSM operator. The GSM operator may use his own trunks or those of other GSM networks to bypass much of the fixed network, subject to whatever national regulations apply. If the mobile is roaming, the route is determined by the network that is setting up the call, not by the home network of the subscriber. The HLR of the home network does not normally play a direct part in the interactions on call set-up.

If the subscriber is roaming, the network that sets up the call makes a call record and bills the subscriber's own network periodically. The subscriber does not need to make any prior arrangements with the network that he is visiting. He is billed by his home network for the call at a rate based on, but not identical to, the rates charged by the network that he actually used.

Incoming calls When an incoming call is made, the network on which the call originates examines the number dialled. If it recognises the number as a GSM network to which it has a connection, it will route the call to the Gateway MSC (GMSC). Because the fixed network may not know the locations of the HLRs in the GSM network, it will normally choose the nearest GMSC. If there are no direct connections to the GSM gateway, the call will be routed via other networks.

For international calls there are two options: the call can be routed either to a fixed network in the destination country, or a signalling interaction can commence with a GSM gateway in the originating country, leaving the choice of traffic route to be determined later. In this second case, the originating network must be able to recognise the GSM network identity of foreign networks. The purpose of this second case is to provide optimal routing for the call if the called mobile is roaming away from its home network.

The GMSC interrogates the HLR of the called mobile either directly if the HLR is on the same network or via another GSM network (see Figure 12.7). The HLR is identified from the called number.

The HLR returns a Mobile Station Roaming Number (MSRN) that has been given to it by the VLR of the MSC at the location of the mobile. The MSRN is a 'directory number' used for identifying a fixed point at an MSC in the GSM network. The MSRN is sent from the HLR to the GMSC to tell it where to route the call to. If the

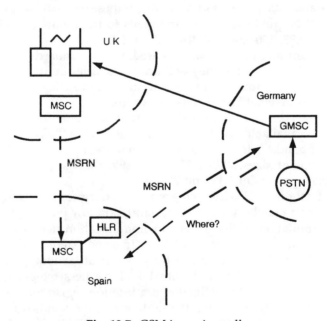

Fig. 12.7 GSM incoming call

mobile is roaming, the MSRN will be a number on the network on which the mobile is roaming. The GMSC then determines the routing and sets up the call to the MSC to which the mobile is logged on. That MSC knows the location area of the mobile and instructs all the base stations in that area to page the mobile. The mobile responds, an authentication procedure is normally conducted, and the call set-up is completed.

The routing taken by the call may be quite different from that of the signalling interactions. For example, a call from Germany to a Spanish mobile that is roaming in the UK may be routed directly from Germany to the UK by a GMSC in Germany, even though the initial signalling interactions were with the HLR in Spain.

When a mobile is out of reach of his home network, it will roam onto whatever network is available. All classes of service offer this facility. If more than one network is available, the question arises as to which network will be used. The GSM specifications say that the user should have manual control of this choice, but in practice it is likely that an order of preference will be programmed into the mobile. This may be done by the dealer for the subscriber and lead to some interesting negotiations between network operators and dealers.

12.5.3 Personal Communication Networks (PCN)

As a result of the rapid growth in the use of analogue cellular radio, the UK Government has chosen to stimulate the market by issuing licences to Mercury PCN, Unitel and Microtel for Personal Communication Networks (PCNs) which will bridge the gap in the market between GSM networks, as currently envisaged, and the universal mobile telecommunications system, UMTS, which RACE is developing. The standards (known as DCS1800) for these services have been prepared in ETSI, and will be published as amendments to the following GSM standards (ETS 300 022, 300 033, 300 034, 300 044 and 300 045).

Personal communication networks are based on the standards and design work carried out for GSM except that they will use frequencies in the 1.7–1.9 GHz band and employ microcells with sizes ranging from 100 m to 1 km (radius) in areas of dense traffic and somewhat larger cells elsewhere.

The smaller cell sizes are necessary to ensure that adequate capacity can be achieved from the available spectrum but they also have the advantage that they allow low power handsets to be used, and the handset sizes should be comparable to the early Telepoint sizes (about $200 \, \text{cm}^3$). Two classes of handset have been defined with peak powers of 1 W and 250 mW.

Some of the features of GSM which are unnecessary for long range operation, such as equalisation may be simplified or omitted in low power handsets.

During the formulation of the personal communication network concept there was extensive debate about the role of DECT. To some extent DECT was seen as a possible alternative to GSM because of the obvious attractions of a handset that could be used both with PCN and with DECT cordless telephones and PBXs. However, DECT was not fully defined and a location registration scheme would have had to be developed to enable the updating of the handset location so that incoming calls could be handled. Also the absence in DECT of any form of equalisation limits the useful range to about 300–400 m, which is significantly less than a low power handset could otherwise achieve and would severely limit coverage during the introduction of the network. Adding equalisation would be a very major change to DECT. Nevertheless it is possible that some PCN networks will eventually provide DECT service capabilities in some urban areas and DECT handsets may incorporate message pagers to enable customers to receive call-me-back requests.

Many of the developments that will form part of PCN may also be undertaken by the GSM operators at 900 MHz and the services at different frequencies will compete with each other to a significant extent.

The fixed network parts of PCN will be based on the GSM standards except that the PCN networks will be allowed to share infrastructures in areas of the country designated by the Director General of Telecommunications. The purpose of this is to help them to establish coverage rapidly and reduce their investment costs so that they can compete more effectively with the GSM operators.

The principal method for infrastructure sharing between the three networks is called national roaming which is defined in the DCS 1800 standards and is based on GSM roaming with some minor modifications. This method is easy to implement but it restricts the services available to those of the provider of the network in the shared infrastructure region. Different regions would be allocated to different operators. Final decisions on the roaming arrangements have yet to be taken.

In the Spring of 1991, Mercury and Unitel announced that they would implement a method of network sharing called parallel network architecture in order to reduce fixed network costs. This method is best thought of as two logically separate but physically integrated networks, and it has the advantage that it does not constrain competition in services. Both networks were to share a common backbone of MSCs and BSCs with software partitioning in an intelligent network structure. Base station sites were also to be shared but with each network having separate transceivers with separate frequency allocations. Subsequently, in March 1992, the two companies agreed on a full merger of both their PCN networks and services.

12.5.4 Universal Mobile Telecommunications System (UMTS)

As a part of the European Commission's RACE programme a study is being made on the design of a universal mobile system for introduction towards the end of the 1990s using part of the 1.7–2.3 GHz band, probably 1.9–2.1 GHz. The aim of UMTS is to produce a personal communicator that will replace both GSM and DECT. The service will probably be in two parts, one aimed at voice and low to medium speed data, and the other at broadband services for which the use of millimetre waves (30 GHz upwards) is being studied. Progress in this area of the RACE programme is slow because the GSM development programme is currently using most of the available R & D resources, but ETSI is forming a new group to cooperate with CCIR to undertake work in this area.

12.6   DIGITAL SHORT RANGE RADIO (DSRR) SYSTEM

In mid-1991, ETSI published the draft of a standard (I-ETS 300 168) for a private Digital Short Range Radio system (DSRR) for single frequency simplex operation in a single 2 MHz band (933–935 MHz), or two-frequency semi-duplex operation in the bands 933–935 MHz and 888–890 MHz. Unlike the other systems described here, DSRR is designed for communication directly between handsets without signals having to pass through a base station, although if two bands are available, a fixed repeater may be used to improve the coverage of an area. DSRR is intended for use at sports events, fetes, building sites, factories, etc.

DSRR uses a distributed self trunking system (dynamic channel allocation) whereby the handset that initiates the call selects an unused voice channel and sends a signal on the control channel which is received by the called party and which contains the number of the handset which it is calling and the number of the proposed voice channel. Once a handshaking procedure has been completed satisfactorily, the handsets switch to the voice channel. The voice coding system uses the full rate GSM codec.

The handsets are specified for a maximum transmitting power of 4 W using GMSK modulation. The channel separation is 25 kHz, giving 79 channels (2 control and 77 traffic).

12.7   MARITIME SERVICES

The main long-distance international maritime services operate in the mf and hf bands, with two international distress frequencies, one for Morse code at 500 kHz, and one for telephony at 2812 kHz. Shorter range international maritime services operate at both VHF (156–157.5 MHz ship Tx, 160.6–162.1 MHz Coast Tx) and UHF (around 460 MHz). These services provide coverage in coastal areas. A maritime business service also operates around the UK at UHF. The telephony services at VHF and UHF have channel spacings of 25 kHz and use frequency modulation.

The development of satellite services has made a very significant impact on maritime communications, and most larger ships that travel long distances now use satellite services.

12.8   CONCLUSIONS

Mobile radio systems have developed rapidly during the last ten years as a result of developments in technology and the introduction of competition in the provision of services. Considerable further developments are in train and the costs of mobile communications should continue to fall steadily. The fall in prices for mobile voice communications is expected to make mobile communications more of a direct competitor to fixed services by the mid-late 1990s.

Within mobile systems, the development of software is becoming a major component, if not the major component, of the costs both of the fixed networks and the mobile terminals. It will therefore be interesting to see the effects of this phenomenon in the services and apparatus markets.

The steady fall in prices is expected to lead some price sensitive customers, who have chosen radio paging or PMR as a lower cost alternative to cellular, to switch to cellular voice services. Nevertheless radio paging and PMR are expected to maintain a significant market share for at least the next ten years.

With the exception of radio paging, the main growth hitherto has been in voice services, however, interest in mobile data services is growing rapidly and should provide a substantial market during the 1990s.

The change from one generation of service to the next will be particularly interesting in the area of cellular radio because of the long time and huge investment needed to provide good coverage. Consequently there may be difficulties in introducing a new service such as GSM in countries where the existing analogue networks are well established and where they offer services that are quite adequate for the majority of customers.

12.9 BIBLIOGRAPHY

D. M Balston (1989) Pan European cellular radio. *Electronics and Communications Engineering Journal*, Jan/Feb, 7,13.

G. Calhou (1988) Digital cellular radio, *Artech House*.

R. Dettmer (1989) Trunked private mobile radio, *IEE Review*, June, 223–226.

R. J Holberhe (1985) *Land Mobile Radio Systems*, Peter Peregrinus.

J. D Parsons and J. G. Gardiner (1989) *Land Mobile Radio Systems*, Blackie & Halstead Press.

13 SATELLITE COMMUNICATIONS

13.1 INTRODUCTION

In this chapter we discuss the main technical aspects of both satellites and Earth stations. We then consider the role of satellites in communications in general terms, and the four main service areas. Finally we discuss the problems of frequency coordination in the geostationary orbit and some new ideas for non-geostationary satellites.

13.2 TECHNICAL ASPECTS OF SATELLITE COMMUNICATIONS

13.2.1 Introduction and system design

Satellites provide communications by receiving radio signals transmitted by one Earth (or ground) station and transmitting them back at a different frequency to another Earth station. Communications satellites normally operate in a geostationary orbit, which is an equatorial orbit with a period of 24 hours such that the satellite is always in the same position relative to a point on the Earth's surface, so that Earth stations do not need to move their antennae appreciably to track the satellite.

Because the signals have to travel to the satellite and back and because the geostationary orbit has a radius some six times that of the Earth, transmission via satellite takes appreciably longer than transmission by terrestrial means, and a single satellite hop takes about 250 ms, the exact figure depending on the relative locations of the Earth stations and the satellite. This delay is quite noticeable in voice telephony where the speech is interactive, and it also necessitates the use of echo control in speech circuits.

Communications systems which use satellites need to be designed as a whole to minimise the cost of the complete system i.e. the satellites, the launchers and the Earth stations. The principal part of the design is the RF link budget between the satellite and the Earth stations. This aspect of the design is much more critical than in most terrestrial radio systems because the cost implications of each decibel are much greater, and because there are normally no highly variable factors such as multipath to take account of, the only significant variable is the effect of precipitation (rain, snow) which is the subject of careful study to obtain statistics on the probabilities of fades greater than specified levels.

The design will include a trade-off between:

- the satellite EIRP (equivalent isotropic radiated power) which is proportional to the RF power density which would be received by an Earth station antenna);
- the size of the Earth station antenna and their number;
- the geographical coverage required on the Earth's surface; and
- the choice of frequency.

The cost per unit of communication capacity for the satellite will depend on the RF power which the satellite needs to transmit because the satellite has to obtain its power from solar cells which convert solar energy into electrical energy. The cost of the whole satellite in orbit depends strongly on the mass of the satellite because the cost of launching a satellite is typically 80% of the cost of manufacturing the satellite, and the cost of the apogee boost motor within the satellite is a significant proportion of its cost. Consequently satellite designers are continually trying to reduce the mass of the subsystems and to find ways of generating and using RF power more efficiently.

13.2.2 Launchers

A satellite launch can be considered to have three phases (see Figure 13.1). The first phase takes it into a low Earth orbit with a period of about 1.5 h. From that orbit the second phase takes the satellite into a highly elliptical transfer orbit with an apogee (highest point) of 35 400 km (22 000 miles), which is the height of the geostationary orbit. The third phase extends the elliptical orbit into the circular geostationary orbit.

The first phase can be accomplished by an unmanned rocket launcher or by the US Space Shuttle. The second phase is normally accomplished by the same rocket launcher, but in the case of the Shuttle a separate perigee stage is needed. Unlike the Shuttle, Ariane launches the satellite straight into transfer orbit, combining the first two phases into one. The third stage is accomplished by a rocket motor called the apogee boost motor which is usually integral with the satellite itself and can be either a solid motor or a liquid motor.

The cost of a launch is typically 80% of that of the satellite and the failure probability of launchers is typically in the region 5–10%. Consequently it is common practice to insure the launching of satellites, and to take contingency measures such as procuring a spare set of those parts with the longest lead times.

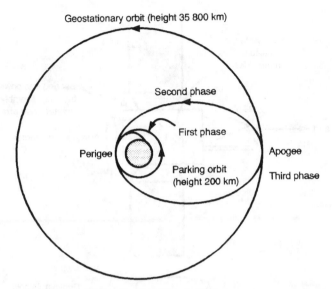

Fig. 13.1 Satellite launch system

Although launcher technology is improving, the rate of improvement is not high and no significant breakthroughs such as the development of a new low mass high energy fuel are foreseen.

13.2.2 Satellite platform

There are two basic designs of satellite platform and Figure 13.2 shows the basic designs. Spin-stabilised satellites (Figure 13.2b) provide their main stabilisation by the body spinning, although the antennas and part of the communications payload, which are connected to the body by a bearing, remain stationary. Spin-stabilised satellites look like a large cylinder with the antenna dishes at one end.

Three-axis-stabilised satellites (Figure 13.2a) obtain their main stabilisation from momentum wheels within the main body of the satellite. Such satellites look like boxes with two wings (the solar arrays).

The main parts of the satellite platform are:

The body

The body or structure which is normally made of aluminium alloy or a carbon composite such as Kevlar.

The motor

The apogee boost motor which takes the satellite from transfer orbit to geostationary orbit is either a liquid or solid fuel motor. Because of the quantity of propellant

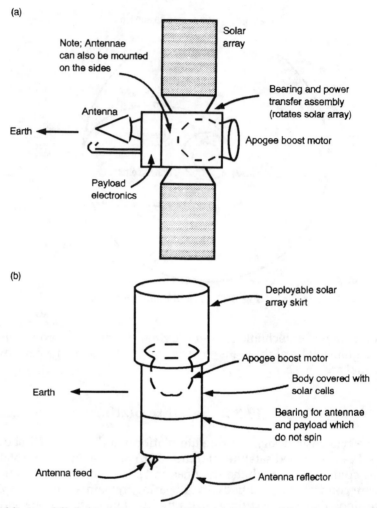

(a)

Note; Antennae can also be mounted on the sides

Solar array

Antenna

Bearing and power transfer assembly (rotates solar array)

Earth ←

Apogee boost motor

Payload electronics

(b)

Deployable solar array skirt

Apogee boost motor

Body covered with solar cells

Earth ←

Bearing for antennae and payload which do not spin

Antenna feed

Antenna reflector

Fig. 13.2 Highly simplified diagrams of a satellite: (a) 3-axis stabilised satellite; (b) spin-stabilised satellite

needed, the motor accounts for between one third and one half the mass of the satellite.

The attitude and orbit control system

The attitude and orbit control system which uses a set of orthogonal thrusters to manoeuvre the satellite into the correct position and maintain it there. In the case of a spin-stabilised satellite it also controls the spinning. The system is controlled by an on-board computer and telemetry from the ground, and the thrusters normally use a liquid propellant such as hydrazine. A number of experiments have been made with ion thrusters which use electrical energy, but ion thrusters are not likely generally to replace liquid fuel thrusters in the foreseeable future. One interesting technique that has produced some fuel savings for three axis stabilised satellites is solar sailing where the angles of the solar arrays relative to the Sun's rays (which exert a small pressure) are controlled to provide small torques.

In addition to manoeuvring the satellite into the correct position and attitude in the orbit, the attitude and orbit control system has to maintain the satellite in that attitude and position, by counteracting the effects of the forces and torques which would disturb the satellite. The main forces and torques are:

(a) a gravitational pull along the orbit towards either of two potential wells (countering this force is called east–west station keeping);

(b) a torque which arises because the Earth is not a true sphere which tends to displace the orbit, producing a daily figure-of-eight movement around the desired orbital position;

(c) a force and a torque produced by the pressure of the Sun's rays on the satellite.

The power subsystem

The power subsystem consists of the solar arrays (groups of solar cells) which convert solar energy into electrical energy, batteries which are used to store electrical energy during launch and during the twice yearly sequences of eclipse (when the satellite is hidden from the Sun by the Earth for a small part of its orbit), and controllers which control the charging of the batteries.

In a spin-stabilised satellite, the solar cells are located on the outside of the cylindrical body of the satellite, whereas on a three-axis stabilised satellite there are two solar arrays which extend outwards, one from each side of the satellite. These solar arrays are deployed after launch. The solar arrays consist of large numbers of solar cells which are semiconductor diodes which produce a small current in the presence of sunlight. These solar cells are covered by a thin layer of glass to protect them. In the past the cells have been based on silicon semiconductors but gallium arsenide is likely to give more efficient energy conversion although the material itself is much more expensive. Developments are also taking place in the glasses used for cell covers which will lead to small but worthwhile savings in mass.

The most commonly used battery technology in the last decade is nickel cadmium but this technology is now being replaced by nickel hydrogen although further developments of nickel cadmium may be possible in the future.

13.2.4 Communications payload

The communications payload consists of antennas and transponders. The transponders receive the signals transmitted from Earth, downconvert them to an intermediate frequency or to baseband, upconvert them to the transmit frequency, and transmit them at an appropriate power back to Earth. Figure 13.3 shows a diagram of a simple transponder.

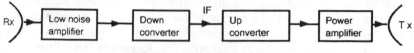

Fig. 13.3 Simple transponder

The communications payload of a satellite may provide one or more different services. Most early satellites were dedicated to providing one particular service but more recently some satellites have been designed to provide combinations of services such as fixed and mobile, or fixed and broadcasting services.

Each service is designed to work in a particular pair of frequency bands, one band being used for the uplink and the other for the downlink. These frequency bands are allocated by CCIR and some pairs are allocated exclusively to satellites whilst others are shared with terrestrial services such as microwave links with the result that the locations of the Earth stations have to be chosen carefully to ensure that there is no interference. The satellite bands are listed in Table 13.1.

The antennas are designed to illuminate those areas of the Earth where the service is to be provided. Some services such as maritime mobile or aeronautical may require illumination of the whole hemisphere of the Earth visible from the satellite. Other services require illumination only of a small area, and antennas can be designed only to illuminate areas no larger than a medium-sized English county.

Because of the high cost of a satellite, the communication system has to be designed to make the most efficient use of its radio transmissions, i.e. radio signals should be transmitted only where they need to be received. Thus antenna beams need to be shaped to the requirements of the service to be provided.

Most terrestrial antennas use a single feed and a single reflector. However, because many satellite antennas need to provide carefully shaped or contoured beams it is common practice to use multiple feeds to provide one or more beams from the same reflector. Probably the most extreme examples of this technique are the INTELSAT VI satellites which have over a hundred feeds sharing the same reflector.

A further improvement can be achieved by using very narrow spot beams which illuminate only the area for which the traffic is destined, not the whole area covered by the service. This type of system requires the payload either to switch rapidly between beams if the down link spot beams are fixed, or to steer the spot beams rapidly from destination to destination.

One way to provide reconfigurable and rapidly steerable beams is to use an antenna made of a phased array of small flat (patch) transmitters instead of feeds and a reflector. The advantages of this design is that each patch can be fed with RF power by a small semiconductor and the phase of each patch can be controlled electronically, allowing rapid steering of the beam. Array antennas of this type are likely to be used first for mobile services at L-band.

To obtain a very small coverage, a high gain antenna is needed, and this is achieved by using either a large reflector or a high frequency or a combination of both. The size of the reflector is limited by the size which can be accommodated within the shroud or payload bay of the launcher unless an unfurlable antenna is used, but it is difficult to design an antenna that will unfurl to give the surface

Table 13.1 Frequencies allocated to commercial satellite communications in region 1 (Europe) up to 35 GHz.

Frequencies (GHz)	Link	User	Service
1.530–1.544	down	INMARSAT	Maritime mobile
1.544–1.599	down	INMARSAT	Mobile (distress and safety)
1.545–1.599	down	INMARSAT	Aeronautical mobile
1.6265–1.6455	up	INMARSAT	Maritime mobile
1.6455–1.6465	up	INMARSAT	Mobile (distress and safety)
1.6465–1.6605	up	INMARSAT	Aeronautical mobile
2.500–2.690	down		Sound broadcasting
3.400–4.200	down	INTELSAT	Fixed
4.500–4.800	down	INTELSAT	Fixed (planned)
5.725–7.075	up	INTELSAT	Fixed (planned 6.725–7.025)
10.70–11.70	down	various	Fixed (planned 10.7–10.95 and 11.2–11.45)
11.70–12.50	down	various	TV boradcasting (planned DBS)
12.50–12.75	down	various	Fixed (small dish)
12.75–13.25	up	various	Fixed
14.00–14.80	up	various	Fixed
17.30–18.10	up	various	TV broadcasting (planned DBS)
17.70–20.20	down	experiment	Fixed
20.20–21.20	down	experiment	Fixed and mobile
22.55–23.55	inter satellite		
27.50–30.00	up	experiment	Fixed
30.00–31.00	up	experiment	Fixed and mobille
32.00–33.00	inter satellite		

accuracy which is needed to produce the correct phase across an aperture greater than the 3–4 m which can be carried within Shuttle or Ariane. A further limiting factor is the accuracy with which an antenna can be pointed. Notwithstanding these problems, there is interest in developing unfurlable antennas to generate extremely narrow spot-beams.

Communications payloads can be very complex with tens of transponder chains and several antenna systems. Because of the problems of reliability it is normal to provide switches which can alter the connections between transponders and antennas if the needs of the service require it or if certain transponders fail.

In services other than TV broadcasting, satellite transponders will normally carry signals from a number of different ground stations. These signals will share the transponder either by frequency division multiplexing or by time division multiplexing. Older and simpler systems use FDM, in which case it is necessary for the transponder to have a high degree of linearity to ensure that harmful intermodulation products are not created which would interfere with the wanted signals. If TDM is used, this problem does not occur and amplifiers can be operated at correspondingly higher powers; however, the use of TDM requires complex time adjustment algorithms in the Earth stations to ensure that signals from different Earth stations do not overlap at the satellite.

A new technique called Satellite Switched Time Division Multiple Access (SSTDMA) is being developed to give flexibility in the operation of payloads with multiple spot beams. In this technique an uplink will carry transmissions which are to be routed to a number of different downlinks which use different spot beams. The transponder will downconvert the uplink transmissions to an intermediate frequency or to baseband and switch each part of each frame to the upconverter and power amplifier that are feeding the spot beam for the required destination. The transponder will repeat its switching cycle for each frame, and the switching pattern can be controlled from the ground using telemetry. Thus any given uplink station may have channels through a single transponder to a number of different downlinks and connection arrangements may be reconfigured.

A further development would be an on board processor or call routing apparatus where the satellite would receive a switching signal, e.g. a form of CCITT No. 7 and would route calls from the uplink spot beam to the downlink spot beam in accordance with the instructions in the signalling channel instead of in accordance with a pattern defined by the telemetry. The satellite would then be analogous to a digital exchange, whereas with the SSTDMA described above it is analogous to a reconfigurable digital cross-connect.

The RF power sources used in the transponders are currently either travelling wave tubes (TWTs) or solid state devices. TWTs are normally used where RF powers in excess of 10–20 W are required because solid state devices of that power are not available and the use of too many RF combiners is inefficient. However, array antennas require a number of separate lower power sources and so array antennas and solid state devices make a combination that is likely to be used increasingly in the future.

In the past, satellites have always operated in isolation. However intersatellite links are being developed to enable signals to be sent from one satellite to another. These links, which would probably operate at frequencies above 20 GHz, will enable satellite systems to provide communications between points on the Earth which cannot be seen simultaneously by a single satellite. They will also enable different services to be interconnected, e.g., a mobile service uplink provided from one satellite to be connected to a fixed service downlink provided from another satellite. However the greatly increased distances which the signals may have to travel will increase the delay and be a distinct disadvantage for telephony.

13.2.5 Satellite reliability

Geostationary satellites have to be designed to work without maintenance. (The possibilities for using the manned Space Shuttle to maintain satellites refer to satellites in low Earth orbit, not geostationary orbit.) Consequently a satellite has to be designed for very high reliability. This reliability is achieved by the following:

(a) duplication of the key elements of most subsystems and flexibility to reconfigure connections to switch out failed components (e.g. TWTs) or assemblies, and switch in spares;

(b) use only of components which have undergone rigorous testing of their suitability for use in space, which means that the technology of satellites is usually a few years older than the latest technology used in terrestrial systems, and this is particularly noticeable in the computing technology used;

(c) use of an extremely detailed quality control system which enables the source of each component or material to be traced in the event of a fault and so ensures that lessons can be learned from each failure;

(d) testing of all the main subsystems in simulated space conditions such that all the components are 'burnt-in', i.e. used for sufficiently long to ensure that the higher probability early failures occur before launch and the faulty components are replaced.

All these procedures add considerably to the cost of satellites. Satellite designers continually have to exercise judgement on the extent of testing required.

13.2.6 Satellite lifetime

The lifetime of a satellite is determined by

1. the probability of failure of the various key equipments and subsystems;
2. the fuel available for attitude control;
3. the degradation in efficiency of the solar arrays because the cells and cell covers slowly deteriorate;
4. the degradation in performance of the rechargeable batteries.

Satellites are normally designed for a lifetime of 7–10 years which is an appropriate period in view of the 30–36 months which it takes to build a satellite. Lifetimes of 10–15 years would be possible but anything over 10 years is beyond the horizon of most investors and so there is not much demand for longer lifetimes.

13.2.7 Earth stations

The term 'Earth station' covers any terrestrial transmitter or receiver that communicates with a satellite. The term ground station is normally used for Earth stations on land, and ship Earth station for ones on ships.

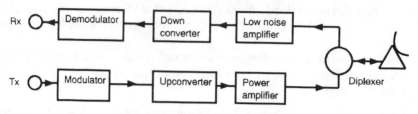

Fig. 13.4 Typical Earth station

Figure 13.4 shows a diagram of a typical Earth station. The baseband signal is modulated, upconverted, amplified and transmitted from an antenna. The converse applies to reception.

The antenna design is very important because whilst gain must be maximised for a given size, it is also essential to achieve low sidelobe levels to ensure that interference to neighbouring satellites is at an acceptably low level.

The other critical item is the low noise receiver whose performance is a major factor in the link budget for the communications system. (The link budget for the downlink from the satellite to the Earth station is normally more demanding than that for the uplink.) In more expensive earth stations, it is common for the low noise amplifier at the front end of the receiver to be cooled to very low temperatures to reduce thermal noise. In some designs, the amplifier is mounted next to the feed horns of the antenna to reduce loss and additional noise.

When the antenna is large and the beam width narrow, the Earth station is fitted with a control system to enable it to track the small perturbations in the orbital position of the satellite.

13.3 THE ROLE OF SATELLITES

The role of satellites in communications is determined by the differences in their inherent characteristics from those of line communications or terrestrial radio. Because satellites provide radio communications to a large area they are well suited for broadcasting and mobile services and a single satellite can replace a whole network of terrestrial radio stations. Line communications are less suitable for broadcasting and are unsuitable for mobile services.

Furthermore because satellites are capable of providing line of sight communications they can use relatively high frequencies and therefore can provide high capacity links over a wide area. In contrast, terrestrial radio systems are limited to either low capacity wide area services or high capacity short distance services.

Satellites can provide communications to any location where there is a ground station and therefore they are well suited to providing communications across terrain where the political situation or the geography makes the use of cable or microwave links difficult. In addition, ground stations can often be installed quickly since the equipment can be flown in, and this feature makes satellites suitable for providing communications at short notice or for limited periods in the case of special events or in the aftermath of disasters.

Having examined the role of satellites in theory, we now consider the main current satellite services.

13.4 SATELLITE SERVICES

13.4.1 Fixed satellite services

Fixed satellite services are services between Earth stations that are fixed or transportable. Because of their high cost, satellites were used initially only for long-distance international telephony or TV distribution, but as the cost has reduced, they are now used for a much wider range of services.

Fixed satellite services are, however, facing increasing competition from cable systems, where developments in fibre optics are producing far more substantial reductions in costs than are possible with satellites, thus in the longer term satellites will tend to be used only in situations that exploit their special characteristics.

The two main organisations that provide satellite capacity for fixed services are INTELSAT and EUTELSAT, which are described in the following sections.

INTELSAT

The early developments of satellite communications were led by INTELSAT, which was formed in 1964, and is an international organisation with some 120 member countries. INTELSAT has the task of providing international fixed communications, primarily for telephony but more recently for TV distribution and business services.

INTELSAT is funded by PTTs and provides satellite capacity to the PTTs. The PTTs are responsible for running their own ground stations and providing the complete communications service to their customers. INTELSAT enjoys a form of international monopoly under which member countries are not allowed to operate other satellite systems that could cause economic harm to INTELSAT. This monopoly caused difficulties for the creation of EUTELSAT (see Section 13.4.1, p.211) to provide fixed services within Europe, and has been the cause of continuing problems and discussions especially during the 1980s when commercial organisations such as PANAMSAT established alternative systems to carry transatlantic traffic. With the change in policy, in many countries, towards communications, there is likely to be a change to the monopoly aspect of INTELSAT's constitution.

INTELSAT provides capacity for a wide range of services from major international trunk connections through to VSATs and domestic services. The main services carried are listed in Table 13.2.

INTELSAT has defined a number of different Earth station standards for use with different services. These standards are shown in Table 13.3. Developments in technology over the last 20 years with a corresponding redesign of link budgets, have led to significant changes to the older of these standards, for example the diameter of the Standard A Earth station antenna has been reduced by a factor of two from around 30 m producing substantial cost savings.

Table 13.2 INTELSAT services

Frequency division multiplexed, frequency modulation (FDM/FM)
Companded FDM?FM (CFDM/FM)
Low density telephone—VISTA
International leased and occasional use TV
Single channel per carrier (SCPC)
SCPC access with demand assignment (SPADE)
Time division multiple access and satellite switched TDMA (SS/TDMA)
INTELSAT Business Services (IBS)
Intermediate Data Rate carriers (IDR)
Very small aperture terminal system—INTELNET
Domestic leases and Planned Domestic Leases (PDS)

Table 13.3 INTELSAT Earth station standards and uses

Type	Frequency band (GHz)	Diameter (m)	G/T (dB/K)	Use
Std A	6/4	15–18	35	Main telephony, TV, FDM/FM, CFDM/FM, TDMA, SCPC, SPADE, IDR, IBS, VISTA
Std B	6/4	10–13	31.7	Thin route telephony & TV, CFDM/FM, SCPC, SPADE, IDR, IBS, VISTA
Std C	14/12/11	11–14	37	Main telephony, TV, FDM/FM, CFDM/FM, IDR, IBS
Std D	6/4	4.5–6	22.7–31.7	VISTA
Std E	14/12/11	3.5–10	25–34	IBS, IDR
Std F	6/4	4.5–10	22.7–29	IBS, IDR
Std G	6/4 + 14/12/11	several	several	INTELNET, international leases
Std Z	6/4 + 14/12/11	several	several	Domestic leases

INTELSAT has now operated six successive series or generations of satellites beginning with Earlybird which was launched in 1965. Each generation has been larger than its predecessor, giving considerable economies of scale in the cost of a telephone circuit. The current system uses a combination of INTELSAT V and INTELSAT VI. Five INTELSAT VIs have been produced and the first entered service in 1990, but the second was the victim of a launch failure.

INTELSAT is currently in the middle the procurement of its seventh generation, which breaks the trend in that it is smaller than the sixth generation because it is designed to replace INTELSAT V satellites in operational roles which do not justify the larger INTELSAT VI. Table 13.4 shows the main characteristics of the more recent satellites.

INTELSAT is now facing rapidly growing competition from international fibre optic submarine cables especially on routes such as those across the North Atlantic where the volume of traffic is high, and also from other satellite systems, therefore some reduction in the size of its business is to be expected.

Table 13.4 Comparison of INTELSAT satellite characteristic

	V	VI	VII
Launch mass	1832 kg	4170 kg	3810 kg
Solar array power	1290 W	2200 W	3970 W
Operational life	7 years	13 years	15 years
Capacity in bearer circuits for Std A & C	13 000	24 000	17 000
C-band			
Hemi. coverage antennas	2	2	2
Hemi. Transponders	10	12	10
Hemi. EIRP	29 dBW	31 dBW	33 dBW
Zone antennas	2	4	4
Zone transponders	8	20	10
Zone EIRP	29 dBW	31 dBW	33 dBW
Global coverage antennas	1	2	2
Global transponders	3	6	6
Global EIRP	23.5 dBW	26.5 dBW	26.5 & 29.5 dBW
Ku-band			
Spot beams	2	2	3
Spot transponders	6	10	10
EIRP	41.1/44.4 dBW	44–47 dBW	45–48 dBW

Hemi. = Hemispherical: dBW = dB relative to 1 W

EUTELSAT

EUTELSAT was formed as an interim organisation in 1977 and became a permanent organisation in 1985. Its basic role is to provide fixed satellite services in Europe, although it has recently added a land mobile service. Its structure is modelled on that of INTELSAT, with capacity sold through the national signatories, although this arrangement is due to change under pressure from the Commission. At the time of writing, EUTELSAT had 28 members but the figure is growing following the political changes in eastern Europe.

Initially EUTELSAT was expected mainly to provide telephony services between European countries, but with the improvements in cables, EUTELSAT has carried much less telephony traffic than expected. The areas that have provided most traffic are TV distribution and broadcasting and latterly services to small business terminals (VSATs).

The political changes in eastern Europe are leading to new opportunities for fixed satellite services and VSATs as a rapid means of installing a communications infrastructure. Satellites will be used extensively for this purpose in eastern Germany, and this application will generate additional traffic for EUTELSAT.

In addition to these fixed services, in 1991 EUTELSAT commenced a data communications and position fixing service for land mobiles. This system operates in the 12/14 GHz band with a combination of TDMA and spread spectrum such that the signals can share transponders with TV signals. The mobiles are fitted with

Table 13.5 EUTELSAT satellites

	EUTELSAT–I	EUTELSAT–II
Launch mass	1160 kg	1800 kg
In orbit mass	550 kg	866 kg
Power	900 W	3000 W
Lifetime	7 years	7–10 years
Frequency bands	14/11, 14/12 GHz	14/11, 14/12 GHz
Number of transponders	14 (10 simultaneous)	16
Transponder power	20 W	50 W

Table 13.6 EUTELSAT Earth station types

Type	Frequency band	Diameter (m)	G/T (dB/K)	Use
Standard	14/12/11 GHz	11–13	37	QPSK, TDMA, TV-FM, FDMA
TV-only	14/12/11 GHz	5.5	30.5	TV-FM, FDMA
Business	14/12/11 GHz	5.5	29.9	SCPC-PSK, FDMA
		3.5	26.9	
		2.4	22.9	

G = Gain: T = Temperature in K

terminals based on the US Omnitrac design with a small antenna that produces a fan shaped beam with sufficient vertical beamwidth not to require any pointing adjustment in elevation. The antenna is however rotated in a horizontal plane to track the satellite.

EUTELSAT originally procured five ECS satellites (also known as EUTELSAT-I) from the European Space Agency and three of these satellites are still in operation. Six second generation satellites (EUTELSAT-II) have been procured of which three had been launched by spring 1992. One will be kept as a spare. All the spacecraft are three-axis stabilised. Details of these satellites are given in Table 13.5

The EUTELSAT standards for Earth stations are shown in Table 13.6.

13.4.2 Mobile satellite services

Mobile satellite services developed more slowly than fixed services because of technical constraints and because of the high cost of a satellite which can transmit sufficient power for reception by a mobile earth station. The early Earth stations used for fixed services had antenna dishes with diameters of the order of 30 m which is clearly impracticable for a mobile. However, by the early 1970s satellites had become sufficiently large and receiver electronics sufficiently sensitive for telephony communications to ships with antenna diameters of 1–2 m. These developments led to the formation of INMARSAT, the International Mobile Satellite Organisation which provides most mobile satellite services.

INMARSAT was constituted by 26 governments in 1979 to have a monopoly in the provision of maritime satellite services. It now has some 64 members and its terms of reference were widened in 1985 to include aeronautical services which began in 1989. Also in 1989, further amendments were made to allow INMARSAT to provide land mobile services. INMARSAT is funded by PTTs who provide the services to customers by using the INMARSAT satellites.

The formation of INMARSAT resulted from an initiative by the International Maritime Organisation (IMO) to examine the possibility of developing a global satellite system for shipping. INMARSAT continues to work very closely with the IMO and INMARSAT's facilities will provide a major part of the Global Maritime Distress and Safety System (GMDSS) which is to be implemented in 1993. In addition to normal communications, INMARSAT provides a relay facility for Emergency Position Indicating Radio Beacons (EPIRBs) which enable distress situations to be located accurately.

The following frequencies, in L-band, are allocated to mobile satellite services, following the 1987 World Administrative Radio Conference (WARC) on mobile satellite services:

- 1.530–1.544 GHz   Satellite to ship or land mobile (only the upper part is for land mobile)

- 1.544–1.545 GHz   Satellite to mobile distress channel

- 1.545–1.599 GHz   Satellite to aeronautical mobile

- 1.6265–1.6455 GHz   Ship or land mobile to satellite (only the upper part is for land mobile)

- 1.6455–1.6465 GHz   Mobile to satellite distress channel

- 1.6465–1.6605 GHz   Aeronautical mobile to satellite.

The INMARSAT system provides satellites over the Atlantic, Indian and Pacific Oceans. In the earlier years the system was composed of a combination of MARECS satellites leased from the European Space Agency, maritime payloads on INTELSAT V satellites, and capacity on MARISAT satellites from Comsat General of the USA. In 1990 the first of at least five INMARSAT-2 satellites built by British Aerospace was launched which provides much higher capacity to accommodate the steadily growing traffic. Yet larger INMARSAT-3 satellites have been ordered from GE-Astro for launch in 1994.

The INMARSAT satellites communicate at C-band (4/6 GHz) with Coast Earth Stations located in a number of different countries. These Coast Earth Stations connect the calls into the fixed public networks. Ships and aircraft using the INMARSAT telephony service can be dialled directly from ordinary telephones.

INMARSAT has defined three main Earth station terminal types to be used by mobiles.

- INMARSAT A is the larger with a dish of typically 1 m diameter that is stabilised to point continuously at the satellite. INMARSAT A is available in three classes (Class I—telephony and telegraphy, Class II—telephony and shore to ship telegraphy, Class III—telegraph only). The telephony facility can also be used for data via a modem.

- INMARSAT C is much smaller, with a fixed antenna, and can be installed easily on vehicles or very small vessels. It provides data communications at 300 or 600 bits/s depending on the satellite that it is used with.

- INMARSAT M, currently under development, will provide low bit rate encoded telephony at 6.4 kbit/s and data at 2.4 kbit/s. It is expected to be ready for service at the end of 1992.

Two types of aeronautical terminal are available; one provides telephony, telex and data, the other provides only telex and data.

13.4.3 Broadcasting

Broadcasting satellite services have also developed more slowly than international telephony services because they too have needed small earth stations. The use of satellites for TV broadcasting began in North America during the early 1980s and they are now used widely throughout North America. Its growth has been facilitated by the US open skies policy of minimising the regulatory restrictions on the use of satellites.

The development of satellite broadcasting within Europe has been more problematical. The 1977 World Administrative Radio Conference (WARC) prepared a plan for the orbital positions and frequencies to be used for direct broadcasting by satellite (DBS). Each European country was allocated five channels and the link budgets showed that the larger countries such as France and Germany would need satellite transponders with some 150–200 W of power for reception with 90 cm dish antennae and low cost receivers, much greater than the 20–30 W used for fixed services and difficult to design with the necessary reliability.

France and Germany formed a joint project to develop national TV broadcasting satellites (TDF1 and TVSAT respectively) using these frequency allocations, whilst the UK, Italy and other countries worked within the European Space Agency to develop a demonstration satellite with a DBS capability called Olympus. At the same time various commercial initiatives began. The situation was complicated by discussions about the allocation of the various national channels and by a long debate about the choice of a new TV broadcasting standard based on the IBA MAC system (see Section 14.4) for use with DBS.

Whilst the problems of DBS continued, receiver technology improved and it became possible to provide coverage of a substantial part of western Europe using the fixed services transponders on the INTELSAT and EUTELSAT satellites. This solution was cheaper and offered the advantage of better international coverage and so TV broadcasting began in practice by these means, and an independent satellite system called Astra was launched to provide additional capacity.

Satellites are likely to be used increasingly in two roles; they can complement cable systems, by providing programmes to the head ends of cable systems, and they are also well suited to serve homes directly, particularly in areas where cable systems are not available. The development of high definition television and digital television will provide further opportunities for satellites because they may require bandwidths that cannot be provided easily within existing terrestrial off-air broadcast networks.

13.4.4 VSATs

One possible application for satellites which first attracted attention during the early 1980s was the provision of business communications within a company using small dish antennae sited on the company's premises. In the USA, these services achieved only very modest success initially; however, recently the Very Small Aperture Terminal (VSAT) services have begun to grow more rapidly. The initial services were probably too sophisticated and expensive, whereas VSATs have found their niche in the market by avoiding complexity, concentrating on data and exploiting the broadcasting (point to multipoint) capability of satellites.

In Europe, until recently, there have been no real opportunities for entrepreneurs to experiment with such services because of the monopolies of the PTTs, although this situation is now changing. In November 1990, the European Commission issued a Green Paper on Satellite Communications setting out its policy:

- to liberalise the supply and operation of both Rx only and Tx/Rx Earth stations (subject only to approval and any necessary licensing procedures);

- to allow unrestricted access to space segment capacity on an equitable, non-discriminatory and cost orientated basis;

- to ensure freedom for direct marketing of space segment capacity (e.g. from EUTELSAT) to service providers;

- to ensure interoperability through transmission standards.

These provisions will lead eventually to an open European market equivalent to that in the USA where there are no restrictions to prevent private companies from offering services.

13.4.5 Other services

The use of satellites to provide temporary services, for example to support special events, e.g. Olympic Games and special conferences, and to provide communications in emergencies has not yet developed to any appreciable extent but may develop further in conjunction with other small dish fixed services.

One application which is beginning to attract interest is the combination of position locating services (or navigation) and messaging services for mobiles such as long-distance trucks.

13.5 GEOSTATIONARY ORBIT

Because of the need to avoid interference, satellites in the geostationary orbit which use the same frequencies need to be spaced sufficiently far apart that Earth stations are able to discriminate between different satellites. Spacings of 1.5–6° are typical depending on the service. Locations in the geostationary orbit are therefore a finite resource.

Particular arcs in the orbit are required for certain services and the arc over the Atlantic Ocean is in the highest demand being used both for fixed transatlantic services and for maritime mobile services.

Larger satellites are generally more cost effective than smaller ones because the overhead elements are shared more widely. Satellites are now capable of carrying transponders which make use of all the available spectrum in any given service band and therefore the possibility of services which use different bands being provided on the same satellite offers further gains in efficiency, and some multiservice satellites have been built. However, in practice the scope for multiservice satellites is severely limited by the practical problems of collaboration between the funding, procurement and running of different services and the problem of finding free orbital slots in the various frequency bands for the different services. Thus unless there is more central planning of the use of the geostationary orbit, which is unlikely, the size of the frequency bands imposes a limit on the size of satellites.

13.6 NON-GEOSTATIONARY SATELLITES

So far this chapter has referred only to geostationary satellites. Communications can also be provided by non-geostationary satellites which have a lower but not necessarily circular orbit. Because such satellites move relative to the Earth's surface, a single satellite cannot provide continuous coverage at any point on the Earth's surface and therefore several satellites in different orbits, or in different positions in the same orbit, are needed for continuous coverage.

During the mid 1980's there was considerable interest in the development of a joint UK–Japanese experimental satellite with a highly elliptical polar orbit which could provide single hop communication between the Far East and Northern Europe for continuous periods of up to four hours. The project however was never completed.

Very recently INMARSAT and Motorola have cooperated to study a system, called Iridium, of some 77 low-Earth orbit satellites distributed across 7 polar orbits to provide continuous global communications in the 1–2 GHz band. The small portable Earth stations would have a wide antenna beamwidth. The satellites would be able to switch and relay signals from the satellite closest to the source to the satellite closest to the destination of each call.

The use of the low-Earth orbit will reduce the delay to a level much more acceptable to voice traffic than that for a geostationary orbit, and it will also enable frequencies to be reused in different orbits, except near the poles. Each Iridium satellite will cost less than a geostationary satellite with economies of scale providing considerable savings, but many more satellites will be needed than with a geostationary system to provide global coverage. The overall comparison with a geostationary system will be complex and the results are awaited with interest.

13.7 CONCLUSION

Satellites have an important role to play especially in the provision of services which exploit their fundamental characteristics such as mobile services and broadcasting.

Geostationary satellites have played a valuable role in the provision of international fixed telephony services but despite developments such as spot beams and on-board switching their use for such services is likely to diminish in the face of competition from optical fibre submarine cables especially on heavily used routes, because the overall rate of development of the cost effectiveness of satellite services is much slower than that of optical fibre systems. For satellites an improvement factor of two might be achieved in about four years but for optical fibre an improvement factor of one to two orders of magnitude might be achieved in the same period. Geostationary satellites also have the disadvantage of the unavoidable delay which is a distinct nuisance in telephony.

The development of satellite services within Europe has been retarded by the existence of PTT monopolies, but with the removal of many of the restrictions as a result of action by the European Commission, innovation in service and further growth in VSAT systems is to be expected.

13.8 BIBLIOGRAPHY

B. G. Evans (1987) *Satellite Communication Systems*, Peter Peregrinus.

14 CABLE AND MICROWAVE TV DISTRIBUTION

14.1 INTRODUCTION

For most countries in which Cable TV (CTV) is available the network is exclusively used for TV services but in the UK and in the USA the legal framework provides the environment for sharing networks which were initially installed purely for cable TV with telephony and other services and for allowing cable TV operators to install new multiservice networks. In the following the emphasis is strongly on the evolution of CTV networks and services towards the provision of other services rather than the extension of CTV technology in isolation.

Cable systems can provide several different forms of TV service. The simplest is a broadcast service where any subscriber can receive any channel. Slightly more complex is a system whereby individual customers can subscribe separately to extra channels in addition to receiving a basic selection which includes the terrestrial broadcast channels. A further refinement is pay-per-view, where customers pay to view specific programmes whose times are scheduled in advance. Finally there is a system whereby video tapes can be selected at any time from a storage system. The ease with which these different services can be made available depends on the technology of the cable system.

The general consensus is that an advanced cable system should be capable of delivering at least 30 different TV channels as well as sound channels, data and telephony, although only two or three TV channels need to be capable of being received simultaneously by any one subscriber.

In this chapter, after briefly describing the competitive situation in the UK and USA, we shall describe the main forms of cable systems, including their topology, the form of TV transmission and the choice of transmission medium (copper or fibre). We then discuss how additional services such as telephony can be made available on the same system, and the possibilities for microwave and millimetre wave technology in TV distribution networks. We conclude with an overview of the future for cable.

14.2 CABLE TV IN THE UK

At present in the UK, cable TV services are not so well established as they are in some other European countries or in the USA, and they are not growing as rapidly as a number of other communication services. The reasons for this situation are historical. Cable services were introduced originally by entrepreneurs, who were often TV dealers, who wanted to provide services in areas where direct radio reception was inadequate. The result was a modest number of fairly small systems. These systems could have developed by offering services which were not available off-air, such as services with a special local flavour, but they were generally restricted in their growth by regulations which prevented them from distributing more than one extra channel and restricted in the insertion of local material.

This situation began to change in the early 1980s when the Government undertook to issue a number of new wide area cable franchises for systems which would deliver typically 10–30 channels and endeavoured to encourage the use of new high speed technology which would provide a local infrastructure capable of carrying a wide range of services in addition to television. In particular, franchises were permitted to offer telephony provided that they did so in conjunction with either British Telecom or Mercury, but only Mercury became involved and then to a very limited extent. Since 1991 the situation has been further liberalised and operators who are licensed under Section 73 of the Broadcasting Act 1990 (and these include the cable TV operators) are able to operate on similar terms to BT, Mercury and Kingston Communications, but it is too early to say what the effect is likely to be although the regulatory changes have stimulated interest and investment. Despite considerable initial enthusiasm for cable, some of the companies who were awarded franchises have not developed their system as rapidly as expected and some not at all. However, in the year to October 1991 the number of subscribers to multi-channel networks grew by 89% to reach a figure of 220 000. Recently too BT has shown a willingness to provide interconnection services.

During the 1980s two other important developments have taken place which compete with cable. Firstly satellite television broadcasting has commenced and the sale of satellite receiving equipment is growing, and secondly some 50% of households have acquired video recorders and have ready access to video tapes from local hire shops. Thus cable systems are facing strong competition. There is however a cogent argument that satellite television will help cable in the short term by stimulating the production of more TV channels which cable can also distribute. Wherever cable is available it can compete effectively with satellite because cable can make all satellite channels easily available without the need for one or more receiving dishes. The competition from video is mostly from customers who want to access new films but are otherwise happy with ordinary TV.

14.3 CABLE TV IN THE USA

Cable TV systems started in about 1948/49 to serve areas where off-air television reception was poor. In order to compete in areas in which off-air reception was satisfactory, CTV companies began to offer more channels and a greater choice of programme material. The Federal Communications Commission (FCC) began to introduce limited jurisdiction in 1962. In 1985 there were over 4700 CTV systems serving 23 million customers, i.e. about 1 in 4 households, and offering 20–30 channels with premium pay-per-view and subscription services. By 1989 about 50% of households were being served.

The 1984 Cable Communications Policy Act allowed CTV operators to provide other services, such as telephony, but largely outside a regulatory framework and not under FCC control. Under the 1984 Act telecommunication and broadcasting companies are not allowed to own CTV companies. This has put the CTV companies in a strong competitive position and there are numerous complaints both from customers (e.g. of high prices and poor technical quality) and competitors that they are abusing their position. In some areas they have what amounts to a telecommunications monopoly.

Currently the whole situation is the subject of widespread debate with particular reference to what new statutes and regulations might be introduced in preparation for the introduction of broadband services and the replacement by optical fibre of much of the coaxial cable which is nearing the end of its service life. Many of the major telecommunication companies are obtaining CTV experience abroad in order to be in a better position to compete if and when they are allowed to do so. It is too early to say what the likely outcome will be.

14.4 TYPES OF CABLE SYSTEM

In this section we consider different types of cable system from the point of view of the delivery of TV services. In Section 14.5 we consider the provision of other services. There are no standard types of cable systems and so we need to look separately at the issues of topology, TV standards and transmission, transmission media and system design. Finally we consider the role of satellite master antenna TV systems (SMATV) which are miniature cable systems.

14.4.1 Cable network topologies

Two terms 'tree and branch' and 'switched star' are frequently used to describe cable systems and they are shown in Figure 14.1. However, their use can be a little misleading. There are two aspects to cable topology:

- the layout topology of the ducts and cable runs;

- the distribution topology of the electrical and switching arrangements.

These aspects are independent and so a switched star system distribution topology can use a tree and branch cable layout. The terms 'tree and branch' and 'switched star' when used without qualification usually refer to the distribution topology or electrical paths.

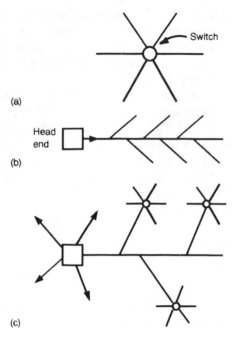

Fig. 14.1 Cable topologies: (a) star topology; (b) tree and branch topology; (c) distributed star

The layout topology is determined by the positions of houses and roads and will normally be based on a tree and branch layout although there may be small stars at the ends of some of the branches, particularly in housing estates and small closes.

The distinction in the distribution topologies is important because it influences the services which can be offered. The tree and branch topology is the older topology and is a system whereby many customers are connected to the same transmission medium, as shown in Figure 14.1(b). Thus all the channels are delivered to each premises all the time and the number of channels available to a subscriber is determined by the capacity of the transmission medium, although of course two cables or fibres can be used instead of one if extra capacity is required. In a tree and branch system, therefore, there needs to be equipment on the subscriber's premises to select the channel to be watched. A number of separate cables may radiate out from the head end in different directions without altering the basic tree and branch nature of the system.

In a switched star topology (Figure 14.1a), customers have separate electrical or optical access to the head end and only those channels which they have selected are transmitted to them. The head end therefore incorporates a switch, usually a space switch, which operates in response to signals from the customer. Such signals may be sent on the same transmission medium used for the TV distribution or on a separate medium which could even be the local telephone network. Thus the capacity of the transmission medium can be lower than for tree and branch and the potential choice of programmes can be higher. In a switched star the existence of separate transmission paths to each subscriber helps in the provision of two way services because there is no problem of multiple access control in the upstream (subscriber to head end) direction. A further advantage of a switched star is that in the provision of subscription channels and pay-per-view it is more fraud resistant because the complete repertoire of programs is not available on the customer's premises.

The choice between tree and branch and switched star is not clear cut, and many systems can use a hybrid topology where the main trunks are tree and branch (all channels are distributed) and the local distribution is a switched star. It is also possible for there to be some switching at the head end as well as at an intermediate point, and such systems are called distributed stars. At present switched star technology appears to be more expensive than tree and branch and many of the new cable systems in the UK are still favouring tree and branch.

14.4.2 Television standards and methods of modulation

Colour television pictures are composed from the three colours red, green and blue. In the current UK standard for 625 line colour television in the UK, which is called PAL or CCIR System 1, the basic colours are mixed into three components for transmission known as Y, U and V, such that the Y or luminance component taken on its own gives satisfactory black and white pictures. The following brief description of the modulation scheme omits some of the finer detail. The two chrominance components (U and V) are amplitude modulated in phase quadrature onto a sub carrier at 4.43 MHz before the whole vision signal with a base bandwidth of about 5.5 MHz is amplitude modulated with vestigial sidebands (AMVSB) onto the vision carrier. The sound signal is frequency modulated onto the sound carrier which is 6 MHz above the vision carrier. Because of the filtering for the vestigial sidebands, the channel spacing is only 8 MHz. This signal format was chosen to be compatible with black and white television but it has the disadvantage that the luminance and chrominance signals overlap, giving rise to cross colour effects which disturb reception of images of stripes or checks.

For satellite broadcasting, because transmitter power on the satellite is at a premium, it is necessary to use frequency modulation instead of amplitude modulation to achieve a better signal to noise ratio for a given carrier to noise ratio (see Section 7.4). However, with an FM signal the demodulation process concentrates the noise in the high frequency parts of the passband, where it interacts with the chrominance signal in a way that affects large areas of saturated colour particularly badly. To overcome this problem, the IBA developed an entirely new system of transmitting television pictures called Multiplexed Analogue Component (MAC), which has aroused widespread interest in Europe as well as the UK.

In the MAC system, the three signals (sound/data, chrominance and luminance) for each line of the television picture are transmitted sequentially using time division multiplexing, with the analogue chrominance and luminance signals being compressed in time. Sound and data are carried digitally in packets of 751 bits at an instantaneous bit rate of 20.25 Mbit/s, which gives an average rate of about

3 Mbit/s. There are three versions of the basic MAC structure: C-MAC, D-MAC and D2-MAC.

In C-MAC the time division multiplexing is carried out at RF after modulation, the analogue signals using FM or AMVSB and the sound/data signals using PSK modulation. C-MAC with FM gives the lowest error rates for a given carrier to noise ratio of all the MAC versions. With FM, the total bandwidth is approximately 27 MHz and with AMVSB it is about 15 MHz. C-MAC can carry eight high quality sound channels (15 kHz) in addition to the picture.

D-MAC is identical to C-MAC except that the time division multiplexing is carried out at baseband, which reduces the bandwidth of the AMVSB modulated signal to 10.5 MHz and also increases the bit error rate for a given carrier to noise ratio compared to MAC. The unmodulated bandwidth of D-MAC is 8.5 MHz. The D in D-MAC stands for duobinary, which is the three level digital coding used for sound/data.

D2-MAC is similar to D-MAC except that the sound data rates are halved to 10.125 Mbit/s instantaneous or about 1.5 Mbit/s average, giving only four high quality sound channels in addition to the picture. The reduction in the sound rate reduces the unmodulated bandwidth to 5 MHz and the bandwidth of a signal using AMVSB to 7 MHz, which fits many continental cable systems.

At present a considerable amount of effort is being devoted to the development of an international standard for high definition television (HDTV) with a little more than 1000 lines per frame and an uncompressed bandwidth of about 40 MHz or up to 300 Mbit/s. With complex signal processing to compress the signal these figures could be reduced by a factor of about four, and the use of compression is likely especially for satellite broadcasting.

In the UK Sky TV has established a strong market position using the PAL system and this must inevitably delay the widespread adoption in the UK of the more technically advanced systems for some time to come.

Where coaxial copper cable is used it is normal practice to employ AMVSB modulation because of its low bandwidth; however, where optical fibres are used and bandwidth is less critical (even if the capacity of a fibre which can be used cost effectively is exceeded an additional fibre can be used) frequency modulation is normally preferred at present because it gives an improved signal to noise ratio for a given carrier to noise ratio compared to amplitude modulation. Nevertheless, in the future the development of amplitude modulated distributed feedback (DFB) lasers may lead to the use of AM for optical fibre systems. Although people commonly associate optical fibre with digital transmission, and although PAL TV signals are encoded digitally (normally at 34 Mbit/s) for distribution on national trunk networks the use of digital transmission in cable systems is some way off because of the much higher bandwidth required and currently because of the significantly higher cost.

14.4.3 Transmission media

The choice of a cable TV transmission medium lies between copper coaxial cable and optical fibre. All the older cable systems use coaxial cable. At present the costs of cable and fibre alone are broadly comparable, but the costs of optoelectronic equipment to work with fibre are generally higher than those of electronic equipment to work with cable. These costs are influenced by the current very low volumes of equipment for fibre systems. Other significant factors are that optical fibre has much lower attenuation and can be used without repeaters or amplifiers, whereas cable has significant attenuation and requires amplification about every kilometre. Inherently fibre has a higher bandwidth than cable. It also occupies less physical space within ducts.

Optical fibre is well suited to use for high capacity trunks from the head end to the local distribution or switching point in a distributed star configuration, and will be used increasingly for cable trunks as costs fall. (The AM DFB laser developments referred to in the last section will help to reduce costs.) Copper is well suited to tree and branch systems because signals can easily be tapped off the cable for transmission to each house, and as a transmission system it is cheaper than fibre for short links from a local switch to the subscriber's premises. However, techniques have now been developed for splitting off optical signals from fibres and these techniques could be used for a tree and branch or unswitched form of distribution which British Telecom Research Laboratories have termed a passive optical network, and their Broadband Passive Optical Network (BPON) is described in Section 16.2.3.

The cost-effective use of fibre in the local network depends to a large extent on the availability of good low cost connectors, which is an area that has caused problems since the early days of fibre, but considerable progress is now being made.

14.4.4 System design

We have already pointed out that there is no standard form of cable system design comparable to the relatively standardised approach to local telephony using copper pairs.

The mature cable technology at present is a tree and branch system using copper coaxial cable. The TV channels are amplitude modulated (normally AMVSB) and frequency division multiplexed. Amplitude modulation is used because of the bandwidth limitations of coaxial cable (frequency modulation and digital coding and modulation would take up too much bandwidth). The carriers used for the TV channels can be in either the VHF range 45–225 MHz or the UHF range 470–890 MHz. VHF has the advantage that the cable losses at those frequencies are lower than at UHF and so the distances between repeaters can be significantly greater. If the transmissions on the cable use the same UHF frequencies as off-air transmissions then the signals can be fed directly into TV sets without conversion. Systems can use VHF for distribution and provide upconversion to UHF either on the customer's premises or locally in the street. When 'sound radio' channels are provided, they are normally distributed as VHF FM signals which can be fed directly into radio tuners.

It would be unusual to design a pure switched star system where several thousand subscribers were all served individually from a large switch at the head end.

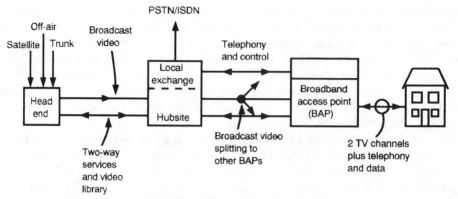

Fig. 14.2 Broadband integrated distributed star

The current economics of switching and transmission for star systems suggest that the optimum design is for 200–300 customers to be served from each wideband switch. One design is British Telecom's Broadband Integrated Distributed Star System (BIDS) see Figure 14.2 where all the channels are broadcast on fibre from the head end via an exchange hub site where telephony can be added to a broadband access point which serves about 200 customers. These broadcast transmissions use frequency modulation with frequency division multiplexing and there is capacity for 16 channels. In addition to the broadcast channels, capacity can be provided on other fibres for two-way channels and for video library channels which are used for one customer at a time. At the broadband access point the channels are downconverted through a space switch with two outputs per customer. They are then upconverted to carrier frequencies of 175 and 225 MHz for transmission by fibre or cable to the customers premises, where they will be converted to a form suitable for input to the TV set (AM UHF or video at baseband).

Radio is an important alternative to the use of cable or fibre for local distribution and the use of millimetre waves is described in Chapter 7. The choice of millimetre waves as opposed to microwaves is the result of pressures for other uses for microwaves.

14.4.5   Satellite Master Antenna TV systems (SMATV)

SMATV is a system whereby satellite transmissions are received at a small Earth station with one or more dish antennas of 1–3 m diameter and distributed by a simple cable system over a small area which is much smaller than would normally be served by a cable system, typically perhaps a block of flats or a small housing estate. The advantage of SMATV over individual domestic satellite reception is that every house does not need its own dish, better quality signals can be obtained, and simultaneous access to more than one satellite is possible if multiple dishes or multiple feeds are used at the head end. The advantage over larger cable systems is that SMATV can take full advantage of small areas of high density housing, but this does not imply that similar antenna cannot be used to provide satellite channels on large cable systems.

There has been considerable debate, without clear conclusions, as to whether SMATV would help or hinder the development of larger cable systems. Its advocates claim that it will stimulate the cable market because SMATV systems could be integrated later into larger cable systems. Its opponents claim that it would cream off the most lucrative areas.

14.5 SERVICES ADDITIONAL TO TELEVISION

So far we have focused on downstream services particularly TV. At present the provision of additional two-way services such as telephony and Viewdata is rare but in the long term it will be important for cable systems to be able to provide a full range of services. In general two-way services have low bandwidth or bit rate requirements compared to TV and little demand is foreseen for high bandwidth two-way services to the home. Owing to the advances in coding, even good quality video telephony can be provided in bandwidth equivalent to a few digital telephony channels. Three main options are available for providing additional services.

- Use a separate system of cables in the same ducts. This obvious low technology solution has considerable economic attractions at the present time and is being used by at least one UK cable company to offer telephony in conjunction with Mercury.

- Use separate fibres within the same cable form.

- Combine the additional services with the downstream services on the same fibre using wavelength division multiplexing or frequency division multiplexing to separate services. Within a two-way service, because there is a common bearer, time division multiplexing is the most likely choice as a means of channel separation but frequency division multiplexing is a possible alternative to TDM. Channels would be allocated for the duration of a call rather than permanently. For example, with Telephony over Passive Optical Networks (TPON) (see Section 16.2.2) telephony will use TDM at one wavelength and TV transmissions may in a future enhancement use FDM at a different optical wavelength.

14.6 MICROWAVE AND MILLIMETRE WAVE FIXED SYSTEMS

Microwave links at 4–6 GHz and 10–12 GHz are widely used for fixed point-to-point transmission with fairly large numbers of channels multiplexed (normally FDM) onto the one carrier. This area of microwave technology is not developing very rapidly because the tendency is to replace these links with optical fibres, but at millimetric wavelengths there is an application which is likely to develop quickly and that is TV distribution over relatively short distances.

Satellites are ideal for broadcasting TV channels to small fixed (60 cm to 1 m diameter) domestic dishes but the number of channels available from a single satellite is limited. To receive a greater number of available channels it is necessary to be able to steer the dish to different satellites or to use multiple dishes or dishes with several adjustable feeds. Satellites are most likely to be used by viewers in areas with low population densities. In areas with high population densities, cable TV is able to offer greater choice at lower cost, although the extent to which cable and satellite compete depends on whether the same TV channels are available on both systems.

Millimetre wave radio links would be able to fill in the gap between satellite and cable TV by distributing large numbers of TV channels over short distances to receivers in homes. Microwave frequencies at 2.5 GHz or 12 GHz could be used but there would be problems of coordination with other services and the amount of spectrum available is very limited. Consequently in the UK an initial decision has been taken to use millimetre wave frequencies at 40.5–42.5 GHz for this application; the question of the use of lower frequencies is, however, still subject to review.

A trial at Saxmundham by British Telecom Research Laboratories (BTRL) has shown the feasibility of transmitting good quality TV signals over ranges of up to 3 km using a directly modulated Gunn diode oscillator as the transmitter at a power of 100 mW per channel, and a separate oscillator antenna combination for each channel. The receiver consists of an offset reflector of 15 cm diameter with an integral gallium arsenide monolithic integrated circuit amplifier and downconverter.

At these frequencies, the attenuation due to rain is a particular problem but BTRL calculate that 99.9% availability can be achieved at the maximum range with at least very good (Grade 4) PAL video quality. An advantage in the use of these high frequencies is that the frequency reuse distance is a few tens of kilometres compared to 150 kilometres or more at lower frequencies.

Inevitably the future of this type of distribution system depends very much on the growth in the number of TV channels, which in turn depends more on the availability of viewing material than on anything else, but if demand rises in the way expected, it would seem that millimetre wave distribution systems will have a useful role.

14.7 CONCLUSIONS

The use of cable for telecommunications services other than TV is of interest only in countries, such as the UK, that allow competition in local telecommunications. Some high quality research and development is being devoted to the design of local networks which can carry a wide range of services including TV, telephony and data. At present most new cable systems have tree and branch topologies based primarily on coaxial cable with separate twisted pairs for telephony but during the next decade fibre will gradually displace copper working from the head end towards the subscriber. In the long term, given sufficient manufacturing volume it is easy to see the attractions of a system based mainly or wholly on optical fibres using some passive techniques and possibly some switching, and making full use of wavelength division multiplexing to carry both narrow and broadband services.

The future of switching for broadband services depends on the nature of the services offered and especially on whether or not there is is demand for two way broadband services for which there are very few bullish forecasts at present. Costs will be highly dependent on manufacturing volumes. The switching of narrow and broad band services at the head end of the network is expected to be accomplished separately (even though the transmission and multiplexing may share the same medium) for the next 10–15 years with narrow band services using digital switches similar to those employed generally in public telephone networks while broadband traffic is routed by analogue space switches.

In the short to medium term there is some uncertainty about substantial investment in very new technology by new companies who would have to compete in the TV area with off-air broadcasts, satellite television and videos, and in the telephone area with the established PSTN. At the same time BT as a public network operator experienced in the use of advanced technology is prevented by regulation from providing TV services. With the low volume of manufacturing and the need for a cautious commercial approach, it is not surprising that some new cable companies are choosing more mature technology which benefits from slightly higher manufacturing volumes. On the other hand the recent resurgence in the number of customers is encouraging and perhaps the end of the economic recession will result in substantial growth.

One particular problem which affects cable is that the most lucrative areas for television services (e.g. housing estates and flats) do not coincide with the most lucrative areas for telephony (business areas and industrial estates).

Although it is difficult to predict the future for cable in the UK, and commentators differ in their views, one attractive possibility which could give a considerable fillip to the industry is the opportunity to use millimetric radio for TV distribution from the basic cable arteries in areas where the cost of installing cable to the customers premises is high. Another possibility is to add CT2 or DECT for two-way Telepoint based telephony services.

14.8 BIBLIOGRAPHY

I. Childs (1988) HDTV putting you in the picture, *IEE Review*, July/Aug, 261–265.

DTI (1984) *Specification of Television Standards for 625-line System I Transmission in the UK*, DTI.

EBU Doc. Tech. 3258 (1988) *Specification of the Systems for the MAC Packet Family*.

J. R. Fox and E. J. Boswell (1989) Star-structured optical networks, *Br. Telecom Technology J.*, 7 (2), 76–88.

M. Pilgrim, R. P. I Scott, R. D Carver and B. J Ellis (1989) The M3VDS Saxmundham demonstrator—multichannel TV distribution by mm-waves, *Br. Telecom Technology J.*, 7 (1), 5–19.

M. Pilgrim, R. P. Searle and R. I. P. Scott (1988) M3VDS—the cheapest, quickest and least obtrusive means of providing multichannel domestic TV, *IBC Conference*, Brighton.

D. E Robinson (1989) Fibre optics: the CATV evolution of the 1990's, *Communications Technology*, Dec, 36,95–98.

W. G Simpson (1984) Broadband cable systems, *Br. Telecommunications Engineering*, 3, 6–13.

J. N. Slater (1988) *Cable Television Technology*, Ellis Horwood and Halstead Press.

15 ISDN

15.1 INTRODUCTION

What is ISDN? The initials stand for Integrated Services Digital Network, which implies a separate discrete network, but in practice in many countries ISDN will use the same switches and transmission facilities as the PSTN. The title is therefore slightly misleading and it is not as easy as one would expect to say exactly what ISDN is. Nevertheless under the heading of ISDN some very important standards and service concepts are emerging, which are having a major influence on network design.

To quote CCITT Recommendation I.120, '. . .the main feature of the ISDN concept is the support of a wide range of voice and non-voice applications in the same network'. In pursuit of this goal, CCITT has prepared a number of ISDN Recommendations for concepts, services, interfaces and signalling. In particular, they have defined basic and primary rate access methods which are capable of providing both circuit and packet switched service access via the same physical exchange line. Such access will involve common channel signalling facilities, which enable sophisticated new services to be offered to the subscriber. One of the key design objectives was to define a basic (144 kbit/s) access method which could be provided on the existing copper pairs in the local area network.

15.2 HISTORICAL PERSPECTIVE

In practice the development of ISDN has been rather slow. Basic rate access has only recently been made available in those European countries which are most committed to ISDN, although a form of primary rate access (2.048 Mbit/s) has been available for a little longer in some countries.

The concept of ISDN has been under development since the late 1970s during a period in which the value of a rigorous and methodical approach to standards writing has been recognised. Consequently ISDN involves far more standards which deal with service and access concepts than earlier work in CCITT.

In this section, we will attempt to introduce the basic concepts so that the reader can interpret 'ISDN-speak', we will describe the basic and primary access methods and we will outline the work to date on broadband ISDN.

15.3 SERVICES AND SERVICE DESCRIPTIONS

ISDN services are known generically as 'telecommunication services' which title is not to be confused with 'teleservices'.

There are two types of service. A bearer service is a service for the transfer of information between two network access points, or network to terminal interfaces. For example a switched 64 kbit/s circuit is a bearer service. A teleservice is a description of a complete service provided to a user (rather than to a terminal) by both the network and the terminals. For example voice services and facsimile services are teleservices. It is interesting to note in passing that the distinctions in UK licences between voice and data are teleservice distinctions, whereas distinctions between circuit switched and packet switched services are bearer service distinctions.

Within the realms of teleservices and bearer services, the services consist of basic services for straight information transfer, and supplementary services which provide additional functionality, although it is not necessarily true to say that a supplementary service is a value added service in the sense of the legal or licence definition. Basic services may exist on their own, whereas supplementary services exist only in conjunction with basic services.

Figure 15.1 shows the interrelationship of these terms, and Tables 15.1, 15.2 and 15.3 list some of the different bearer services, teleservices and Supplementary services. Not all the definitions of the services listed are complete. Tables 15.9, 15.10, and 15.11 in Section 15.7 list the European standards. Generally far more work has been done on bearer services than teleservices, and further work is needed on the definition of some more complicated supplementary services.

ISDN services are defined rigorously and the service descriptions themselves follow a methodically organised structure.

Service descriptions are given in three stages:

- *Stage 1* Describes the service from the user's point of view.

- *Stage 2* Describes the functional network aspects including an intermediate view of what happens at the user-to-network interface and at interfaces between exchanges.

- *Stage 3* Describes the network implementation aspects including the switching and service nodes, and the protocols and formats.

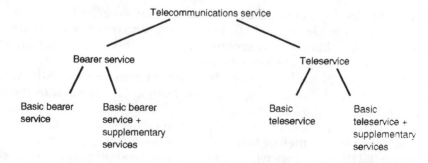

Fig. 15.1 Interrelationship of 'bearer services', 'teleservices' and 'supplementary' services

Table 15.1 ISDN bearer services

(A) Circuit switched mode Description	Comment
64 kbit/s, unrestricted	For circuit switched data
64 kbit/s, for speech	May involve conversion to analogue or lower bit rate; echo cancellling will be used if necessary
64 kbit/s, for 3.1 kHz audio	
Alternative speech/64kbit/s unrestricted	For transfer of data during a speech call
2 × 64 kbit/s unrestricted	
384 kbit/s unrestricted	
1536 kbit/s unrestricted	For the USA
1920 kbit/s unrestricted	
(B) Packet switched mode Virtual call and permanent virtual circuit	
Connectionless	
User signalling	

Table 15.2 ISDN teleservice

Description
Telephony
Teletex
Telefax 4 (Fax Group 4)
Mixed mode document transfer
Videotex
Telex

Table 15.3 ISDN supplementary services

Description

Number identification:
- Direct Dialling In
- Multiple Subscriber Number (assigns more than one number to an interface)
- Calling Line Identification Presentation (gives the called party the calling party's number)
- Calling Line Identification Restriction (allows the calling party to prevent his number being presented)
- Connected Line Identification Presentation (gives the connected party's number to the calling party)
- Connected Line Identification Restriction (allows the connected party to prevent his number being presented
- Malicious Call Identification
- Sub-addressing

Continued overleaf

Table 15.3 *(continued)*

Description

Call Offering:
- Call Transfer (allows user to transfer an active call)
- Call Forwarding Busy (call are forwarded to another number when user is busy)
- Call Forwarding No reply (calls forwarded when no reply)
- Call Forwarding Unconditional (calls forwarded in all circumstances)
- Call Deflection (for further study)
- Line Hunting (assigns single number to group of interface)

Call completion:
- Call Waiting (notifies user of incoming call)
- Call Hold (allows call to be interrupted and re-established)
- Call Completion to Busy Subscribers (for further study)

Multi-Party:
- Conference Calling
- Three Party Service (allows a user to make a second call during an existing call and switch between the calls while preserving the privacy of each)

Community of interest:
- Closed User Group (restricts members of a group to calls to other members, but specific member can have additional facilities)
- Private Numbering Plan (for further study)

Charging:
- Credit Card Calling (for futher study)
- Advice of Charge
- Reverse Charging (for further study)

Additional information transfer:
- User to User Signalling (allows a user to send a limited amount of information over the signalling channel in association with a call)

Each stage is further divided into several steps, and the formulation of the description at each stage and step is described in three different levels (Note levels here are used in different sense to that of layers in the ISO 7-layer model.) Full details are given in *CCITT Recommendation I.130*.

15.4 USER ACCESS

Two forms of user access have been defined to date—basic rate access and primary rate access.

Basic rate access is a form of access that is designed to give the user two 64 kbit/s channels called B channels and a single 16 kbit/s channel called a D channel, making a total of 144 kbit/s. The use of the B channels is unrestricted and independent but typically they may be used for circuit switched services, permanently assigned connections and packet switched services, whereas the D channel is used only for packet switched services and signalling between the user and the network. Basic rate access was designed to be capable of being provided over the existing copper pairs in the local network and yet give the user a significantly greater range of services than was possible with the analogue PSTN.

Primary rate access is access at 2 Mbit/s, which provides 30 B channels at 64 kbit/s, each plus a D channel at 64 kbit/s (higher than the D channel rate for basic rate access). Primary rate access can also provide channels at an intermediate rate of 384 kbit/s called H0 or a single channel of 1920 kbit/s called H12. (There is also a USA equivalent called H11 at 1536 kbit/s.)

It is important to understand that these channels are access channels over which services (teleservices, bearer and supplementary) are provided. The channels are not themselves services and they do not extend unaltered through the public network. The nature of the channels does, however, limit the services which can be provided on them. For example, whereas the B channel can provide both circuit switched and packet switched services, the D channel can provide only packet switched services because it also carries signalling in packet form.

A reference configuration has been defined for the user–network interface which is shown in Figure 15.2. Each box in the diagram has different functions.

The NT1 box contains the network termination functions which would normally be found in the box on the wall provided by the operator of the public network. These functions include termination of the line transmission, maintenance and performance monitoring, transfer of power to the user's apparatus, OSI layer 1 multiplexing, transfer of timing and contention resolution. (*Note*: for an introduction to the concept of layers see Appendix 1 on OSI.)

The NT2 box, if present, contains layer 2 and 3 protocol handling and switching functions, switching and concentration, as well as maintenance and some layer 1 functions. The NT2 can best be thought of as a PABX, a local area network or a terminal controller. NT1 and NT2 functions can be combined in the same physical equipment although in the UK and other European countries the supply of the NT2 functions is being liberalised which means that the NT1 and NT2 are more likely to be split.

The TE1 box represents terminal equipment which connects to the network at the S reference point which is an ISDN interface. The TA box represents a terminal adapter which converts from the S reference point to the R reference point where there is a non-ISDN interface such as an analogue telephony interface or a modem interface.

The descriptions above explain the functions represented by the boxes in the reference configuration and the reference configuration applies at both basic rate and primary rate. We will now consider the interfaces at the reference points.

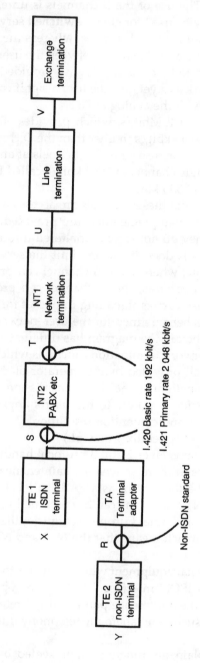

Fig. 15.2 ISDN reference model

Notes:
(1) Bearer services are provided at S and/or T
(2) Teleservices are provided at X and/or Y
(3) Other CCITT services are provided at R

15.5 INTERFACES

15.5.1 Interface at the U reference point

The interface at the U reference point is the interface between the equipment containing the NT1 functions and the exchange line to the public exchange. The U reference point is not defined or referred to in CCITT Recommendations but it has been defined in the USA. We use the term here as a useful shorthand.

Although the U interface is not defined, because the form of the signals at the U interface needs to be matched to the physical characteristics of the exchange lines, which differ from country to country, CCITT Recommendation G.961 contains general requirements for a digital transmission system for basic rate ISDN on metallic local lines, and contains six annexes giving detailed definitions of alternative transmission systems:

- MMS 43, a modified monitoring code with echo cancelling, where 4 bits are mapped into 3 ternary symbols with a line rate of 120 kbaud;
- 2B1Q, a four-level code with echo cancelling, where two binary bits are are mapped into a single quaternary symbol with a line rate of 80 kbaud;
- AMI, alternate mark inversion with echo cancelling and a symbol rate of 160 kbaud;
- AMI with alternate interleaved transmissions from both ends (ping pong) and a line rate of 320 kbaud;
- binary biphase code using echo cancelling with a line rate of 160 kbaud;
- SU32, a substitutional unconditional 3B2T code with echo cancelling and a line rate of 108 kbaud.

Within Europe, there is a possibility of a European standard being produce based on either the MMS43 or the 2B1Q system. MMS43 is used in Germany and France, and 2B1Q is used in the UK.

The information rate in each direction at this interface is approximately 160 kbit/s and special line codes rather than binary codes are used to match the spectrum of the signal to the frequency response of the line. It should be noted that the information bit rate at the U reference point is normally lower than that at the S or T reference point. Both contain the same 144 kbit/s of the Band D channels but local signalling (e.g. activation/deactivation which are not on the D channel) and framing are handled differently.

The form of the primary rate interface is normally either a screened four-wire pair or an optical fibre cable. Again special line codes will be used.

15.5.2 Interface at the S and T reference points

The S and T reference points are the source of some confusion. Strictly speaking, S and T denote the reference points, not the interfaces. The S point is the point of

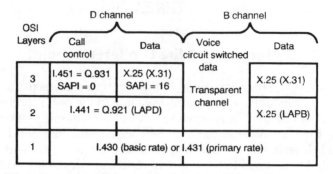

Note: AN X.25 terminal would require an X.31 adaptor to connect to ISDN.
Alternatively, adding the functions of X.31 to the X.25 terminal would convert
to a packet TE 1.

Fig. 15.3 ISDN interface standards

connection of a TE1 terminal, and the T point is the point of connection to the public network terminal, NT1. If no NT2 functions are present, the reference points are coincident. If an NT2 is present the interfaces at both reference points may be identical at layer 1 and layer 2. However, at layer 3 they may differ slightly in that the signalling protocols at the S interface are private network protocols, whereas at the T interface they are public network protocols.

Figure 15.3 shows the structure of the ISDN interface standards and should be referred to in reading the following paragraphs.

Layer 1

Basic Rate

The basic rate interface at layer 1 is defined in I.430. The physical form of the interface is two separate pairs of wires for go and return signals and optional additional pairs for power transfer. The bit rate at this interface is 192 kbit/s which consists of the two 64 kbit/s B channels, the 16 kbit/s D channel and a further 48 kbit/s for framing, octet timing and activation and deactivation of communication between the terminals and the network termination (NT1/NT2). The frame length is 48 bits and the frame duration is 250 μs, see Figure 15.4. A pseudo ternary code is used with no line signal denoting a binary 1 and alternating positive or negative pulses denoting a binary 0. A code violation of the alternating pulses is used to identify the start of the frame.

Where a single terminal (TE1) or terminal adapter (TA) is connected at the S or T reference point, the maximum length of wiring is approximately 1 km, which limit is determined by timing considerations. However, several terminals may be connected at the same reference point using what is called a passive bus. Up to 8 terminals may be connected to a short passive bus of up to 100–200 m length, see

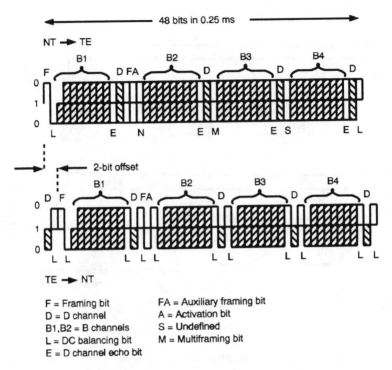

F = Framing bit
D = D channel
B1,B2 = B channels
L = DC balancing bit
E = D channel echo bit

FA = Auxiliary framing bit
A = Activation bit
S = Undefined
M = Multiframing bit

Fig. 15.4 Basic rate frame structures at S and T reference points

Figure 15.5. Alternatively if the distance between the terminals is restricted to 25–50 m, the terminals may be grouped at the end of an extended passive bus of up to 1 km length. These lengths are maximum figures. Practical limits depend on the choice of cable. A protocol operating within the layer 1 frame is used to resolve contention between terminals on the same passive bus. The passive bus can be used for point-to-point or point-to-multipoint (broadcasting NT to TE) communications but it cannot enable more than one point-to-point communications to take place simultaneously. All communications are between the terminals and the NT; the protocols do not allow the terminals to communicate with each other.

Primary rate

The primary rate interface is defined in I.431. Unlike the basic rate interface, only a single terminal or NT2 may be connected at the S or T reference point and the limit to the cable length is determined by loss rather than timing considerations. The physical form of the interface is a balanced $120\,\Omega$ symmetrical pair of wires although an unbalanced $75\,\Omega$ coaxial interface may also be used in the short term. Within Europe, the signal structure is the same as the normal 2 Mbit/s multiplex which is defined in CCITT Recommendation G.703 with a frame length of 256 bits (duration 125 µs) consisting of 32 octets or time slots numbered 0–31. Time slot 0

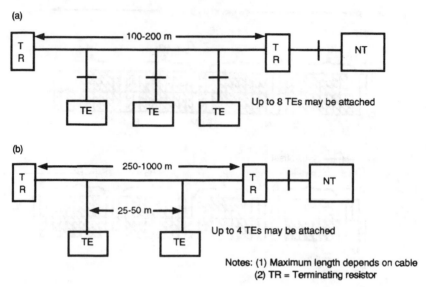

Fig. 15.5 ISDN passive bus (a) short passive bus (b) extended passive bus

carries a frame alignment signal in alternate time slots and time slot 16 carries the D channel with the remaining slots carrying the B, H0 or H12 channels.

Layer 2

The Layer 2 protocols are defined in CCITT Recommendations I.440 (general aspects) and I.441 (detailed specification), but these Recommendations also have the numbers Q.920 and Q.921 respectively. The protocols apply both at basic rate and at primary rate but they apply only in the D channel because the B and H channels are transparent.

The Layer 1 protocols and signal formats described above were concerned with the physical transport of bits across the interface, with the activation and deactivation of terminals, and with the transfer of timing and framing information which is essential for PCM coding on the B channel to be unravelled (i.e. to indicate where each octet or sample begins). The Layer 2 protocols are concerned with the use of the D channel as a data link to connect processes in the terminal to processes in the NT. The Layer 2 protocols therefore provide multiplexing, error control and frame references with respect to each logical communication link.

The format of the Layer 2 signals is an HDLC frame known as LAPD. The format is shown in Figure 15.6. The frame begins and ends with a unique flag and all other bits in the frame are processed by adding bits where necessary and removing them at the receiver to ensure that the bit pattern of the flag is not duplicated. The address field contains two important identifiers—the Service Access Point Identifier (SAPI) and the Terminal Equipment Identifier (TEI).

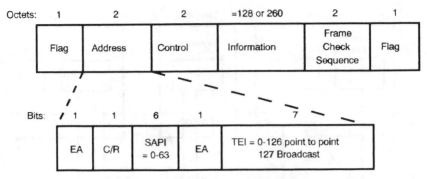

Fig. 15.6 LAPD format (data link layer)

The SAPI is used to identify types of services provided to Layer 3 and can have values from 0 to 63. The value SAPI = 0 is used to identify a frame which is used for signalling and the value SAPI = 16 is used to identify a frame of an X.25 data protocol.

The TEI is used to identify the process in a terminal which is communicating. The TEI can have any value from 0 to 126, allowing up to 127 different processes to be identified. These processes may be spread between the eight terminals which can be accommodated on the passive bus. The value, TEI = 127 is used to identify a broadcast transmission for reception by all terminals. Figure 15.7 shows the use of SAPI and TEI in diagrammatic form.

Layer 3

Layer 3 on the D channel contains any of the following protocols:

- Signalling protocol as defined in CCITT Recommendation I.451 or Q.931 (which is the same) in which case SAPI = 0. The signalling protocol is used to set up and clear calls and to invoke supplementary services.

- Packet layer data protocol as defined in CCITT Recommendation X.25, in which case SAPI = 16.

- Any other protocol to be defined, in which case SAPI is set to the defined number.

The normal format is that each Layer 2 frame carries a single Layer 3 packet and so the Layer 3 frame does not have to be delimited by flags.

15.5.3 Interface at the V reference point

The V reference point has recently been defined, and is currently the subject of standardisation. This reference point exists between the line termination equip-

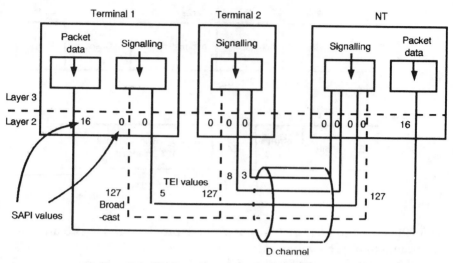

Fig. 15.7 D-channel with SAPI and TEI values

ment at the public exchange end of the local exchange line, and the exchange termination (switch). The purpose of introducing a standard at this point is to enable exchanges from different manufacturers to be used with different line transmission systems including both optical fibre and copper cable.

15.6 PACKET SWITCHED DATA SERVICES

At the beginning of this chapter we pointed out that despite the title Integrated Services Digital Network circuit switched and packet switched services were still provided on separate networks and ISDN so far provides only an integrated form of access. When packet switched services are provided on separate networks, there are two principal ways (in the ISDN context) in which access to public services can be provided.

(a) A packet terminal employing normal D channel signalling sets up a connection using the B channel for access, through the ISDN to a gateway between the ISDN and the packet switched network. The packet terminal then communicates via this connection to the gateway using X.25 to set up a virtual call across the packet network. This is a two-stage call set-up process; the first stage uses ISDN numbering (E.164) to reach the gateway, whereas the second stage employs X.121 numbering to cross the packet network. (See Section 25.3 for numbering details.)

(b) A packet terminal communicates with a packet handler, within ISDN, which is connected via X.75 to the packet network. In this case the user's packet terminal would normally identify the call destination with an ISDN (E.164) number and the packet handler would do the number translation to X.121 for onward routing through the packet network. Although the call set-up is a two stage process, the user is not aware of the second stage in the same way as he is in (a) above.

Access from the packet terminal to ISDN may be either by the B channel but using D channel signalling to request packet transfer mode or wholly via the D channel using SAPI=16.

Because X.25 on its own is not sufficient, in both the cases (a) and (b) above, an existing X.25 terminal must contain the appropriate terminal adaptor, which will include the functions described in X.31, to be able to connect to packet services via ISDN. A packet terminal that incorporates X.31 functions as well as X.25 is called a packet ISDN terminal, TE1.

Existing packet networks are somewhat restricted in terms of both their through-put and the delays which they introduce, because of the need to interrogate packets up to network Layer 3 at each X.75 connection between networks and at each X.25 connection to terminals. In addition the methods of packet access described above violate one of the objectives of ISDN, which is to separate the user plane and the control plane (user information and signalling information) and to provide a wide range of user services using a limited set of connection types and interface arrangements. Thus the ISDN packet service should ideally have the same packet signalling protocols as the other ISDN basic services, but with X.31 and X.25 it does not do this because much of the signalling is contained within X.25.

Consequently CCITT is developing additional packet mode bearer services that meet these objectives. The main such services are called frame relaying and frame switching. Both will use separate standard ISDN signalling to set up a call and the network will then transfer frames or packets until the call is cleared. In frame relaying, the network will transfer frames transparently in both directions, protecting the order, or sequence, of the frames and detecting errors in the frame format (error frames are discarded). This basic frame relay service is called Frame Relay 1. Frame Relay 2 is similar but in addition it provides congestion control at Layer 2. Frame switching provides acknowledged transfer of frames and both detection and recovery from errors as well as flow control. In addition there has been some interest in an X.25 based packet bearer service in which X.25 is simplified at Layer 3 and call control procedures are transferred to the ISDN signalling.

The frame relay concept that originated in this ISDN related work became the subject of intensive development by many manufacturers in 1990/91 because manufacturers realised the potential of this technique to provide a high speed form of packet communications suitable for interconnecting LANs. However, the initial interest is in frame relay as a packet form for use on leased circuits (permanent virtual circuits) not as a public service, although it is expected that standards for public services (switched virtual circuits) will be developed later. These public services are expected to use separate signalling based on Q.931 for call set up and clearing. Frame relay is described in more detail in Section 16.3.4.

15.7 ISDN IN EUROPE

15.7.1 CEPT MOU

In 1986, the European Commission issued a recommendation (86/659/EEC) on the coordinated introduction of ISDN within Europe. In response, the European Tele-

Table 15.4 Minimum set of services to be provided by TOs under CEPT MOU

Bearer services	Supplementary services
Circuit mode 64 kbit/s unrestricted	Calling Line Identification Presentation (CLIP)
Circuit mode 3.1 kHz audio	Calling Line Identification Restriction (CLIR)
	Direct Dialling In (DDI)
	Multiple Subscriber Number (MSN)
	Terminal Portability (TP)

communications Organisations (public network operators) discussed the implementation of ISDN within CEPT, and in April 1989 signed a Memorandum of Understanding on the Implementation of European ISDN Service by 1992.

The MOU covers the implementation of common services and access arrangements, and the development of the necessary standards by the European Telecommunications Standards Institute (ETSI) with common conformance testing procedures to enable the interchangeability of terminals. The target is for services to commence in 1992, but the commitment is to opening service by December 1993 at the latest. The intention was that the necessary standards should be developed by ETSI by December 1989.

Table 15.5 Services to be provided by TOs to European Standards under CEPT MOU

Bearer services	Supplementary services
Circuit mode speech	Advice of charge at call set-up time (AOC-S)
Circuit mode 2 × 64 kbit/s unrestricted	Advice of charge during call (AOC-D)
Packet mode	Advice of charge at end of call (AOC-E)
X.31 case A (B-channel)	COnnected Line identification Presentation (COLP)
X.31 case B (D-channel)	COnnected Line identification Restriction (COLR)
X.31 case B (B-channel)	Closed User Group (CUG)
	Call Waiting (CW)
	Completion of Calls to Busy Subsribers (CCBS)
	CONFerence call add on (CONF)
	Meet Me Conference (MMC)
	Call Forwarding Unconditional (CFU)
	Call Forwarding Busy (CFB)
	Call Forwarding No Reply (CFNR)
	Call Deflection (CD)
	FreePHone (FPH)
	Malicious Call IDentification (MCID)
	SUB-addressing (SUB)
	Three ParTY service (3PTY)
	User–User Signalling (UUS)

The TOs committed themselves to providing the following minimum set of services shown in Table 15.4.

In addition, the TOs agreed to operate the following services listed in Table 15.5 in accordance with European standards.

15.7.2 Open Network Provision (ONP) and ISDN

The European Commission attaches great importance to the introduction of ISDN for the benefit of European industry and commerce and as a means of introducing a common market in telecommunications apparatus. Under the ONP framework (see Section 23.5.3), the Commission introduced a Recommendation on ISDN during 1992. The intention is that ONP conditions are to be applied to both basic and primary rate services at the S/T reference point, at which customer apparatus is attached to the network.

The ISDN ONP Recommendation specifies a minimum set of services that must be offered by 1 January 1994 (see Table 15.6) and lists other services which are to be provided by an unspecified date, (see Table 15.7). It also lists supplementary services that must conform to European standards if provided (see Table 15.8).

Tables 15.9 and 15.10 list the ETSI standards and their current status. The date under the standard number shows the expected date of publication.

Some of the services deserve special comment. Terminal addressing is concerned with the method of identifying particular terminals on an S-bus. At present differ-

Table 15.6 Services to be available by 1 January 1994 under ONP

Access	Bearer services	Supplementary services
Basic rate at S/T	Circuit mode 64 kbit/s unrestricted	Calling Line Identification Presentation (CLIP)
Primary rate at S/T	Circuit mode 3.1 kHz audio	Calling Line Identification Restriction (CLIR)
		Direct Dialling In (DDI)
		Multiple Subscriber Number (MSN)
		Terminal Portability (TP)

Table 15.7 Services to be available by an unspecified date under ONP

Access	Bearer services	Supplementary services	Teleservices
	Circuit mode 64 kbit/s unrestricted on reserved or permanent mode	Call transfer services	Telephony 3.1 kHz
		Call forwarding services	
		Reverse charging	
		Freephone	
	Cricuit mode 2 × 64 kbit/s unrestricted	Kiosk billing or equivalent	
		Closed user group	
	Packet mode over B and/or D channels.	User–user signalling	
		Malicious call identification	
		Network management services	

Table 15.8 Services to conform to European Standards under ONP if provided

Supplementary Services

- Advice of charge services (AOC)
- Number Identification Services (COLP, COLR)
- Call Waiting (CW)
- Completion of Calls to Busy Subscribers (CCBS)
- Conference services
- SUB addressing (SUB)
- Three ParTY services (3PTY)

Table 15.9 Status of standards for bearer and teleservices

Title	Standard	Publication date
1. Bearer services to be available by 1.194		
Circuit mode bearer service 64 kbit/s unrestricted 8 kHz structured	ETS 300 108	1992
Circuit mode bearer service 64 kbit/s 8 kHz structured for 3.1 kHz audio	ETS 300 110	1992
2. Bearer services to be available at unspecified date		
Circuit mode 64 kbit/s unrestricted on reserved permanent mode	No number	Work started
Circuit mode 2×64 kbit/s unrestricted	No number	No work
Packet mode bearer service, virtual call & permanent virtual call on B channel	ETS 300 048	Published
Packet mode bearer service, virtual call & permanent virtual call on D channel	ETS 300 049	Published
3. Other bearer services		
Circuit mode bearer service, 64 kbit/s 8 kHz structured	ETS 300 109	1992
4. Teleservices to be available at unspecified date		
Circuit mode teleservice, telephony 3.1 kHz	ETS 300 111	1992
5. Other Teleservices		
Circuit Mode Telefax group 4	ETS 300 120	1992

ent methods are used in different member states and ETSI has been asked to develop a common method. The network management service is not defined at all at present and is expected to be the subject of an independent study. PSPDN interworking using a single-stage selection process is under study in CCITT and ETSI will prepare appropriate European standards once the CCITT work reaches a conclusion.

In addition to the service requirements given above, ONP will require:

- publication of information on the services;

Table 15.10 Numbers and status of European standards for supplementary services

Supplementary service		Service description	Functional capabilities and information flows	Access signalling protocol
(a) *To be available by 1.1.94*				
Calling line identification presentation	CLIP	ETS 300 089 Published	ETS 300 091 Published	ETS 300 092 Published
Calling line identification restriction	CLIR	ETS 300 090 Published	ETS 300 091 Published	ETS 300 093 Published
Direct dialling In	DDI	ETS 300 062 Published	ETS 300 063 Published	ETS 300 064 Published
Multiple subscriber number	MSN	ETS 300 050 Published	ETS 300 051 Published	ETS 300 052 Published
Terminal portability	TP	ETS 300 053 Published	ETS 300 054 Published	ETS 300 055 Published
(b) *To be available by unspecified date*				
(i) Call transfer services				
Explicit call transfer	ECT	nna 1992/3	nna 1993	nna 1994
(ii) Call forwarding services				
Call forwarding busy	CFB	ETS 300 199 1992/3	ETS 300 203 1992/3	ETS 300 207 1992/3
Call forwarding unconditional	CFU	ETS 300 200 1992/3	ETS 300 204 1992/3	ETS 300 207 1992/3
Call forwarding no reply	CFNR	ETS 300 201 1992/3	ETS 300 205 1992/3	ETS 300 207 1992/3
Call deflection	CD	ETS 300 202 1992/3	ETS 300 206 1992/3	ETS 300 207 1992/3
(iii) Miscellaneous services				
Reverse charging (No standards)		—	—	—
Freephone	FPH	ETS 300 208 1992	ETS 300 209 1992	ETS 300 210 1992
Kiosk billing (No standard)		—	—	—
Closed user group	CUG	ETS 300 136 Published	ETS 300 137 Published	ETS 300 138 Published
User–user signalling	UUS	nna 1992/3	nna 1992/3	nna 1992/3
Malicious call identification	MCID	ETS 300 128 Published	ETS 300 129 Published	ETS 300 130 Published
(c) *Services to use European Standards*				
(i) Advice of Charge Services				
at set-up	AOC-S	ETS 300 178 1992	ETS 300 181 1992	ETS 300 182 1993
during call	AOC-D	ETS 300 179 1992	ETS 300 181 1992	ETS 300 182 1993
end of call	AOC-E	ETS 300 180 1992	ETS 300 181 1992	ETS 300 182 1993

Table 15.10 *(continued)*

Supplementary service	Service description	Functional capabilities and information flows	Access signalling protocol
(ii) Number identification services			
Connected Line Identification	ETS 300 094	ETS 300 096	ETS 300 097
Presentation COLP	Published	Published	Published
Connected Line Identification	ETS 300 095	ETS 300 096	ETS 300 098
Restriction COLR	Published	Published	Published
(iii) Call completion services			
Call waiting	ETS 300 056	ETS 300 057	ETS 300 058
CW	Published	1992	Published
Call hold	ETS 300 139	ETS 300 140	ETS 300 141
HOLD	Published	Published	Published
Completion of calls to busy	—	—	—
subscribers CCBS	1992/3	1992/3	1992/3
(iv) Conference serivces			
Meet me conference	ETS 300 164	ETS 300 165	
MMC	1992	1992	
Conference call add-on	ETS 300 183	ETS 300 184	ETS 300 185
CONF	1992	1992	1992
(v) Other services			
Sub-addressing	ETS 300 059	ETS 300 060	ETS 300 061
SUB	Published	Published	Published
Three party service	ETS 300 186	ETS 300 187	ETS 300 188
3PTY	1992/3	1992/3	1992/3
nna = number not assigned			

- publication of information on the target and actual statistics for the availability, delivery, and performance of services;

- a common ordering procedure;

- a one-stop ordering procedure;

- a one-stop billing procedure;

- fair management of numbering arrangements;

- cost-orientated tariffs with scope for volume discounts (Phase 1 services may be bundled but all other services must be offered separately).

15.8 BROADBAND ISDN

The concept of broadband ISDN is described in general terms in CCITT Recommendation I.211 and it is due to be developed further during the 1989–92 study period.

I.211 envisages two main categories of service—interactive and distribution—with interactive services being subdivided into conversational services, messaging services and retrieval services. The following are some examples:

- Interactive: Conversational: Video telephony
  Video conference
  High speed data;

- Interactive: Messaging: Mail services for films or images;

- Interactive: Retrieval: Film and image retrieval;

- Distribution: Television and audio distribution.

The reference model for broadband ISDN is the same as the model for ordinary ISDN shown in Figure 15.2.

The channel rates envisaged are:

- H21: 32.768 Mbit/s

- H22: an integer multiple of 64 kbit/s in the range 43–5 Mbit/s

- H4: an integer multiple of 64 kbit/s in the range 132–8 Mbit/s

and two user–network interfaces are planned, one at about 150 Mbit/s and the other at about 600 Mbit/s, compatible with SDH.

The intended form of operation uses Asynchronous Time Division (ATD) as described in Section 3.5, which is a multiplexing technique where the information is carried in cells which contain an information field and a header to identify the channel (e.g. Asynchronous Transfer Mode). Cells are assigned on demand and connections are set up for the duration of a call. Signalling and user information are carried on separate channels.

15.9 BROADBAND ISDN AND NETWORK EVOLUTION

The adoption of B-ISDN is bound to be a slow process that keeps in step with the evolution of networks from their current state to one in which broadband services exist as part of a multiservice environment (see Section 16.3.5). Particularly in the context of the RACE project, consideration is being given to how best to implement these changes and while this work is likely to go on for some time (as the CCITT recommendations emerge) the following points can be made:

(a) The Integrated Broadband Communications Network (IBCN) of RACE is based on B-ISDN in terms of facilities and access mechanisms. The objective is commercial introduction by 1995.

(b) Application areas where a potential requirement for broadband services have been identified include:

(i) Publishing—multinational and multimedia with particular reference to high quality image material.

(ii) Financial—arising from the aggregation of many requirements, e.g. voice, data, document and video.

(iii) Healthcare (Telemed)—again arising from an aggregation of requirements.

(iv) Industry—computer aided design and manufacturing information transfer.

(v) In the longer term, home entertainment.

(c) Currently work on B-ISDN is very much tied to the ATM and this is not suited to all applications with high definition television being a particular case where the STM appears more suitable. Of course further development of B-ISDN in this direction is not precluded.

(d) A problem area is network dimensioning for traffic flows. The nature of the bursty traffic associated with video telephony and video conferencing needs to be characterised and measured, and simulation models and/or mathematical theories need to be developed.

15.10 THE FUTURE FOR ISDN

The introduction to this chapter pointed out that ISDN is really a collection of service concepts and network access standards together with CCITT No. 7 as the public network signalling system. In many respects the descriptor Integrated Digital Access (IDA) as used by BT for its early ISDN service trials is a better descriptor because the most important part of ISDN, so far as the customer is concerned, is the means of access to it. The approach of BT and other public network operators is to add ISDN like facilities to existing digital exchanges (e.g. System X, System Y or DMS).

The success of primary rate ISDN for the connection of PBXs to public networks that provide bearer services is almost guaranteed because of the better quality offered by digital transmission, the wide range of supplementary services potentially available and the absence of competition from other digital systems or services. Primary rate ISDN will probably establish itself as the main form of 2 Mbit/s PABX to public exchange access in the medium to long term, i.e. by 1995.

Basic rate access has the potential to provide a wide range of supplementary services but some of these services can be provided by other means, e.g. tone dialling using the digits 0–9 plus * and # on digital exchanges, and it is difficult to predict when and to what extent business and domestic customers will be prepared to pay a premium for such extras. Basic rate does offer 2B+D access over a single copper pair and if the installation and rental were lower than twice the equivalent figure for analogue access, basic rate ISDN would be attractive to customers who want two separate circuits, e.g. one for telephony and one for fax. But initially at least, and purely from the tariff point of view, the situation in some countries favours two analogue lines. Furthermore, because of the high production volumes associated with an established product, simple terminal apparatus for analogue connection is likely to be cheaper. Growth of basic rate ISDN will therefore depend on the ability of industry to develop terminals which exploit features most readily available through ISDN such as integrated work stations for small businesses or 64 kbit/s low bit rate video telephony for business and domestic use. This latter could be a very significant development. Overall the acceptance of basic rate ISDN will certainly lag behind the adoption of primary rate ISDN.

Preparations for broadband ISDN are advancing steadily with the development of the Asynchronous Transfer Mode and the Synchronous Digital Hierarchy and support of the Race programme. The main applications for broadband ISDN are likely to involve the use and manipulation of high quality images.

15.11 BIBLIOGRAPHY

P. Bocker (1988) *ISDN—The Integrated Services Digital Networks: Concepts, Methods, Systems,* Springer-Verlag.

J. M. Griffiths (ed.) (1990) *ISDN Explained: World Wide Applications and Network Technology,* Wiley.

A. M. Rutowski (1985) *Integrated Services Digital Networks,* Artech House.

16 DATA AND MULTISERVICE NETWORKS—LOWER LAYERS

16.1 INTRODUCTION

16.1.1 Scope

The subject of this chapter is networks in the physical sense of their topology and performance characteristics and in OSI terms it is largely confined to layers 1 to 3. The services that may be run over these networks are covered in the next chapter. A network is a means of connecting users' terminals to each other. A network's design must represent some sort of compromise between its ability on the one hand to provide simultaneously every possible interconnection pattern and on the other hand the need to minimise the cost of its provision and maintenance. We are used to the idea of a network being associated with a service; for example, the telephone network, the telex network, the packet switching network and so on. This association is not fundamental and in fact in its early days the telex service shared the telephone network. Today we are moving towards the concept of multi-service networks and a major aim of this chapter is to describe ways in which those networks can be realised. If technology can provide the means by which we can use the same transmission and switching equipment for a multiplicity of services, there should be cost savings compared with the case where separate networks are provided for each major service. Because this chapter is mainly concerned with the future it has to take for granted that the reader has some background knowledge of existing networks; if that is not the case then Chapters 3–5 are recommended as prior reading. Nevertheless the rest of Section 16.1 contains some introductory material which is partly a recapitulation based on the more general coverage of Chapters 3 and 4, but expressed in network terms, and which is specifically directed to the rest of this chapter.

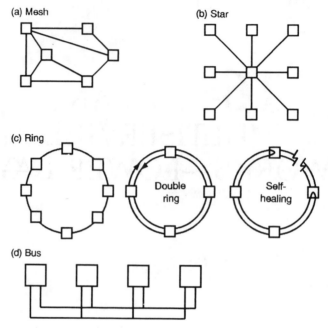

Fig. 16.1 Network topologies

16.1.2 Network topology

There are several different basic topologies or connection arrangements for networks; the mesh, the star, the ring and the bus. These are shown in Figure 16.1. These descriptions normally refer to the nature of the electrical connections rather than to the physical layout of the cables or the layout of the ducts. For example, equipment may be connected electrically in the form of a ring, yet the cable layout take the form of a star.

Network topologies should be chosen to match the operational requirements of the network and to minimise cost. Topology will therefore be heavily influenced by the relative costs of switching and transmission. For example, a star is a good topology where a number of simple equipments need to be connected to a single more complex equipment which will need to participate in most of the communications. A mesh is a good topology where there is a possibility of damage or congestion provided that calls can be rerouted through intermediate nodes. A mesh is also useful if the number of equipments which can be linked for transmission needs to be minimised, e.g. to minimise delay. The PSTN is a combination of star topologies in the local network and mesh topologies in the trunk network.

The realisation that the cost of installing, rearranging and maintaining a star configuration network in an urban area or large building is a substantial proportion of the cost of ownership has tended to concentrate attention more recently on ring and bus architectures and these receive detailed treatment later on in this chapter.

16.1.3 BT's telephony network

Today the BT telephone network represents for the UK by far the largest network in terms of size, traffic and revenue earning. It will be taken as an example of the present situation and its existence must be regarded as a major factor in any evolution towards new network configurations.

Most telephone users are connected to their local exchange by pairs of copper wires which are capable of transmitting the analogue speech signal with an acceptable amount of impairment. A minority of users, largely businesses, have a digital connection to the nearest exchange that provides a digital interface. The network between the user and the local exchange is called the local area network and represents probably the largest single item in the investment inventory of any telephone company.

By the end of the 1990s all exchanges in most of the more developed countries should be digital (or at least provide digital access) but currently some are analogue and some digital.

In addition to the local area and trunk transmission networks there is a junction network that interconnects local exchanges over distances of a up to a few tens of kilometres; these connections are sometimes known as sideways connections. The junction and trunk networks use multiplexing of speech channels on to a common bearer as the means of conveying a large amount of traffic over appreciable distances. The bearer may be a copper pair, a coaxial cable or an optical fibre depending on economic considerations relating to distance and the number of channels.

The telephony network can be regarded as star connected as far as the end terminals to local exchanges are concerned and mesh interconnected at the junction and trunk level. The density of local exchanges in relation to the population that they serve has been the subject of much study in the past and has resulted in the near optimum siting of exchanges in most areas. The advent of relatively cheap broadband technology allows a number of the assumptions on which these older studies were made to be questioned and permits some new thinking about the topology of networks in general.

16.1.4 Wideband and broadband digital networks

The term 'broadband' is defined by CCITT to mean a bit rate in excess of a nominal 2 Mbit/s while the term 'wideband' is deprecated by CCITT but is sometimes used to cover the range between 64 kbit/s to 2 Mbit/s. Narrowband covers 64 kbit/s and below. For simplicity here it will be assumed that broadband includes wideband unless specifically indicated to the contrary.

There are two, not incompatible, reasons why broadband networks are worthy of consideration:

(a) The sharing of one broadband bearer between a number of users, most or all of whom may not individually need the total bandwidth, may offer economic advantages in terms of transmission plant and/or flexibility of service provision.

(b) Individual transmission channels may need to be broadband by virtue of the information that they are required to carry.

Because a user does not normally require to be connected for 100% of the time, the ability to assign bandwidth on demand is a vital requirement in the efficient use of broadband transmission.

Two broadband (or wideband) access means and three area network topologies are to be considered:

1. Integrated Digital Access (IDA)

2. Optical fibre in local area networks

3. Local Area Networks (LANs)

4. Metropolitan Area Networks (MANs)

5. Wide Area Networks (WANs)

IDA means basic rate or primary rate access in the local area using the relevant protocols of the Integrated Services Digital Network (ISDN) as described in Chapter 15. IDA therefore does not require further description here. It does not directly involve changes to network topology and it does not constitute a user service in its own right.

Optical fibre in local area networks is intended to apply to changes to existing networks rather than a completely new network topology in its own right. It does need a fairly detailed treatment however and is the topic of Section 16.2

The three area networks do imply unconventional topologies and the distinguishing feature between them is size; as a rough guide LANs cover distances associated with a single site, say 1 km, MANs cover an urban area, say several tens of kilometres, while WANs cover the world. The distinguishing features within an area network type, irrespective of size, are topology and protocols. However, there is not necessarily a one-to-one correspondence between a particular area network design and the LAN/MAN/WAN nomenclature, some designs are suitable for both LANs and MANs, for example.

16.1.5 Transmission media

Section 16.3 covers in some detail designs for each of the three types of broadband area networks listed above. The main emphasis there is on the network aspects and the transmission medium is a secondary concern; normally the transmission medium will be a land line (e.g. twisted copper wire, coaxial cable or optical fibre) but other media are potentially involved in all types of network and these are very briefly mentioned below, and also in the context of building wiring in Chapter 10.

Satellites are an example of an existing multi-service transmission medium and can be regarded as an adjunct to other networks (particularly WANs), or as in the case of satellite TV, the primary transmission medium of a specialised broadband network. Chapter 13 is concerned in detail with satellite developments.

Cable networks, which today means coaxial cable networks, are usually identi-fied with cable TV but they potentially have a much wider role and one that is pursued further in Part 4. The more specific aspects of cable are covered in Chapter 14, where future developments in topology and technology are discussed.

The term 'wireless' is used here literally to mean without wires and this includes radio (which is the topic of Chapter 12) but it also includes optical communication over a line-of-sight path and acoustic communication, whether sonic, subsonic or ultrasonic. Certainly radio is relevant as a transmission medium in the immediate context but the role of other forms of wireless communication is less certain and is briefly touched upon in Section 16.2.4.

16.2 OPTICAL FIBRE IN LOCAL AREA NETWORKS

16.2.1 Introduction

The local network based on the use of copper pairs has been in existence since telephony began and is the part of the network for which up until very recently there has been very little pressure for change; at least on technological grounds. Optical fibre as a transmission medium is still much more expensive on a circuit-to-circuit basis than a copper pair, with the break-even point in 1989 estimated as being about 3 or 4 to 1 for new cable in an existing duct. There are however a number of advantages of optical fibre which are outlined below:

(a) potentially improved reliability, although this has yet to be demonstrated in a practical situation;

(b) much greater transmission capacity, although the exploitation of the spare capacity has hardly begun;

(c) low attenuation and freedom from crosstalk.

In advance of an established need for very broadband services the high capacity of optical fibre can be used in another way; that is by sharing a reduced number of very high capacity circuits between a number of users. Clearly such sharing must entail multiplexing techniques. There are a large number of ways in which the local area network might be reconfigured and here we shall concentrate on one of these which is based on a star configuration. Configurations based on buses and rings are covered in Section 16.3, although the technology considered there is not necess-arily optical.

There are two main ways in which the local area network might evolve:

1. from the existing telephony network;

2. from the cable TV network.

The second alternative normally involves a double or distributed star configuration with active nodes and is given further consideration in Chapter 14. What follows is strongly influenced by BTRL's work on what they call Telephony on a Passive Optical Network (TPON) and Broadband Passive Optical Network (BPON). For a comparison between the passive and active approaches see Ritchie (1989).

16.2.2 TPON

The local area network uses optical fibre but is still based on a star topology. Instead of there being one point or arm of the star for every customer there are fewer arms, but up to about 120 customers are connected to each arm using two stages of passive optical splitting. The optical fibre splits at the street cabinet and splits again at the distribution point or footway box; the final connection can then be either the continuation of the fibre or a copper pair. The former is called Fibre To The Home (FTTH) and the latter Fibre To The Curb (FTTC) (US spelling of kerb). Figure 16.2 shows the basic arrangement; it is as the name indicates entirely passive and all the information for all the customers on an arm is time division multiplexed and broadcast over the optical fibre to be received at all the termination points. Information as received at a termination will consist of a number of channels each destined for a particular customer; selection then takes place and customers only have access to the channel or channels allocated to them. Because all the information sent on each arm is receivable at each termination point, the network remains a virtual star even if topologically it has moved in the direction of a tree and branch.

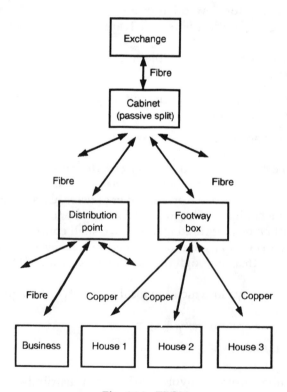

Fig. 16.2 TPON

Signalling on the exchange side of TPON was DASS2 and is now the BT version of ISDN access (Section 5.4.1 and 5.4.2) which is converted at the TPON interface to a form suitable for delivery to network terminals. Once a connection is established, information for a specific recipient is recognised by its position in time in the multiplex. Synchronism is controlled from the head end and the distribution of down stream information is in the form of a conventional TDM frame. The transmission of up stream information from customer to the exchange also uses TDM but is complicated by the variable time delays and amplitude levels inherent in the different path lengths. The customer terminal has to be told at what time to transmit so that all information arrives at the head end in the correct relative time relationship. Optical splitters separate the upstream and downstream information at both ends; such splitters are the optical equivalent of an electrical hybrid.

The number of customers on each arm is likely to be of the order 30–120, a limit set mainly by reliability rather than economic considerations. There is an ability to allocate transmission capacity as customers' needs require; the total bit rate of 20 Mbit/s can be allocated with a granularity of 8 kbit/s. Current estimates suggest that the cost of TPON equates to the cost of new copper lines when each optical termination provides a capacity of 3 to 4 lines and so TPON could be used economically in residential areas if several individual customers were connected by copper from an optical termination in a distribution box in their street.

A version of TPON using frequency, as opposed to time, division multiplexing is under consideration.

16.2.3 BPON

By adding another level of multiplexing, based on wavelength division (see Section 2.1.7) it is possible to extend TPON to provide BPON a broadband passive optical network, (see Figure 16.3). TPON continues to use the 1300 nm band while the additional BPON services use the 1550 nm band. The 1550 nm band is split into four wavelengths each of which can be further sub-divided by using time or frequency division multiplexing. One of the four wavelengths is, however, reserved for maintenance. The remaining three could be used for example for TV services.

There are two versions of BPON: analogue BPON and digital BPON. Analogue BPON uses frequency modulation because amplitude modulation puts severe constraints on the power budget. Using FDM, up to 32 conventional TV channels or alternatively the appropriate smaller number of MAC or HDTV channels can be carried at each optical wavelength. Digital BPON needs technology that still has to be developed whereas analogue BPON technology exists and it has to be established whether and when digital BPON will become cost effective. Looking even further ahead to a time when the SDH multiplex hierarchy discussed in Sections 3.4 and 16.3.5 is used in asynchronous transfer mode, information will be carried on a broadband local area network and the acronym APON (Asynchronous Passive Optical Network) will apply.

Under the RACE programme (see Section 16.3.5) the Optical Line Outlet (OLD) concept (Vandenameele and Evers, 1990) is employed for the early integration of

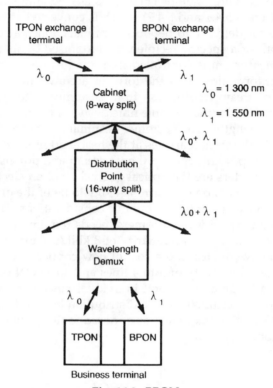

Fig. 16.3 BPON

CATV and ISDN networks. The 800 nm band is allocated to CATV and the 1300 nm band to ISDN.

For the distribution of a large number of television or video channels in the local area a purely passive network may not provide the best solution and it seems likely that some at least of the broadband services are more economically realised if amplification is employed. The use of erbium-doped optical fibre amplifiers, as discussed in Section 2.1.8, results in optical power levels of the 100 mW order at the head end of the network and makes possible the distribution of, for example, 160 digital PAL TV channels to over 7000 customers over distances of up to 30 km. Thus the high cost of the optical amplifier can be spread across a large population of customers.

16.2.4 Rival technologies to optical fibre in the local area network

The main rival to optical fibre, apart from coaxial cable which is dealt with above, is radio. Radio is a serious competitor, particularly for the tails of the local area network, and is covered in Chapter 12.

There are two less serious competitors, again for the tails of the local network (e.g. on the customer side of the street cabinet), which are a line-of-sight optical link and a highly directional acoustic link.

Line-of-sight optical communication links are used in some specialised applications but problems of fading in heavy rain, obstruction of the line-of-sight path by large vehicles and new buildings are potentially fairly serious impediments. The equipment is unlikely to be much less obtrusive or cheaper than microwave radio and hence it would be rash to be optimistic about the future for line-of-sight optical transmission links. An advantage compared to radio is that it it does not need a spectrum allocation.

Ultrasonic acoustic transmission is a theoretical possibility; a high acoustic frequency is the obvious choice if the size of equipment is to be kept within reasonable bounds. There are several unknown quantities, that of interference from and to an acoustic link being perhaps the major factor. While there is radio spectrum available it would seem unlikely that the pressure towards an alternative acoustic solution will be great enough for the necessary research to be undertaken.

16.3   AREA NETWORKS

16.3.1   Introduction

Area networks are categorised in terms of increasing area of coverage as local, metropolitan or wide, but as we shall see there is no hard and fast demarcation between the three and the indications are that, particularly for high speed networks, the basic concepts at least are tending to merge. In fact even computer backplane interconnection standards are showing signs of sharing common ground with LANs although that is not a topic that is treated here. Figure 16.4 attempts to put some of the area networks covered in this section into perspective by showing them on a chart of bit rate against distance.

Existing networks, particularly in the local area, are based on star topologies and a star topology uses more cable or wire in connecting a number of network terminals to a central switching point than is needed if all those terminals share a common ring or bus. In the past the technology has not been available to allow the transmission medium to be shared economically, but that situation is fast changing and the cost of installing, rearranging and maintaining cables and wires is becoming increasingly important not only for building wiring as we saw in Chapter 10 but generally.

Hence most emerging local area networks are based on the bus and the ring topologies. When it comes to the question of how the network behaves in the presence of equipment failures the distinction between a bus and a ring may become significant, but in most other contexts it is not and for simplicity we will talk about rings, bearing in mind that what is said also applies to buses.

With a ring all information is potentially available to all the nodes (terminals) connected to the ring although only the node(s) to which the information is addressed will accept it. Unlike the telephone network, no dedicated path need be

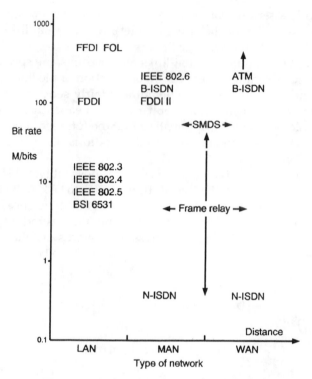

Fig. 16.4 Area network—application chart

allocated for the duration of a call between terminals that are involved in a trans-action or a conversation, and this suggests the packet mode of working is very appropriate. Because the ring is in effect a single transmission path most of the complexity lies in the rules or protocols for access to the ring. The information-carrying capacity of the ring depends on these protocols as well as on the ring's inherent transmission bandwidth, but the different protocols constitute the major distinguishing feature between the various ring standards.

There is always a limit to the size of a ring, which can be expressed either in terms of the number of nodes attached to it or in terms of the circumferential path length. This limit is determined by one or more of the factors, delay, jitter and the reliability of what is essentially a serial arrangement.

Because, owing to contention in access, rings can have a high and variable access time, or latency, they are better suited to data than they are to speech. Delay and variations in it do not matter much or at all in most data transmission applications, whereas with speech delay and particularly variations in it are more objectionable. Most early rings were designed for data, with speech as very much a secondary consideration. More recently, the need to provide multi-service networks has led to approaches which permit the satisfactory conveyance of both types of information.

With some designs the physical transmission medium is discontinuous at each node and the data on the ring are regenerated there. Provision may be made for the automatic by-passing of a node on detection of failure. However, it is not essential for a node to introduce a total discontinuity into the transmission medium and for example a design using an opto-coupler might remove a fraction of the power on the ring at each node with the large remaining fraction continuing on its way. Power is then inserted at any node by means of an optical combiner.

16.3.2 Local Area Networks (LANs)

There are many LAN network designs but the discussion here will be confined mainly to those that have received or are under consideration for international standardisation. There many proprietary LANs, most of which are for personal computer networks, for example ARCNET and PC-Network, but to treat these is difficult if only because there is a tendency for proprietary products to be upgraded and redesigned. We therefore concentrate on those that are standardised, and Table 16.1 summarises the key characteristics of the five best-known arrangements.

The transmission media standards or recommendations (OSI level 0) for the LANs in Table 16.1 are dealt with in Section 10.3.1, and in this chapter the concern is mainly with behaviour at OSI layers 1–3.

Jitter is produced by each repeater on a ring and for most repeater designs accumulates in proportion to the square root of the number of repeaters. The actual number of repeaters depends on whether each station acts as a repeater (i.e. interrupts the continuity of the transmission medium) and on whether the distances between stations are so great that true repeaters are needed. This may limit the number of stations to a maximum value that is less than that implied by the system addressing capability and the figures given in Table 16.1.

The major features of the LANs listed in Table 16.1 are described below as are, briefly, the properties of register insertion rings and several newer designs that are candidates for standardisation.

Table 16.1 LAN characteristics

Standard	Access protocol	Speed (Mbit/s)	Distance (km)	Max. nodes*	Comments
IEEE 802.3 ISO 8802/3	CSMA/CD Bus	⩽ 10	2.5	500	Ethernet
IEEE 802.4 ISO 8802/4	Token Bus	⩽ 10	100s m	250	
IEEE 802.5 ISO 8802/5	Token Ring	⩽ 10 4 typ.	m	250	
BS 6531 ISO 8802/7	Slotted Ring	10	10	254	Cambridge Ring
FDDI	Token Ring	100	100	500	Fibre optic

* These are design limits only; practical considerations may limit the number of nodes to appreciably less than the stated figure.

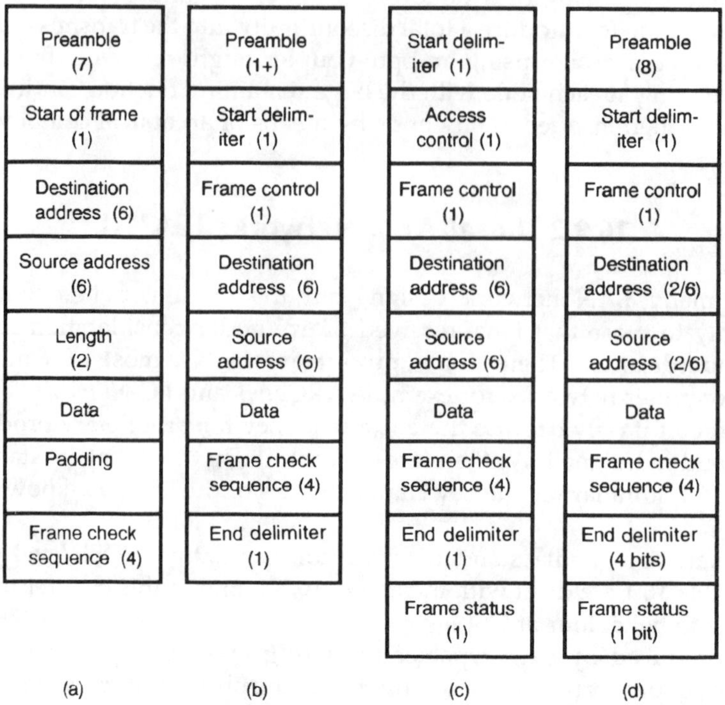

(a)	(b)	(c)	(d)
Preamble (7)	Preamble (1+)	Start delimiter (1)	Preamble (8)
Start of frame (1)	Start delimiter (1)	Access control (1)	Start delimiter (1)
Destination address (6)	Frame control (1)	Frame control (1)	Frame control (1)
Source address (6)	Destination address (6)	Destination address (6)	Destination address (2/6)
Length (2)	Source address (6)	Source address (6)	Source address (2/6)
Data	Data	Data	Data
Padding	Frame check sequence (4)	Frame check sequence (4)	Frame check sequence (4)
Frame check sequence (4)	End delimiter (1)	End delimiter (1)	End delimiter (4 bits)
		Frame status (1)	Frame status (1 bit)

Fig. 16.5 IEEE 802 Frame Formats; (a) IEEE 802.3 Ethernet; (b) IEEE 802.4 Token Bus; (c) IEEE 802.5 Token Ring; (d) FDDI

IEEE 802.3—Ethernet

The basic Ethernet protocol is called Carrier Sense Multiple Access with Collision Detection (CSMA/CD). Only one station or node has access to the bus at a time. A bus busy state is detected through the carrier sense function. In the absence of carrier any node can attempt access; if more than one does so at the same time, then a collision will be detected and all parties back-off and try later, but with different delays in order to try to avoid further collisions. This is a widely used system for LANs (not only Ethernet) but will not perform well with high traffic or with long bus delays. It is intended for packet working and is unsuited to speech even when speech is packetised. To ensure that a collision can be detected the time taken to transmit a packet must be longer than the time taken for the header of the packet to transit the bus. This makes the design a compromise between packet length, packet speed and bus length.

The Medium Access Control (MAC) frame (see Figure 16.5a) is the packet format, which includes an overhead for framing, addressing, length indication and check sequence. Frames with a length below the minimum (64 bytes at a 10 Mbit/s data rate) require padding. The maximum message content is 1500 bytes.

Stations are connected to transreceivers and the transreceivers, which present a high impedance to the bus, are teed onto the bus via an attachment cable (see

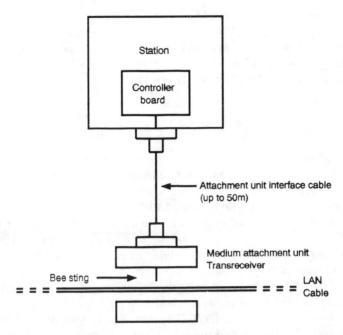

Fig. 16.6 Connection of a CMSA/CD station to the LAN cable

Figure 16.6). There are rules for the minimum distance between teeing points (see Chapter 10). One type of teeing using a 'bee-sting' connector does not require the bus cable to be cut.

Ethernet is the most widely accepted and used of the LAN systems and finds its main field of application in the interconnection of personal workstations.

IEEE 802.4—Token Bus

When the bus is free a token message passes along it and any station can seize the token, hold it and insert a message onto the bus. When the message has passed round all the stations (which are logically in a ring), having been read by the recipient in the process, the originating node recognises its own message, removes it from the bus and re-inserts the token. Hence the larger the loop delay, which depends primarily on the number of stations, the less traffic the bus can carry. Because the token bus does not waste time with collision avoidance it might be expected to work at higher traffic levels than the CSMA/CD system.

With a token-passing system a maximum delivery time can be guaranteed (which is not he case with CSMA/CD) and this makes the system suitable for manufacturing applications. (See under MAP in Section 17.6.3.)

The frame format (see Figure 16.5b) includes a frame control field which makes the distinction between tokens, LAN management information and user data, together with a priority indication for user and management messages. There are the usual overheads for framing, addressing and checking. User messages must fit into an overall frame size of 8191 octets.

A considerable amount of the system complexity lies in the management of failed nodes and the removal and insertion of stations from the bus. The priority scheme works by queuing messages at nodes in priority order.

IEEE 802.5—Token Ring

The principle of operation is very similar to the token bus. When the ring is free a token message passes round it and any station can seize the token, hold it and insert a message onto the ring. A transmitted message passes from station to station until it reaches the recipient. The recipient adds a receipt marker and re-inserts the message on the ring where it continues in the same direction until it gets back to the sender. The sender recognises the message as its own and removes it from the ring and re-inserts the token. There is a delay at each station while the message is read, hence the more stations there are, the larger the loop delay and the less traffic the ring can carry. Like the token bus, the token ring does not waste time with collision avoidance and for equality of data rates it might be expected to work at higher traffic levels than the CSMA/CD system and at similar traffic levels to the token bus. One station is assigned the role of active monitor and is concerned with error recovery and other management procedure.

The frame format is shown in Figure 16.5c. A token packet consists only of the start delimiter, access control and end delimiter fields. The access field contains a priority indicator. The data field contains user information or management packets. There is no minimum or maximum size for the data field but there is a maximum time during which a station can access the ring. However, there is a requirement that the system must work with a frame length of 133 octets.

As with the token bus much of the complexity lies in the management of start-up, failure, and reconfiguration. The active monitor is responsible for ensuring that messages only go once round the ring and for ring management generally. Any station can assume the role of active monitor and failure of the active monitor requires a procedure for establishing another station in that role.

IBM were the originators of the token ring and it might therefore be expected to find similar applications areas to those of Ethernet, e.g. in the inter-connection of personal computers. There is controversy over the relative merits of Ethernet and token ring and very often commercial considerations are decisive.

IEEE 802.9—Multimedia Interface

IEEE 802.9 defines a multimedia access unit that is connected via two UTP cables per station to a number of stations. The total bandwidth is either 4 or 20 Mbit/s which can be divided into (1) the ISDN channel structure, i.e. two 64 kbit/s B-channels and one 16 kbit/s D-channel, (2) IEEE 802 series LAN channels and (3) channels for circuit switched traffic. The multimedia access unit simplifies the work station interface by reducing the number of cables between it and the outside world; it could perhaps become obsolete if ATM takes the place of the current multiplicity of transmission networks.

Fibre Distributed Data Interface (FDDI)

FDDI is a standard being developed by the American National Standards Institute Committee ASC X3T9.5 but is also being standardised by ISO in their 9314 series. Figure 16.5d shows the frame structure. It is a token passing arrangement and differs from the IEEE 802.5 only in that the token is not held until the message goes round the ring but is released immediately by being attached to the end of the transmitted frame or frames. If the transmitted message does not come back to the sender, then an alarm is raised and recovery action attempted. The early release of the token and its higher data rate would appear to be the primary reason for the greater range of FDDI compared with the 802.3/4/5. The pure transmission delay, i.e. excluding nodes and repeaters, of a 100 km FDDI ring is 500 μs compared with 10 μs for a 2.5 km 802.3.

FDDI II is a version of FDDI in which the 100 Mbit/s bandwidth is split up into sixteen 6.144 Mbit/s isochronous* channels, giving in effect sixteen independent slower speed rings. (Further discussion on FDDI II is in Section 16.3.3 below.) It is noteworthy that 6.144 Mbit/s is a multiple of both the European and North American primary data rates. FDDI II can interface to the SDH (see Section 3.4) although its data rate is less than the 155 Mbit/s of STM-1.

FDDI follow-on LAN (FFOL) is an extension of FDDI-II which is being put forward for standardisation. It will work at speeds of 600 Mbit/s plus, it is compatible with the synchronous digital hierarchy and will transport ATM cells as well as FDDI I and FDDI II.

Fibre channel

Fibre channel is in the process of being defined for point-to-point connection at multiple data rates ranging from 100 Mbit/s through 200 Mbit/s and 400 Mbit/s to 800 Mbit/s. The transmission medium is single mode fibre up to 10 km, multi-mode fibre up to 2 km or coaxial cable up to 50 m. Crossbar switches are being considered for the higher data rates and ring configurations at the lower rates with broadcast hubs at all rates. There are three classes of service: Class 1—dedicated connection; Class 2—frame switched; and Class 3—connectionless or datagram.

Slotted rings

The slotted ring to BS 6531 differs fairly radically from those described above in that the transmission on the ring is divided up into a number of synchronised time

---

* Isochronous is defined by CCITT to mean that the time scale is such that the intervals between significant instants either have the same duration or durations that are integral multiples of the shortest duration. In other words the bit rate on each synchronised channel is a sub-multiple of the 100 Mbit/s (nominal) clock rate.

slots which can be of length 40, 56, 72 or 88 bits as preassigned. Any station can fill the first available free time slot on the ring and there is no need for tokens or collision avoidance. The number of messages that can coexist on the ring clearly goes up as the size of the ring, and delay round the ring does not affect throughput. However, messages must be of a fixed length and because each must contain address and format information there is a fairly high overhead, particularly if a large amount of data is to be transferred in one transaction. Each node on the ring introduces a delay of three bits which is necessary to determine whether a slot is available or not. The message is marked by the recipient and continues on back to the originator who removes it but is not allowed to use that time slot again immediately.

Of the LANs discussed here, apart from FDDI II, which is specifically intended for speech, the slotted ring is undoubtedly the most suitable ring for speech and its suitability can be increased if certain time slots are allocated to speech and perhaps assigned to a particular terminal pair for the duration of a conversation.

The BS 6531 slotted ring appears not to have received widespread acceptance, its use being confined mainly to computer networks in UK universities. There are, however, two further developments of slotted rings, both based on UK work that are worthy of a brief mention:

1. Orwell is a BT development in which a node can use any free slot that passes it and must then back off for a fixed period to let other nodes have fair access to the ring.

2. The Cambridge Backbone Network (CBN) is a simple slotted ring with four time slots each of 256 bits. The objective is high speed operation at 1 Gbit/s.

Register insertion

Although it appears not to be the subject of standardisation, the register insertion ring should be mentioned. Each node contains a register which is loaded as a parallel operation with typically a message of some 40 bits. If there is no message passing the node, the register is transmitted serially. Any message coming up behind is shunted serially into the register. Unpredictable delays can result which become longer as the ring becomes more heavily loaded.

Cyclic Reservation Multiple Access (CRMA)

This is an IBM proposal for a 1 Gbit/s plus LAN/MAN in which capacity is allocated by a head-end station on the basis of demand forecast by all stations. If the network overloads, stations are told to back-off temporarily. This is the first indication of a move towards a centralised control mechanism for LANs and it will be interesting to see if it continues.

Wireless LANs

Wireless LANs (WLANs) are LANS using radio rather than wire as the transmission medium. WLANs (also known as CLANs; C for connectionless or cordless) are currently slower and more expensive than wired LANS. They find their major application in situations, such as in the retail trade, where computer terminals are frequently moved because of changes in floor plan. There are several proprietary systems of which NCR's Wavlan, Motorola's WIN and a California Microwave product using frequency hopping are perhaps the best established. The IEEE committee 802.11 has started work on standardisation. Frequency spectrum allocations in Europe and the UK have still to be established.

A number of new products working at higher data rates, up to 15 Mbit/s, are anticipated and the use of spread spectrum techniques may help both from the point of view of spectrum allocation and privacy. Motorola's Altair design operates at 17 Gbit/s, at which frequency transmission can be confined to a very limited area.

16.3.3 Metropolitan Area Networks (MAN) and the interconnection of area networks

A metropolitan area network can be a network in its own right with individual user nodes or it can be a means of interconnecting LANs or it can fulfil both these roles. There is a considerable overlap in network technology between MANs and LANs and looking further ahead between MANs and some of the new technologies discussed in Section 16.3.5 (Wide Area Networks). We start by taking the relevant LANs from the previous section in a comparison with a network specifically intended for a MAN role. Because the network interconnection role is usually a key feature of MANs we go on to consider interconnection of LANs, MANs and WANs in general.

MANs

On the assumption that a MAN will in general need to carry speech and data traffic, the two main contenders for the role are the FDDI II and the IEEE 802.6 standards. Their salient features are summarised in Table 16.2.

The operation of FDDI is outlined briefly in Section 16.3.2. The IEEE 802.6 is also called a dual bus QPSX (queued packet synchronous switch) MAN. The IEEE 802.6 uses a protocol called Distributed Queue Dual Bus (DQDB); it has a 125 µs frame time with mini-packets containing speech or data and the data rate of 150 Mbit/s is directly compatible with SONET (see Section 3.4). IEEE 802.6 is sometimes called 'broadband ISDN' but this is a misnomer because it confuses an implementation standard with a service. Broadband ISDN might well run on an IEEE 802.6 bus, however.

The transmision medium for IEEE 802.6 is a standard public network transmission system such as CCITT G.703 or G.707-9, or in the USA ANSI DS3.

Table 16.2 MAN standards

	FDDI II	IEEE 802.6
Protocol	Token Ring	Distributed Queue Dual Bus (DQDB)
Speed	100 Mbit/s	150 Mbit/s
Max nodes	500	Not specified
Compass	100 km	Not limited
Max. packet length	4472 octets	39 octets

Both the FDDI II and IEEE 802.6 arrangements are capable of handling speech at 64 kbit/s using assigned (circuit switched) channels for the duration of a call. Speech delay on MANs is not therefore likely to be too significant a problem. Bandwidth can be provided in sub-multiples (down to 8 kbit/s) and multiples of 64 kbit/s. Channel bandwidth can be assigned for the duration of a call and 64 kbit/s channels could for example be combined for the duration of a video call. Operation in a packet mode is also common to both approaches.

The differences between FDDI II and IEEE 802.6 arise largely from the backgrounds of their instigators. FDDI has a computer manufacturing background, hence the emphasis on long packets and optimisation for data rather than speech transmission. IEEE 802.6 has been instigated under strong influence from telecommunication administrations and hence there is a greater emphasis on speech as the short packet length exemplifies.

Another major difference is in failure procedures. In an FDDI II system with duplicated rings, if one ring fails either the traffic carrying capacity is halved or the system capacity is limited to that of one ring under normal operation (i.e. the other ring is kept idle). Under failure the halves of the ring are joined on both sides of the break making a single ring of nearly twice the length. With the IEEE 802.6 system the buses under no fault conditions are configured as in Figure 16.7a with the bus ends colocated. If a break occurs, Figure 16.7b, then the function of the ends is transferred to the nodes on both sides of the break and the buses at the original end-point are joined, (see Figure 16.7c). Thus the buses continue to work at the

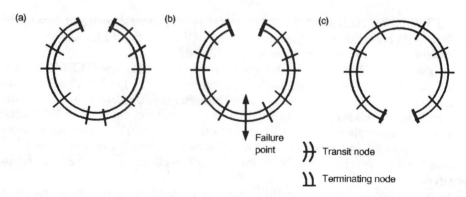

Fig. 16.7 Dual bus failure modes; (a) normal operation; (b) failure; (c) bus reconfigured

same capacity, each bus being fully used both before and after failure. The end-of-bus points do, however, need additional equipment, and this additional equipment must be provided at all nodes that are capable of taking over the end-point role in a failure situation.

IEEE 802.6 does not specify a transmission medium, while FDDI II is based on coaxial cable. There is, however, an FDDI version called FDDI III under consideration that is based on the use of optical fibre for transmission.

Although IEEE 802.6 does not specify a maximum number of nodes nor a maximum distance it is thought that in practice considerations of reliability, delay and jitter may not make all that much difference to the ultimate size of the MAN between the two designs.

The FDDI standard is, at the time of writing, more advanced than the IEEE 802.6 and it is an open question as to whether there is a large enough window for the adoption of the FDDI II standard before the IEEE 802.6 or perhaps the RACE WAN design (see below) takes over. In 1991 BT was close to carrying out trials on its Switched High speed Data Service (SHDS) which in turn is based on the Bellcore Switched Multimegabit Data Service (SMDS). These high speed data services are based on asynchronous transfer mode operation (ATM) (see Section 16.3.5).

Although they may not claim to be WANs there are a number of emerging optical fibre networks, e.g. those provided by BT and Mercury in the central London area, which are moving in that direction. BT call their network a Flexible Access System (FAS) because it provides access to all the major network services as well as providing private network facilities. FAS is currently a single star network for business customers with 25 or more lines. A version for lower traffic customers would need to be based on a double (or distributed) star (see Section 2.3.4) with primary fibres providing access from a central service point to a local access point from which fibre or copper pairs would constitute the final link to the customer.

Area network interconnection

Area networks can be connected by a bridge or a gateway and the distinction is illustrated in terms of the OSI model in Figure 16.8. Figure 16.8a shows a bridge and relay; note (Figure 16.8b) how OSI layer 2 is split into sub-layers for Link Layer Control (LLC) and Media Access Control (MAC). Splitting the link layer in this way is peculiar to LANs and was introduced to provide a common network–link layer interface independent from the specific type of IEEE 602 series LAN implemented at the MAC sub-layer. Figure 16.8b shows a gateway and relay which involves the network layer. In general bridges are used to connect LANs of the same type and gateways to connect dissimilar LANs, LANs to MANs, LANs to WANs or MANs to WANs. Routers are bridges or gateways in which the connections are not one-to-one and there is a choice of route on at least one side of the interface.

As Figure 16.5 shows, there is commonality of addressing between the frame formats of the IEEE LANs so that routing can take place through bridges without reference to the LAN type. A single bridge between a LAN and elsewhere is straightforward but additional complexity arise when there is a multiplicity of

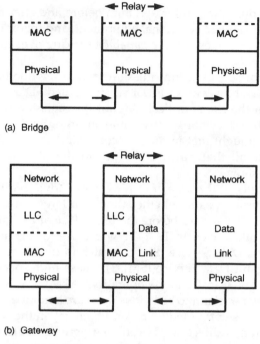

(a) Bridge

(b) Gateway

Fig. 16.8 OSI model for bridges and gateways

bridges. The IEEE 802.1 standard addresses the problem of managing multiple bridges, for example when there is for security reasons two bridges between two LANs.

A gateway as illustrated in Figure 16.8b implies that there are common protocols at both ends above the network layer. If this is not the case then protocol conversion is required and this can lead to much additional complexity, resulting in proprietary products and currently no known standardisation. The ISO 7-layer model does allow for packet length conversion although this can be time consuming.

If the primary function of a LAN or MAN is to connect other LANs or MANs then the preferred choice is probably token ring (IEEE 802.5) or FDDI.

16.3.4. Frame relay

Introduction

As part of the development of ISDN, work began in CCITT on the definition of forms of packet mode communications where the control plane and user plane are properly separated (i.e. signalling information from user information) and not combined as in X.25. This resulted in the concept of a simple packet mode access to ISDN (see CCITT recommendation I.122), known as frame relay or Frame Mode Bearer Service, where the packet or frame is defined at Layer 2 and there is no protocol at Layer 3. (There is an introduction to this topic in Section 15.6.)

This work, which started quietly in CCITT, rapidly became the subject of intense interest and activity as manufacturers realised the potential of frame relay for providing efficient high speed data communications on leased circuits (permanent virtual circuits) for applications such as the interconnection of LANs.

A Frame Relay Forum was formed in January 1991, with members mainly from equipment suppliers, to coordinate developments, accelerate the standards making process (which continued in the formal standards bodies) and effect cooperation on interworking and conformance testing. To date the Forum has proved to be very effective and the model is being copied in that a Forum for ATM has recently been established.

The term 'fast packet' is sometimes used to describe frame relay, but the same term may also be used for cell switching technologies such as ATM. Cell switching technologies are characterised by short fixed length cells.

Technical description

Frame relay is a variable frame structure of up to 4096 octets defined at Layer 2. Figure 16.9 shows the frame structure. In addition to the data carried, there is an address field, which is used for a logical channel number, bits to indicate the existence of congestion, and a frame check sequence to detect errors.

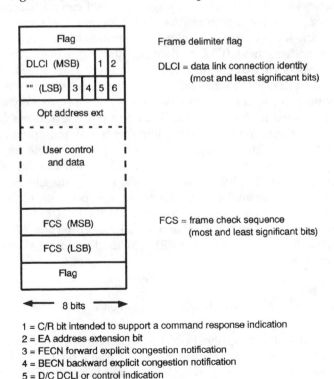

1 = C/R bit intended to support a command response indication
2 = EA address extension bit
3 = FECN forward explicit congestion notification
4 = BECN backward explicit congestion notification
5 = D/C DCLI or control indication
6 = EA address extension bit

Fig. 16.9 Frame relay frame

This frame structure applies to a link layer (Layer 2) and may be used either for communications across a leased circuit or for the access link to a frame relay network service. The frame structure could be used between nodes within a frame relay network but such a network would normally need additional functionality to handle events such as node failures and so proprietary protocols would be used internally.

The service provided across a frame relay network is connection oriented. There is no error correction: if an error is detected, the frame is discarded. Error correction would need to be applied on an end-to-end basis by the communicating applications. The absence of error control makes frame relay well suited to LAN interconnection because LAN protocols include mechanisms to recover from errors or data loss. However, the transmission medium must be good enough to make the probability of an error in a frame low, otherwise throughput is seriously affected.

The FECN and BECN flags (see Figure 16.9) for congestion control operate differently and more simply than the window system on X.25. The flags are simply means of indicating that congestion is present, or imminent, so that the end-point system can reduce the traffic rate. Frames may be marked by an end system as 'discard eligible' so that they will be discarded first in a situation of severe congestion and overflowing queues.

To date the operation of a switched network service (switched virtual circuits) (sometimes referred to as 'frame switching') has not been standardised; the only frame relay standards are for individual links (permanent virtual circuits) and currently network services are collections of permanent virtual circuits with no switching. The address field (DLCI) identifies the permanent virtual circuit (or connection). When switched frame relay services are introduced, they will use separate signalling frames, containing numbers in accordance with the CCITT E.164 numbering plan, to set up and clear the connection. At network nodes the number will be related to the route and DLCI of the frame address field.

The simplicity of the frame relay protocol makes it easy to implement and run at high speed. 2 Mbit/s is currently the maximum speed possible for an X.25 type controller chip. For frame relay the HDLC protocol, similar to X.25, allows peak bit rates up to 45 Mbit/s or 34 Mbit/s (i.e. H.22 and H.21 in the European and North American plesiochronous multiplexing hierarchies respectively).

Public frame relay services began to become available at the end of 1991 with offerings from AT&T and BT. Initially access is likely to be at a 64/56 kbit/s rate and statistical multiplexing techniques will provide the traffic balancing necessary for an economic realisation.

Frame relay applications

Frame relay is best suited to applications that generate bursts of data that must be delivered quickly. The main initial application for frame relay is LAN interconnection, for example interconnecting the LANs in the different offices of an international company. If such connections are not provided by other LANs or a MAN but by an X.25 packet switched network, then long delays can occur when a large burst of data is generated. Depending on the traffic, such arrangements could be

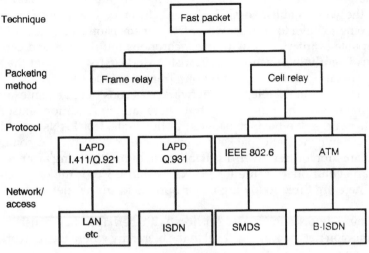

Fig. 16.10 Fast packet techniques

replaced by frame relay operating over say 2 Mbit/s leased circuits or a public frame relay service. Until switched virtual circuits become available, the public service will consist of a number of permanent virtual frame relay circuits between up to 1024 terminal points (limited by the 10 bit field of the DLCI).

Frame relay can be used to interconnect or interwork with data network other than LANs and can provide an attractive method of interworking with other data networks such as ATM carrying B-ISDN, X.25 and SMDS.

Frame relay can be used for voice but steps have to be taken to reduce the variability from frame to frame of the transmission delay. This would most easily be done by adding an additional controlled delay at the receiving terminal. However, delays of the order necessary would make echo cancellation necessary in most cases and it is not expected that voice will be a major traffic component for frame relay.

Frame relay is but one of a number of 'fast packet' techniques and Figure 16.10 puts frame relay into a context which includes a number of the other such techniques which are discussed elsewhere in this section.

16.3.5 Wide Area Networks (WAN)

Introduction

Chapters 4 and 5 deal with transmission (including SDH and ATM) and switching respectively, but here we bring both technologies together and consider them in the context of wide area bearer networks with the emphasis very much on the evolutionary impact of the Synchronous Digital Hierarchy (SDH).

The situation is evolutionary because today's digital transmission network is based on the well-established CCITT plesiochronous hierarchy (see Figure 3.2) which is going to have to co-exist with SDH transmission (see Figures 3.5 and 3.6). SDH is capable of interfacing with plesiochronous multiplexes and carrying their traffic transparently in Synchronous Transfer Mode (STM) while at the same time providing operation in the Asynchronous Transfer Mode (ATM).

In Section 16.1.2 the BT telephone network was considered as an example of an existing circuit switched network, but there are in addition existing packet switched networks. We have considered in the context of LANs and MANs how future circuit switched and packet switched information can be combined on a single transmission bearer given a suitable form of multiplexing arrangement. This is clearly an evolutionary route for WANs too but it is not the only possible way forward. There are three following major options for future networks:

(a) Separate networks for major services with perhaps shared use of the SDH transmission network at any level in the hierarchy where the network topology is the same.

(b) Separate switching and expeditiously combined transmission but with a common control structure. This was the philosophy adopted in the early stages of defining what was to become System X.

(c) Completely common and integrated transmission and switching from the user's network interface throughout the network.

The last of these strategies is the one currently favoured (although not the only possibility) and, in the next section but one, we go on to consider the European initiative called R & D in Advanced Communications technologies in Europe (RACE), which is an EC sponsored research and development programme for an Integrated Broadband Communication Network (IBCN). RACE involves cooperative work by administrations and manufacturers on the definition of a future European network.

Concurrently with RACE, CCITT is in the process of defining broadband ISDN, or B-ISDN, which is a set of services and a set of access protocols rather than a network (see Section 15.8 for details of B-ISDN). The RACE network will certainly support B-ISDN as well as a range of other access means and facilities. The RACE design is currently being strongly influenced by B-ISDN protocols and a lot of the detailed design issues relating to B-ISDN are being resolved following the 1984–88 CCITT plenary, and are being published as they become ready. Rather different approaches are being taken elsewhere, particularly in the USA and Japan, although SDH and B-ISDN are a common theme throughout.

PDH, SDH and ATM—the impact on switching

During a transitional period switches may have to interface with:

(i) the existing Plesiochronous Digital Hierarchy (PDH);

(ii) the Synchronous Digital Hierarchy (SDH) in the Synchronous Transfer Mode (STM);

(iii) the Synchronous Digital Hierarchy (SDH) in the Aysynchronous Transfer Mode (ATM);

(iv) cell based (or Asynchronous Time Division, ATD) transmission in the Asynchronous Transfer Mode (ATM).

These possibilities are illustrated in Figure 16.11. Switches under (i) are existing designs, typically Time–Space–Time (TST) or Space–Time–Space (STS) structures, which will remain as part of the 'old' network as long as that network exists but there is no reason why such switches should not interface to the SDH in STM mode at the 2 Mbit/s level in the multiplex hierarchy. In some cases network switching nodes will be required to deal with STM and ATM traffic over an SDH interface; in other situations all the traffic will be ATM and will be via either an SDH interface or via the special cell-base ATD interface yet to be defined by CCITT.

Although SDH transmission is not an essential part of the ATM approach, the universality of SDH and its flexibility to handle a variety of formats and to provide so efficiently the drop and insert function is clearly a distinct advantage and it is most unlikely that they will develop other than in sympathy.

With ATM, user information is broken down into fixed length cells, the number of such cells per unit time reflecting the instantaneous bandwidth requirement and

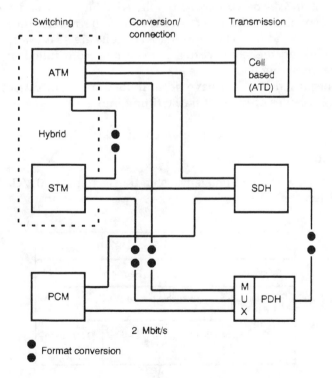

Fig. 16.11 Possible interconnections between switching and transmission

the variation in this number of cells per unit time the bursty nature of the traffic. Network bandwidth might be allocated in sub-multiples or multiples of 64 kbit/s channels based on the average cell rate for a particular service. For an essentially random process, e.g. packets in transaction processing, the average bandwidth may be very small. By taking account of the statistical properties of the (bursty) traffic, transmission time slots can be used more efficiently with ATM (compared with synchronous arrangements) at the expense of a finite probability that data will be lost. (Although ATM is asynchronous in terms of its cell rate, a framing reference is of course needed to identify the start of each cell.)

Each ATM cell has a header and an information field; the primary purpose of the header is to identify the connection number for the sequence of cells that constitute a virtual channel for a specific call. With existing switching in plesiochronous networks there is no header and there is a fixed relationship between a connection and a transmission time slot number. With STM the equivalent of a header is in the form of a pointer within the SDH multiplex and for a connection the time slot is not fixed, although a fixed bandwidth is allocated for the duration of the call.

In an SDH multiplex ATM cells and their headers are contiguous bits in the bit stream whereas STM information at, say the 2Mbit/s level, is interleaved and distributed throughout the frame. This means that incoming ATM information can be switched on the fly as it is received but a frame store is necessary before STM information can be switched. For this reason switches working in a mixed ATM—STM environment tend to have separate switch blocks for each transfer mode.

STM information can be carried across an ATM network and vice versa. For example an incoming STM-1 octet stream can be be re-formatted into 48 octet blocks and a header added. The resulting ATM cells are sent transparently across an ATM network on the same virtual path to be reconstituted in another STM network at the receiving end.

Services transported by ATM have been divided into four classes as shown in Figure 16.12. Examples of types of traffic fitting into each class are:

- Class 1: Voice, video conferencing video telephony and teleservices;

- Class 2: Moving image transfer;

- Class 3: Moving and still image transfer, transactions, CAD/CAM, and block data transfer;

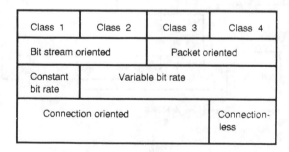

Fig. 16.12 ATM classes

- Class 4: As Class 3.

One might expect a migration of traffic over time from classes 1 and 2 to 3 or 4.

The RACE network

RACE is similar to OSI in that it is pursuing an open network policy in which the general architecture and interfaces are defined but implementations can take any form within these general constraints.

ATM is the long-term aim under RACE but contrary to early assumptions the synchronous and asynchronous strategies are not mutually exclusive and, as indicated above, switches can be designed that are partitioned into sections, one section dealing with ATM cells and the other dealing with synchronous circuit switched information.

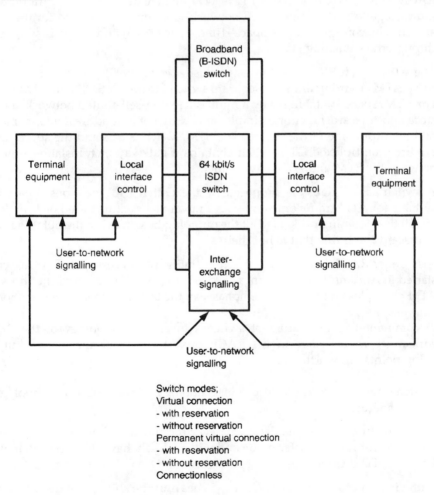

Fig. 16.13 Conceptual configuration of the RACE network

Figure 16.13 summarises the potential connection capability of the RACE net-work. Full reservation of resources will guarantee a fixed network delay and this is clearly important for speech and speech with STM is not a problem. If speech is carried in ATM it should be given priority over data to reduce delay variation although paradoxically it does not matter too much if a few speech samples are lost. In connection mode speech sample will always arrive in the correct order (a virtual path through the network is reserved) but this is not necessarily so in connectionless mode where some re-ordering will be required. In the connection-less mode all the routing information is contained within the cell which is then exactly equivalent to a packet in packet switching. (See Section 8.2.4 for connec-tion–connectionless mode definitions.)

The following are examples of some of the work going on under RACE on switch design:

• RACE BLNT (Fox *et al.*, 1990) is a local switch handling ATM traffic only; multi-stage folded and unfolded networks are being considered for the space switch. The configuration is Space–Time–Space (STS) and traffic is self-routed through the switching stages.

• The ATMOSPHERIC project's design (Fisher *et al.*, 1990) works in a combined STM–ATM environment with separate switch blocks for STM and ATM traffic. For ATM, space switching is by a multi-stage non-self-routed network and the store and forward function is implemented by content addressable memories on the output side of the switch. The STM switch is STS and the space stages are time multiplexed. The switch contains a gateway or translation function between the STM and ATM environments.

• A switch called Gauss is proposed by the Dutch PTT (de Vries, 1990); it is suitable for ATM traffic only and uses a novel architecture based on broadcast-ing all the incoming packets to all the outgoing links leaving the outgoing links to discard the traffic that is not theirs.

• Also under RACE is the Berkom demonstration (Armbruster, *et al.*, 1940) which started in Autumn 1989 in Berlin. This does not use ATM standards but does use ATM principles and there is an emphasis on the transfer of video information.

Any comments on the nature of switches in and the topology of the RACE network must at this stage of the RACE project be rather speculative but the following points are worth making.

(a) Switches will be based on electronic technology initially; optical technology is not yet mature enough.

(b) Switching of synchronous data, whatever the transmission environment, will use a synchronous transfer mode and will probably have its own switch block based on TDM principles and separate from any ATM switching.

(c) With virtual path reservation, ATM switches must be able to handle queues and priority may need to be given to some types of traffic. If speech is handled using virtual circuits, it will need to be given high priority in order to minimise and equalise delay. Switching may well involve multi-stage self-steering networks.

(d) The use of remote statistical multiplexers and the adoption of MANs points to a concentration of traffic more towards the periphery of the network, avoiding the need for the traffic to be concentrated at the existing local exchange and for a proportion of the traffic to go through that exchange at all. The flexible drop and insert capability of STM enables traffic to be easily demultiplexed at any level in the hierarchy. Hence there may be a tendency for the distinction in terms of design between local switches and higher order switches to disappear. Exchanges will get larger and probably fewer in number.

(e) It is considered that the mesh interconnection of major switching nodes will remain a feature of the network for some time to come. The reason for this belief is that the investment in the existing topology, through the geography of transmission plant, is very large and the plant (particularly optical fibre) is suitable for use in the RACE context anyway. When this spare capacity is exhausted it may well be that the new transmission plant is organised to be more ring like and less mesh like but route diversity is always likely to be necessary in the interests of reliability and this will limit the degree to which pure rings can be allowed.

Evolutionary approaches in North America and Japan

The Bellcore approach (Albanese and Jaggernauth, 1990) in the USA puts greater emphasis on an evolution from interfaces with MANs based on IEEE 802.6 through coexistence with B-ISDN towards a largely B-ISDN solution. Clearly this involves establishing interworking protocols between these two types of interface. The concept is called Switched Multi-megabit Data Services (SMDS) and operation (unlike the European solution) is in connectionless mode. Early SMDS architectures confine interworking to established LANs, particularly IEEE 802-3 (Ethernet) FDDI and IEEE 802.6. Network interfaces are based on IEEE 802.6.

In Japan the emphasis is on B-ISDN but with perhaps a more distributed network architecture than is apparent under RACE (Tanabe *et al.*, 1990).

16.3.6 Overall broadband network design

Figure 16.14 illustrates what a complete broadband access network might look like in the local area and contains all of the types of network discussed above. Additionally it includes an integrated services private branch exchange (ISPBX) which is a PBX having integrated services internally and, via ISDN, access to the network. LANs and ISPBXs overlap in function and are therefore to some extent in competition with each other. The interface to the existing 64 kbit/s network carries ordinary speech circuits (POTS = plain ordinary telephone service) and ISDN channels.

Transition from the existing network to a broadband replacement will not happen abruptly and a transitional strategy is necessary. A broadband network is unlikely

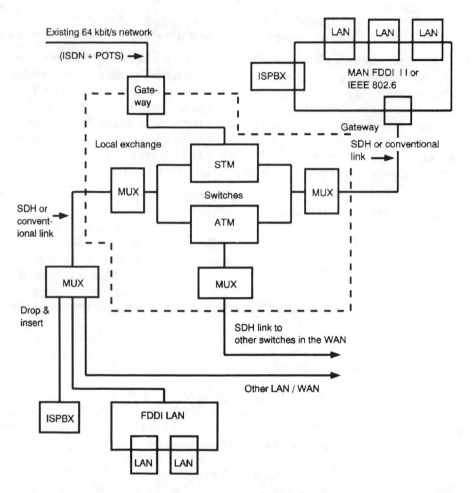

Fig. 16.14 Broadband network example

to be realised before the turn of the century at the earliest, by which time contemporary networks will be all digital. Because subscribers to the new broadband services will still need to communicate with subscribers to the older narrow band network services, there must be gateways between networks and a gateway to the PSTN is shown in Figure 16.14 As the subscribers to broadband services increase there will a migration on to the new network and the gateway traffic will decrease. With SDH it is possible to provide gateways at any level in the switching hierarchy, not just at the local exchange level as shown.

16.3.7 Global gigabit networks

Advances in fibre technology and in particular the availability of optical amplifiers make possible the concept of a global gigabit optical network. In early 1992 a research project with US government funding was being put in place to make this a reality for the USA, and if that can be done the rest of the world is not such a big step thereafter.

16.4 CONCLUSIONS ON BROADBAND AND AREA NETWORKS

16.4.1 The local area

We have considered the adoption of optical fibre in the local area network mainly in the context of the BT TPON—BPON designs, and Europe in general is proceeding along rather similar lines while in Japan and the USA there is perhaps more emphasis on the use of amplifiers and also more emphasis on bus and ring structures as opposed to the star arrangement favoured by BT. Nevertheless there is considerably common ground and a good consensus that a substantial amount of optical fibre will be installed in the local area by the end of the century.

There is potential for downstream broadband traffic in the form of TV, and optical fibre can carry such services readily and will do so unless prevented by regulation. There is increasing interest in HDTV and beyond that what is called a 'wall-to-wall visual experience'. Capacity for such services should also be available on fibre. Broadband traffic in the upstream direction implies a need for switching on a point-to point basis, rather than the selection process needed for downstream TV, and hence the conclusions in the next section below are relevant.

16.4.2 The public switched network

For point-to-point connections a fairly clear picture is emerging as to how broadband networks are likely to develop over the next 10–15 years. Nevertheless speech will still be the major class of traffic in the contexts of MAN and WAN, if not LAN, during that period and will continue to exert a strong influence on network design through the restrictions it places on delay and variations in delay.

The increasing use of digital processing, particularly in the speech coding process, is giving rise to a demand for increased end-to-end network delay. It remains to be seen whether this requirement will be met by:

- (a) relaxation by international agreement of the current delay limit recommendations;

- (b) increased use of echo cancellers; or

- (c) digital processing solutions not involving substantial delay.

Any new network will be introduced as an overlay and will have to interwork with existing networks for many years to come and this places limitations on what can be done generally and especially in the context of speech.

LAN technology is well established, though the trend towards higher speeds will continue and FDDI I appears to have a well assured role here. The situation with regard to WANs is far less clear. Do WANs really need a separate technology (e.g. FDDI II), or does the more advanced LAN technology, or the kind of technology emerging for WANs in the RACE context provide a solution or solutions? Is there an economic justification for new network topologies (e.g. rings), particularly in areas of high terminal density, bearing in mind the existing investment in ducts and the need for an evolutionary approach to the introduction of a broadband network, and how is this situation affected by the increasing trend towards liberalisation? We do not pretend to have answers to these questions and there is probably more than one answer depending on whereabouts in the world you are; more work must be done before anyone can reach sensible conclusions.

The pressure for new multiservice networks comes largely from business and one might expect a demand for networks using perhaps B-ISDN and ATM principles to come from large businesses, either in the form of private networks (virtual or otherwise) leased from PTOs or by direct use of a public service or both.

The investment in RACE is considerable and increasing and there is strong political and economic pressure to find common European solutions to new generations of telecommunication systems. The particular solution of SDH for transmission and hybrid ATM/STM for switching seems to be a promising one. It is encouraging that Europe is taking the lead in the standardisation of B-ISDN.

Finally it is worth re-emphasising that the introduction of broadband networks does not depend wholly on the demand for broadband services. For example the wide area technology of SDH transmission can be used in any context and ATM switching can be used initially for an overlay for narrower band services and subsequently for the gradual introduction of broadband services. Thus the introduction of this new technology is more certain than the widespread adoption of broadband services such as video conference.

16.5 BIBLIOGRAPHY

C. Albanese and J. Jaggernauth (1990) Evolution of the network supporting SMDS to B-ISDN, *Integrated Broadband Services and Networks*, IEE Conference Publication Number 329, October, pp 79–84

H. Armbruster *et al*, (1990) Phasing-in the universal broadband ISDN: initial trials for examining ATM applications and ATM systems, *Integrated Broadband Services and Networks*, IEE Conference Publication Number 329, October, pp 200–5.

R .E. G. Back (1988) British Telecom's future network,' *Br. Telecommunications Engineering*, 7, 82–88.

C. Caves. (1987) FDDI II: a new standard for integrated services high speed LANs, *Network 87*, OnLine Publications, pp 245–256.

R .J. F. de Vries (1990) Gauss: a single-stage ATM switch with output buffering, *Integrated Broadband Services and Networks*, IEE Conference Publication Number 329, October, pp 248–52.

D. W. Faulkner and D. I Fordham (1989) Broadband systems on passive optical networks, *Br. Telecom Technology J.*, 7, 115–122.

D. G. Fisher (1989) The European route to integrated broadband communications, *2nd IEE National Conference on Telecommunications*, York, April, pp 1–5 (CPN-300).

D. G. Fisher *et al.* (1990) An open network architecture for integrated broadband communication, *Integrated Broadband Services and Networks*, IEE Conference Publication 329, October, pp 73–8.

J. E. Flood and P. Cochrane (1991) *Synchronous Higher-Order Digital Multiplexing*, Chapter 10, Transmission Systems, Peter Peregrinus.

P. Food (1989) FDDI II: integrated services on high speed LANs, *ISDN* 1988, OnLine Publications, pp 303–312.

A. L Fox *et al.* (1990) RACE BLNT: a technology solution for the broadband local network, *Integrated Broadband Services and Networks*, IEE Conference Publication 329, October, pp 47–57.

C. E Hoppitt and D. E. A. Clarke (1989) The provision of telephony over passive optical networks, *Br. Telecom Technology J.*, 7, 100–14.

R. W. Newman, Z. Budriks and J.L Hullett (1988) The QPSX MAN, *IEEE Communications Magazine*, 26 (4), 20–28.

S. O Hara (1984) The evolution of the modern telecommunication network, *The Mildner Lecture*, University College, London.

W. R Ritchie. (1989) Overview of local access development, *Br. Telecom Technology J.*, 7, 7–16.

W. Scott Currie (1988) *LANs explained, Ellis Horwood*.

M. Sexton (1989) Synchronous networks—Sonet and SDH. The changing face of telecommunications transmission, *IEE Professional group E7 Colloquium*, 16th Jan pp 6/1–6/7.

S. Tanabe, T. Suzuki and K. Ohtsuki (1990) A new distributed switching system architecture for B-ISDN, *Integrated Broadband Services and Networks*, IEE Conference Publication Number 329, October, pp 258–63

J. Vandenameele and G.Evers (1990) IBCN Introductory Steps: overlay networks and physical integration, *Integrated Broadband Services and Networks*, IEE Conference Publication 329, October, pp 29–34.

17 DATA AND MULTISERVICE NETWORKS—UPPER LAYERS

17.1 INTRODUCTION

Although sometimes the lower layers of the network are service dependent, the upper layers are much more so and the stress in this chapter is, in contrast to Chapter 16, on network services rather than on network technology. The emphasis is on services that have recently been introduced or services that will become available in the short to medium term. The emphasis is also, to a much greater degree than elsewhere in this book, on the UK because treating other countries' services at second hand is not likely to result in a very accurate picture and would make Chapter 17 unwieldy. In Part 4 an attempt is made to bring everything together and we shall present an overall picture of what the user will see up to the early part of the next century.

Communication outside earshot might be divided into a number of broad categories:

(a) Real-time interactive communication as exemplified by voice conversation over a telephone network, voice messaging (for example a telephone answering machine or the centralised equivalent of the same) video telephony and (video) conferencing. Voice services are dealt with in Section 17.3 and video services in Section 17.4.

(b) Postal services for the conveyance of written material. These are only of interest in so far as they are a source for substitution by electronic messages. There are currently what might be termed hybrid arrangements where part of the transport process is electronic and part mechanical; for example telegrams and facsimile post. These are obviously ripe for a completely electronic solution but they do not currently represent a very significant proportion of the total traffic and will be largely ignored.

(c) Electronic means of conveying text and graphical information in a conversational or a non conversational mode. This includes existing addressed services such as telex, teletex, electronic mail or E-mail, analogue and digital facsimile (or 'fax'), information retrieval, electronic data interchange and transaction processing (booking or reservation services). This is the area which is covered in Section 17.5.

(d) Data entry and computer to computer bulk data, or file, transfers and what started as largely one-way services such as Prestel (videotex) are covered in Section 17.6.

(e) Non addressed broadcast services such as radio and television, however distributed, are not covered. Addressed broadcast services, e.g. paging, which have already been considered in some detail in Chapter 12, are considered very briefly in Section 17.5, as is mobile radio in Section 17.3

17.2 TRAFFIC POTENTIAL

To some considerable extent the importance of a service, or a potential service, must be related to the traffic it conveys, and in this context it is perhaps useful to give an overall picture of where such traffic is to be found. Table 17.1 below is based on data for the USA produced in 1970 (Hough, 1970). The concern here is not so much with the accuracy of Hough's 1990 prediction (the video telephone figure is clearly a major overestimate) but with the orders of magnitude associated with each category of data. Table 17.1 expresses activities in transactions/year and bits/year. Bits/year corresponds to traffic/year when transactions are point to point but traffic/year will be much greater than bits/year when the information is broadcast over a network.

Table 17.2 gives similar information for the UK but for a limited set of activities only and takes account of the broadcast nature of much of the material in arriving at the traffic figure. (It is assumed that the broadcast traffic is downloaded as trunk traffic to a number of store and forward switching centres and distributed as local traffic from there.) It must be emphasised that Tables 17.1 and 17.2 are purely indicative of what is potentially available in terms of information transfers and it will be well into the next century before more than a very small fraction of this potential becomes telecommunication data traffic. In any case the availability of information in electronic form may well transform the way in which we use information. Who reads more than about 20% of their daily paper, for example?

A number of points can be highlighted from this analysis:

1. Telephony will be the dominant type of traffic for some time to come.

2. While videophone could become dominant there are no signs of this happening and it would be rash to suggest a date when it will.

3. The distribution of printed material, such as newspapers, could represent a very significant source of traffic but we are a long way away from the date when this will become a reality.

4. A lot of the printed material could be transmitted at night, thus representing a source of traffic and revenue suited to a time when the network is not busy with more immediate traffic such as voice.

Table 17.1 Information services for the whole of the USA ranked in accordance with magnitude of originated traffic

Services	Transactions/year in 1990	Conversions factor	Bits/year
1. Telephone	482×10^9 calls	360s/call 64 kbit/s	1.1×10^{19}
2. Video-telephone	1×10^9 calls	360s/call 6.3 Mbits/s	2.3×10^{18}
3. Mail	100×10^9 letters	3×10^5 bits/letter	3.0×10^{16}
4. Television	72×10^3 hours	3600s/hr 64 Mbits/s	1.7×10^{16}
5. Remote library browsing	20×10^6 accesses	200p/access 3×10^5 bits/p	1.2×10^{15}
6. Interlibrary loans	100×10^6 books	10^7 bits/book	1.0×10^{15}
7. Facsimile transmissions	25×10^6 cases	10p/case 3×10^6 bits/p	7.5×10^{14}
8. Medical lit. search	200×10^6 searches	30p/srch 3×10^4 bits/p	1.4×10^{14}
9. Checks & credit trans	340×10^9 items	50 char/item 8 bits/char	1.4×10^{14}
10. Patent searches	7×10^6 searches	6p/srch 3×10^6 bits/p	1.4×10^{14}
11. Facsimile transmission of newspapers	20 newpapers	365 days @ 180×10^6 50p/day	6.6×10^{13}
All other listed items			24×10^{13}

p = pages, char = characters, srch = search, Trans = transfers, Lit. = literature

Table 17.2 Traffic potentially available in the UK to various information services

Information	Transactions/Years in 1990	Traffic (bits/year)		Comments
Voice	2.6×10^{10}	6×10^{17} 1.2×10^{17}	total trunk	Digitised speech
PO Mail	10^{10}	2×10^{15} 8×10^{14}	total trunk	33% fax, 67% e-mail.
Intra-company mail	7×10^9 pages	1.8×10^{13} 5.4×10^{12}	total trunk	100% e-mail
Newspapers (daily)	$12 \times 356 \times 3.3 \times 10^9$	9×10^{13} 1×10^{18}	trunk local	Fax @ 3×10^7 bits/p
Periodicals (weekly)	$20 \times 53 \times 3.1 \times 10^9$	1×10^{14} 4.7×10^{18}	trunk local	Colour fax @ 3×10^{17} bits/page
Periodicals (monthly)	$200 \times 12 \times 7.2 \times 10^7$	2.5×10^{14} 10^{17}	trunk local	3×10^{17} bits/page
Public libraries	$3 \times 10^6 \times 5.9 \times 10^8$	3×10^{15} 6×10^{17}	trunk local	4.3×10^4 new titles year. Fax as above
Literature searches	2×10^7	2×10^{13} 4×10^{12}	total trunk	

Note the assumption about fax and e-mail are made only to arrive at data quantities; it is not intended to imply that this would necessarily be the means of conveyance.

Because fax in the UK currently uses the PSTN, there is no readily available means for sorting speech traffic from fax and thus there is a danger that the growth of what is essentially data traffic could be lost sight of. ISDN, however, is capable of making a distinction between speech and fax and could therefore lead to a separation of their traffic statistics.

What Tables 17.1 and 17.2 do not show is potential for new kinds of information transmission. It is very much more difficult to predict what these might be, but a good example of a new requirement is in the context of Computer Aided Design (CAD) where it is necessary to download a diagram for display on a terminal from a central data base or to transfer it between terminals. With the number of pixels per high resolution display being of the order of 10^6, megabytes of data can be involved per transfer. Nevertheless it will be a long time before such new applications produce data volumes comparable with those required for speech.

17.3 VOICE SERVICES

Currently voice traffic uses the PSTN and the service is sometimes referred to as Plain Ordinary Telephone Service (POTS). There are several ways in which voice services might evolve that would give tangible advantages to the user, and these are discussed in the following paragraphs.

17.3.1 High quality speech

Confining the speech band to a frequency range from 300 Hz to 3.4 kHz results in some loss of quality. With digital encoding it is quite easy either to code speech at 64 kbit/s with an extended frequency range or to code it using a higher bit rate, for example by combining two 64 kbit/s channels to run at 128 kbit/s. There appears however to be very little demand for such a service, at least to the point where many people are prepared to pay a premium for it. However, as described in Section 6.3.5, there is a recent CCITT Recommendation (G.722) which covers coding for a 7 kHz speech band, but as far as is known there are currently no two-way voice services that use it. Whether it would constitute a service in its own right is debatable, because no more is necessary than special codecs in the telephone set and a guaranteed 64 kbit/s end-to-end connection between terminals, which could certainly be provided in the context of ISDN. It might be noted in passing that there are also CCITT standards (G.735, G.737) for so called 'music circuits' running at 384 kbit/s; these are typically used for outside broadcasts and are capable of transmitting sound of a very high quality.

High quality digital speech is finding its first application in the context of voice conferencing and as the sound channel in video conferencing. With video telephony services there might be a case for providing high quality voice to give the service extra cachet.

17.3.2 Reduced bit-rate speech

In Section 6.3 various techniques for reducing the bit rate necessary to transmit speech below the standard 64 kbit/s are described. Private organisations which are leasing digital private circuits are showing increasing interest in using ADPCM at

32 kbit/s to double their circuit capacity, and as the relevant codecs become freely available one might expect interest to extend to lower bit rate still. The ability (in countries where the regulations permit it) for private network operators to re-sell spare capacity is an incentive here.

17.3.3 Telephony features

Included under this heading are such features (or supplementary services) as call transfer, camp-on-busy, call waiting, call forwarding-no reply, calling line identity, etc. Such features are already available on PBXs where they have received a degree of acceptance. To be readily accepted, features have to be easy to use and to make them easy to use means that a special telephone is required, a 'feature phone' as it is sometimes called (see Section 9.2). Some features can be built into the telephone; short code dialling and last number repeat are good examples. Features that require signalling to and from the exchange while a call is in progress present something of a problem with analogue and simple 64 kbit/s digital single channel transmission. This is where the outband signalling capability available on the D-channel of ISDN becomes important. A complete set of telephony features is in a fairly advanced stage of definition by CCITT in the context of the ISDN I-series Recommendations and one might anticipate the introduction of special voice supplementary services when ISDN becomes more widely available. To what extent the user is prepared to pay a premium for such features is again debatable.

Today the number of features available to subscribers depends on the type of exchange to which they are attached, more features being available on the stored programme controlled BT's System X, System Y and Mercury's DMS than on crossbar or Strowger exchanges. The System X and Y exchanges can have added to them all the features being specified in the ISDN context but in order to benefit from them the subscriber will have to have ISDN access rather than a simple single channel analogue or digital connection.

Looking further ahead to a multiservice broadband environment the situation as seen by a voice only subscriber should not look very different from what is achievable under ISDN.

17.3.4 Cordless telephones and telepoint

The use of very short range radio links to avoid the need for a cord between the telephone base station and the handset is well established using first generation cordless telephone technology (CT1). This is analogue technology and uses a frequency of 1.6–1.8 MHz from base station to handset and 49 MHz for the return direction. The second generation, CT2, uses digital technology and operates in the 1.7–2.3 GHz band and is described in some detail in Section 12.4.2. Cordless telephones need to be protected from interference from other cordless telephones working in the neighbourhood and methods of coding and spectrum allocation are used to try to keep such interference to an acceptable level. CT2 offers a considerable improvement over CT1 in this respect. CT1 and CT2 do not necessarily constitute services in their own right, being primarily a means of providing a cordless tail to an existing service.

CT2 technology in the context of Telepoint (see further in Section 12.4.2) does, however, constitute a new service. With Telepoint, base stations are provided with a range of the 100 m order located in a public place, such as a railway station, and are a substitute for public call boxes. Users of the service carry their own transportable handsets and can make calls from the vicinity of a base station. Initially this was outgoing calls only but the service is being enhanced to handle incoming calls at base stations where a called party has logged on. Four Telepoint service providers were licensed in the UK as service providers with base stations connected to the PSTN; however, three have ceased operation owing to lack of demand, but Hutchison, began providing a service in 1992.

The key to the success of systems using CT1 and CT2 technology (which are of considerable complexity) is VLSI and the low cost and miniaturisation that is implied by it. Combined analogue and digital VLSI has still to make a full impact here, and one can confidently predict that this is an area of significant growth; there is both consumer demand and scope for price reductions.

There are several specialised areas where cordless technology is finding service applications in the private sector; of particular note are cordless secretarial systems, key systems and PBXs (see Section 4.6). The major attraction here is the ability to avoid internal wiring. The cost of installing, maintaining and rearranging internal wiring is a very significant item in the cost of ownership of in-house telecommunications equipment (see Chapter 10) and this is the main justification for developing cordless switches.

17.3.5 Mobile radio–personal communications

Two-way mobile radio falls under two main headings:

(a) Private Mobile Radio (PMR), and

(b) cellular radio

PMR (see Section 12.3) in the UK has grown very slowly over the last five years or so largely because the spectrum is limited. In 1988 there were some 440 000 mobiles. The release of some further spectrum in Band III (174–225 MHz) together with the use of advanced trunking techniques has led to major growth in PMR for applications such as vehicle fleets, police, fire service, etc., but cellular is often the preferred solution for smaller companies.

A major stimulus to the mobile market has come from the introduction of two cellular radio services in the UK aimed more at individual motorists rather than fleets. The services are called Vodaphone and Cellnet and work in the 900 MHz band. In 1989/90 there were some 500 000–700 000 mobiles in both these networks, and a high growth rate, but there were signs of congestion developing in parts of the UK. Since 1990 there has been a downturn owing to the recession. Ultimately the limits on further growth are set by the availability of allocated spectrum and the spacing of base stations.

The next generation of mobile radio is being introduced on a Europe wide basis under the title of Global System for Mobile communications (GSM) which is a digital system. (See Section 6.3.9 for the implications of this in respect of speech quality and Section 12.5.2 for a description of the system.)

The personal communication concept moves mobile communications away from a vehicle towards a hand-held pocket-sized transceiver. In 1990 UK licences for what are called personal communication networks (PCN) were granted to three operators—Mercury, Microtel and Unitel. (Since then, Mercury and Unitel first agreed to share their fixed network infrastructure and after a few months merged their operations into one company.) The systems are all based on GSM technology but with some modifications; for further details see Section 12.5.3.

At the end of 1991 Unitel carried out a market simulation trial and reported encouraging results with over 60% of the trialists wanting to keep their phone. As a result the projected market is 1.8 million households and nearly 0.5 million businesses.

The GSM networks will include small cells with a radius of 1 km or less in areas of high traffic density and provide personal communications to low power pocket size handsets. They will thus compete directly with the new PCNs which are licensed in the UK and which are to be based on GSM technology but operate at higher frequencies (see Section 12.5.3). There is already competition developing between PCN and cellular radio as the cellular radio operators attempt to pre-empt the PCN networks by moving into the hand-held portable market. Vodophone launched a trial GSM service at the end of 1991 but lack of availability of hand-held portables was forecast to delay full implementation until early 1993.

PCN and developments in GSM will depend on advances in microcellular technology where cell shapes and sizes will be tailored to the local geography (e.g. a cigar shape will cover a street) and it will be necessary to make innovations in system management and control for incoming calls and for handover between microcells and between macrocells and microcells.

17.3.6 Billing services

The more advanced stored program controlled exchanges are capable of providing itemised billing and we may expect to see a substantial growth in the use of this feature as the crossbar and Strowger exchanges are replaced in the BT network and as Mercury's network, which had this feature from its outset, expands. Not all customers want all their calls itemised, some may wish itemisation to be applied to only the more expensive calls. The ability to obtain itemisation, which in itself may be charged for, on a selective basis is likely to be an attraction and should help to eliminate faults that lead to metering errors and incorrect bills. Some of the more advanced billing concepts are treated in Section 18.3 under Intelligent Network Services.

17.3.7 Special services

Over the last five years or so there has been remarkable growth in special telephony services where the caller may either pay a reduced or increased tariff, with, in the reduced case, the balance being paid by the service provider. Examples of reduced tariffs are the 0800 'freephone' service and services where the caller only pays the local tariff irrespective of distance. Examples of premium services are chat-lines and premium information services. It is necessary to set the special metering rate and in the case of premium services to provide the information from which the service provider can obtain his share of the call charge. To achieve this end, services are provided by BT in the UK on an overlay network called a derived services network, of which there are analogue and digital versions, the digital version being known as the Digital Derived Services Network (DDSN). The volume and range of such services is likely to continue to increase.

17.3.8 Voice messaging and audiotex

Voice communication has the disadvantage that if the called party's phone is unanswered the caller is frustrated in his attempt at communication. As a second best to person-to-person conversation there are two possibilities;

(a) the called party can provide a telephone answering machine on which verbal messages may be left; or

(b) a centralised voice messaging service can be utilised.

The first approach (a) is well known, there are believed to be some 700 000–800 000 answering machines in service in the UK, and little further need be said about them except to note that a new generation of telephone answering machines will use RAM storage in place of tape storage. RAM storage will make message recording and retrieval faster and cheaper because it is not a serial process and will increase the competitiveness of answering machines relative to voice message services. For (b), details of how voice messaging systems work and what equipment is available see, for example, Section 7 of Roberts and Hay (1988).

Voice messaging services have received a degree of acceptance in the USA but they are less prevalent in the UK; they can be provided privately, as an add-on feature of a PABX or centrex, or as a public service. In the UK, BT Voicebank is available as a bureau service. In general procedures involve:

(a) the re-direction of incoming calls to a centralised answering point;

(b) the recording of messages (low bit rate digital encoding techniques are of relevance here); and

(c) the retrieval of messages by the called party.

The main advantage of a voice messaging system compared with most telephone answering machines is that messages can be readily retrieved from anywhere

where there is a telephone. The service is particularly worthwhile for certain occupations, e.g. sales representatives, but is unlikely to appeal to the working or residential populations at large. People, by and large, do not like speaking into silence to a machine.

There is a need, with message retrieval, to signal from the telephone terminal to the central answering point, and while this can be, and is, done by DTMF (dual tone multi-frequency) signalling there would be clear advantages in some form of outband or common channel signalling such as could be made available under ISDN. The diversion of incoming calls to the answering service is not automatic for public services (at least) and this means that the caller has to make and pay for another call to record the message. The difficulty of making the caller pay extra for leaving a message could be circumvented by provision of a supplementary service (employing common channel signalling) that would result in the call being re-routed with the service charge levied on the called party. The person for whom a message is left must remember to interrogate the service to find out whether there are any messages waiting. It is helpful if a message waiting indicator light on the telephone can be used as a reminder but this is a facility easier to provide on a PABX than as part of a public service.

Thus while there is scope in a number of areas for technology that could enhance the attractiveness of voice messaging services—and voice messaging is being introduced as a feature of the X.400 IPMS (see Section 17.5.3, p.307)—one must bear in mind that there is competition from a number of other services, for example, paging services provide an alternative means of contacting people and paging services with a centralised voice mailbox feature provide direct competition. The increase in mobile voice communications generally means that the number of occasions of personal non-availability is decreasing, particularly perhaps during working hours.

· Audiotex is a form of voice messaging and information service where instead of the message being a recording of a voice it is stored in textual form on a data base, and when requested it is retrieved from the data base and a voice synthesizer produces the message in an audible form so that it can be carried on a speech circuit. There are established services in the UK but it is more widely used in the USA.

17.3.9 Effects of further liberalisation

As from 1989 the owners of private networks in the UK were permitted to carry third party traffic (a situation that has applied for some time in the USA under the acronym WATS, Wide Area Transmission Service), therefore the opportunity exists for further competition with the PTOs for voice (and other) traffic. (Note however that, in the UK at least, the private network circuits are still leased from the PTO rather than privately owned.) Because of economies of scale and existing investment on the part of the PTOs, new service providers are likely either to provide particular value added features or to look for niche markets for voice traffic, probably using the infrastructure of existing private networks. With new mobile services or a multi-service network the situation is more open and will tend to lead to more general competition.

Up to 1991 there were only three PTOs in the UK, although certain CTV operators were licensed to carry PSTN traffic; following the White Paper in early 1991 the opportunity is available for anyone who can satisfy the entry conditions to obtain a licence to act as a PTO. A number of companies and consortia are likely to take advantage of the relaxed regulations; in particular utilities (such as British Rail or the electricity supply industry) who own or carry other services over strips of land are in a strong position to establish services quickly. (See further in Section 23.4.)

17.4 VIDEO SERVICES

17.4.1 Video conferencing and video telephony

Video telephones services have been technically feasible since the 1960s, if not earlier, but apart from some demonstrations and rather specialised applications have not been found to be acceptable. The most obvious reason why this is so is that the extra cost does not justify any advantage that a picture might give in addition to sound. The situation with regard to cost is changing for two reasons:

(a) optical fibre transmission is reducing the cost of bandwidth;

(b) sophisticated picture coding techniques are reducing the requirement for bandwidth.

Nevertheless the issue of acceptability to users remains and it is useful to summarise the factors for and against adoption.

Contra

(a) There are several psychological factors such as status, lack of familiarity and privacy that point to a preference for avoiding visual contact in many situations (Dickson and Bowers, 1973), so that even if there were no premium on video telephony it is doubtful if it would be used in every situation. Certainly, the ability to establish audible communication initially with the option of switching to visual by mutual agreement is seen as a pre-requisite for any service.

(b) Avoidance of travel has been cited as a positive aspect of visual telephony but perhaps surprisingly as a survey has shown the majority of people like travelling on business (Johansen, 1984).

(c) Informal opportunities for person-to-person contact are lost if video conferences are substituted for travel.

(d) Video conferences are difficult when the parties involved work in very different time zones

Pro

(a) In the context of meetings:

- avoidance of jet lag;

- saving of time;

- documentary and staff support available locally on call, making for a more effective use of people's time.

(b) In a general context, trials of a videophone services in a closed business environment have shown that most users find the system well worth having (Wish, 1975).

The interest in the use of video telephony for meetings is clearly based on the economics of video conferencing as against travel. The economics of a high quality service are in general marginal as the slow adoption of BT's Videostream and Confravision services illustrates. In 1991 there were about 200 private video conferencing systems in operation in the UK. Growth of the AT&T Picturephone service and other services in the USA is somewhat faster. As it should be in a country of greater distances and population, the market there is about 100-fold larger than in the UK. Video conferencing is the subject of CCITT Recommendation H.261 and as most systems conform to H.261, international video conferences are quite common. The DIDAMES project under the RACE heading is considering video conferencing in the context of IBCN (Schindler and Heidebrecht, 1990).

There is growth in private video conferencing circuits under the stimulus of the reduced bandwidth video coding schemes described in Section 6.3.8, but from personal experience the results are not impressive, being rather like carrying out a meeting from opposite ends of a tennis court on a rather foggy day. However, the speech quality, which it has been argued is more important than the video, is good.

It is received wisdom that while travel is getting more expensive in real terms telecommunications is getting cheaper. Examination of data for the UK shows that the effect has not been large in the past, being of the order 2% per annum in favour of telecommunications between 1972 and 1982. However, the technological changes mentioned above could accelerate this trend which should be larger now, and there is evidence that the Gulf War's adverse effect on travel has provided a stimulus to video conferencing.

If the economics of video conferencing are marginal it would appear that there is little hope that person-to-person video telephony is anywhere near economic. However, there are several factors that favour the person-to-person service:

(a) The expensive cameras, monitors and control equipment normally used for studio quality in a conference are not needed.

(b) A lower frame rate can be tolerated reducing the bandwidth requirement. Systems using bandwidth compression techniques and working at bit rates as low as 64 kbit/s are being consider and 64 kbit/s, 384 kbit/s or 2 Mbit/s system appear quite feasible (see further in Section 6.3.8). This means that video telephony can be considered in the context of ISDN as is indeed the case in the UK, Europe, the USA and Japan at least.

(c) A multifunction display which could double as a PC/word processor and picture terminal. This would enable the cost of terminal equipment to be amortised over both the office documentation and videophone receiver functions. In a residential environment a TV receiver might be made to be dual purpose.

(d) Video cameras are consumer products providing a potential source of available relatively low cost technology.

(e) The variable bandwidth transmission and switching that is feasible under SDH and ATM makes possible variable bit rate picture transmission which as we have seen in Section 6.3.8 has distinct advantages for picture coding.

Products that combine a PC with the camera and other equipment necessary for video telephony are appearing on the market at a price of about twice that of a top-of-the-range PC on its own.

In the context of the RACE project at least there are signs of a revived interest in video telephony, even though it would seem that its added value is not in general perceived as being worth while at other than a fairly marginal cost compared to telephony. Certainly low cost terminal equipment is a key factor in the adoption of video telephony.

However, there are several potential situations where a videophone can provide obvious added value and some of these are:

(a) One party has goods to offer which the other party needs to see. (This is generally a one-way service for which there may be other solutions.)

(b) Documents are under discussion or are being jointly generated.

(c) Visual authentification is required.

(d) For subjective reasons visual contact enhances a social or business contact.

(e) One or more parties is deaf.

(f) A medical consultation.

None of these on its own, or in fact all of them together, is sufficient to justify a public service, but it would not be surprising to find that some of these applications are first realised on private networks leading ultimately to a public service. In public networks ISDN may act as a catalyst but there is also competition, from speech plus fax, for example.

17.4.2 Telesurveillance

Telesurveillance services are relatively well established (e.g. for intruder detection) and, if a low frame rate is acceptable, can use an ordinary analogue telephone line.

With 64 kbit/s digital circuits, frame storage and picture processing it should be possible to increase the frame rate considerably. This is not seen in general as an application putting significant new demands on the capability of any new network compared with what is currently possible with an analogue line and modem.

17.4.3 Subscriber teleconferencing

There is a somewhat grey area between video telephony and cable TV which is exemplified by teleconferencing in which, for example, you might participate remotely in a learned society meeting or small media event such as a concert, small party political conference or trades union meeting. Two-way video would not in general be required but there would be a greater call for two-way speech (to ask questions, take part in a discussion, applaud, etc.). Whether the cable TV companies or the PTOs exploit this market, if in fact there is a worthwhile market to be exploited, remains to be seen; certainly there is the technology available to do it.

17.5 MESSAGING SERVICES

17.5.1 Overview and introduction

We start by considering in general terms what messaging services are; we then go on in Section 17.5.2 to consider, again in general terms, the bearer services necessary to support the specific message and data services covered in the remainder of the chapter. Section 17.5.3 deals with messaging services on an individual basis, a sub-section being devoted to each in turn.

Messaging services—what are they?

The items under consideration here are broader than electronic mail and are termed Electronic Messaging (EM). Under the electronic messaging heading the following are included (note that precise definition of messaging services is difficult and that the boundary between the services in this section and the bulk data transfer services of Section 17.6 is loose and somewhat arbitrary):

- Electronic mail or E-mail;
- facsimile or fax;
- telex;
- teletex;
- Electronic Data Interchange (EDI);
- transaction processing;
- Paging.

Fax differs from the rest of EM because the latter deal in what might be termed 'semantic units' (e.g. characters), whereas fax is like a photocopier and is not concerned with the elemental content of the page being transmitted. The rest of EM, unlike fax, allows the received version of a message to be printed in a different format from the source document. All the EM services are in competition with each other to some extent and jointly in competition with the Post.

It is not the intention here to treat the well-established service in any great depth; the reader is referred to Roberts and Hay (1988) for an introduction to and specific current information on the topics of telex, teletex, facsimile and what the authors call Computer Based Message Services (CBMS) but what is here called electronic mail.

Within the compass of EM it is possible to consider the business and the residential sectors of the market, but the discussion here will be confined mainly to the former on the grounds that this is where the most immediate prospects of growth lie. In the business environment intra- and inter-company communication will in general involve different networks for conveyance and it is necessary to treat these to some extent separately. Currently intra-company communication over private networks represents on average some 75–80% of the total communications in large organisations and in the context of public networks and value added networks the major interest therefore lies in communications between companies.

Perhaps the most primitive form of messaging is a simple paging service which unlike the other messaging services is basically a one way service.

We shall need to consider each of the above-mentioned user-perceived services in some detail, but before doing so it is necessary to discuss briefly the network services that these higher level services may use.

17.5.2 Means of access to messaging and data services

When considering the use of a messaging or data service there may be a choice of bearer service over which the main service runs. The economic effects of such a choice can be considerable, but it does not automatically follow that the cheapest service is the best, because other factors such as error detection will need to be taken into consideration.

Packet switching services

A number of value added service providers sell packet switching services (based on CCITT recommendations X.25 and X.75) of which one of BT's Public Data Network Services called PSS (Packet SwitchStream) is probably the most widely used in the UK but is in competition with an number of Value Added and Data Services (VADS) including (nationally) Mercury's 5000, IBM's Managed Network Service, EDS, Fastrak and Istel. Basic information on how packet switches work may be found in Section 4.4.

In North America, Internet is a collection of more than 2000 packet switched networks linked by TCP/IP protocols (see Section 8.2.5). Internet developed from

ARPAnet which is the US military's wide area data network. There is a move to make the Internet protocols more acceptable internationally by opening participation in their management to an international body and evolving them towards closer compatibility with OSI.

Packet switching implies of course that the data stream must be segmented into packets, but this presents no problem with messaging services, where delay and variations in delay are not a first-order consideration. Currently access to a port in the packet switching exchange can be via an analogue private circuit and modem, via a digital private circuit, or by means of a modem and a dialled-up connection over the PSTN. The choice is largely an economic one; relatively large amounts of data are necessary to justify direct digital access. Note that when the bearer service consists of the PSTN in tandem with a packet switching service, both have to be paid for. In the future Packet Switching Service access as a part of the ISDN will be possible, and this topic is covered in Section 15.6.

Packet service traffic will grow for a good many years to come but with the promise of multi-service networks and apart from frame relay (see Section 16.3.4) there is limited scope for major technological advances in terms of higher data rates and new protocols, although clearly advantage will be taken of technological trends as the scale of existing services increases.

The PSTN and ISDN

The PSTN currently provides a means of access to all the services listed above except Telex. In the longer term ISDN should provide a means of access to all messaging services. The PSTN is not always an ideal means of access and there are problems in using it particularly in respect of fax.

Currently in the UK there is no fax service as such (although in Germany for example there is) and the great majority of fax transmissions are sent over the PSTN using modems at data rates of 2400, 4800, 7200 or 9600 bits/s. The PSTN provides a relatively cheap means of transmission but a number of problems can arise because the PSTN is designed for speech and is unable to distinguish a fax transmission from any other. Some of these problems are:

(a) High error rates can occur due to crosstalk and noise, particularly on long analogue lines and at analogue exchanges.

(b) There is a need for end-to-end signalling for such features as confirmation of delivery for which the in-band signalling arrangements inherent in the use of the PSTN are far from ideal. There is also lack of standardisation so that terminals from different manufactures may have non matching protocols. A standardised outband signalling system should lead to an improved protocol set and interworking capability.

(c) Wrong numbers can provide a considerable nuisance to telephone users.

(d) It is difficult to prevent unsolicited (junk) mail.

The use of ISDN features in a network supporting fax would go a long way towards removing these difficulties.

The PSTN can be used, as mentioned above, for access to packet switching services and for access direct, or via a packet switching service, to Telecom Gold's electronic mail service, but because these services are, unlike fax, distinct from the PSTN as a bearer, the kind of problems that occur with fax are avoided.

ISDN provides access to the PSTN and to PSS and in due course will provide access to electronic mail services. This gives great potential for flexibility for trading, on a transaction by transaction basis, the lower cost and higher error rate of the PSTN against the higher cost but much improved error performance of an X.25 packet switched network.

17.5.3 Specific messaging services

Individual services are now considered under the headings of telex, teletex, facsimile, electronic mail, electronic data interchange, transaction processing and radio paging. There is also something of a digression to consider message switching or store and forward techniques in connection with messaging services.

Telex

Telex is by far the longest established messaging service and is the only messaging service currently established on a world wide basis, having 2 million contacts in 200 countries. In the UK, BT provide their Telex Service and Mercury provide their Link 7000 series of services. Because telex is so well established, a description of it is outside current scope, and it should suffice to make the following predictions:

(a) Inland telex traffic in developed countries will decline over the next decade, owing to competition from fax and electronic mail, but the service will persist.

(b) International telex traffic will continue to grow for several years more.

(c) There will be an increasing use of electronic mail services to access telex services in both directions (incoming and outgoing).

Most telex services throughout the world use circuit switching to connect terminals. This can be advantageous because it is possible, though rare, to work interactively and have a 'conversation' between the operators of a pair of telex terminals but there are a disadvantages in using circuit switching in isolation which are discussed under the message switching heading below. In the computer world special interactive services have been developed for what is broadly termed transaction processing, but these are designed for a conversational mode between man and machine and do not substitute for the situation where two humans need to converse by keyboard in real time. Perhaps the need for a conversational mode of operation is not great enough to justify its perpetuation in new services and the situation is certainly complicated by the various fax bureau services that operate in a store and forward mode. In any case fax allows a conversational mode with freedom of format.

Teletex

Teletex has been established as a superior telex service with an extended character set (310 printable characters rather than 31), higher data rate (240 characters/s compared with 7) and improved layout features enabling messages to be presented in a format approaching that of a letter quality. It has not been widely accepted in the UK where there is no public service as such, although teletex terminals can operate over the PSTN or PSS (but no longer between the PSTN and PSS). The situation in Europe, particularly in Germany and Sweden, is one of much wider acceptance and public services are provided.

X.400 (MHS) recommendations contain a teletex (and telex) interface but it does not necessarily follow that this interface will be fully implemented in all countries. Some UK electronic mail systems will receive incoming teletex only, but it would seem that in most countries teletex has been ousted in general by electronic mail and fax. While CCITT, in representing the telecommunication administrations, is continuing to develop teletex service standards towards mixed mode operation (i.e. teletex + Group 4 fax) it is doing so to some extent in competition with standards emerging from the computer world (see below) and the computer world's standards are receiving the greater acceptance in the UK at least.

Message switching (store and forward)

One of the disadvantages of a telex service based on circuit switching is that if a called party is busy and/or the message is multi-address, repeat attempts have to be made and repeat messages have to be sent. This ties up the outgoing telex line and wastes the telex operator's time. If, instead of being connected to a telex exchange directly, terminals are connected via a message switch, this problem can be solved. Message switching centres (which were first established in the 1960s) read the telex header, store the complete message and distribute it when favourable conditions arise. Advantage can be taken, for non-urgent messages, of lower tariffs that may operate at night. This mode of operation is often referred to as 'store and forward'. Message switching was adopted first in large corporate or Government networks but is now available as a bureau service from a number of service providers.

The tendency is for the store and forward technique to be combined for a number of services, e.g. telex, electronic mail and teletex, giving rise to multi-service or multi-interface bureau services under the heading of Computerised Message Bureau Service (CMBS). For example, a message may originate as telex and be delivered as electronic mail, being handled by one or more bureaux in the process.

Advances in computer technology will obviously lead to hardware cost reductions, but most of the investment is probably in software and a significant part of the cost is in the management of the service. For new messaging services, such as X.400 electronic mail, store and forward techniques are an inherent part of the service which is not designed to work without them. Clearly message switching, store and forward or whatever one chooses to call it is here to stay. The widespread adoption of electronic mail will lead to economies of scale that should in the longer term result in lower costs and benefit the user. The need to interface to telex will eventually disappear but that is unlikely to be the case in this century.

Facsimile (Fax)

Facsimile has a history going back to before World War II, but has only recently become widely available as a means of office document transmission. Fax has gone through various stages of standardisation which are the subject of the CCITT Recommendations in the T series. These recommendations divide fax up into four groups. The earlier Groups, 1 and 2, are largely obsolete and most of the market today uses Group 3. In 1989 the installed base in the UK was estimated as being some 350 000 machines, the vast majority of which were Group 3. Groups 1 to 3 are analogue machines but Group 4 are digital. The acceptance of Group 4 machines awaits the widespread availability of digital access to the network. Group 4 fax is subject to criticism on the ground that error checking is page-based and hence if an error occurs it is expensive and slow to have to retransmit an entire page. There is pressure for a new standard, which at the time of writing, for want of an official name, is being called 'Group 3 at 64 kbit/s'. The proposal is to have an ISDN interface and to provide interworking with conventional Group 3 machines.

Facsimile machines work by scanning the document to be transmitted with a raster, much as a television camera scans the picture it transmits, and this is both a strength and weakness as the following pros and cons illustrate.

The advantage of Group 3 fax is that:

(a) There is no problem with the preservation of the form of the original document (except grey scale and colour) and diagrams are as easy to handle as text.

(b) The use of the PSTN enables a large degree of intercommunication to be established very readily and relatively cheaply.

(c) It is in general cheaper than telex and electronic mail.

(d) It has a very wide range of optional features resulting in a wide choice of products of varying degrees of sophistication and price. (With some disadvantage in that not all of these features have been the subject of standardisation and are not therefore necessarily compatible between manufacturers.)

Its major disadvantages are

(a) Fax documents are not in word processor compatible form but the transmission as facsimiles of documents produced on a word processor or PC can be automated with the addition of special interfacing equipment. Received documents are not machine readable except on an optical character reader (OCR). OCRs are subject to accuracy and other problems (see Section 9.2).

(b) The use of the PSTN as the transmission medium has a number of disadvantages (see Section 17.5.2).

Group 4 fax improves grey scale resolution, allows colour as an option, reduce transmission time and provides a number of other additional features as part of the standard. Group 4 fax must use digital networks and will normally work at a rate of 64 kbit/s. At present Group 4 fax equipment is much more expensive than conventional Group 3 by a factor of about 10 times but the high cost can be recovered in greater throughput if sufficient traffic is available. Thus Group 4 is likely to find its first applications on dedicated inter-company routes or in private networks, where there is a predictably high traffic between specific network nodes.

While there are clearly several advantages in Group 4 fax compared to Group 3, the investment in Group 3 is so high that it may not be easy to supplant. We would tend to agree with the view that if Group 4 fax has a major role it is a mixed media role (e.g. in combination with teletex) but there are other standards for mixed media messaging that originate from the field of office automation (see below), and the indications are that these may find much wider acceptance.

A relatively new service is a fax retrieval by which documents can be recovered from a data base and transmitted to a customer. DTMF signalling provides the means of communicating with the data base for document selection and calling number identification. A service is available from Racal-Vodata in the UK.

Electronic mail

Currently electronic mail or E-mail offers a means of sending textual information between end terminals and does not differ fundamentally from telex and teletext except in so far as a store and forward mode is an inherent part of the service. Messages addresses still have to conform to a standard format but it is a more user friendly format than that of telex and, also compared with telex, there is much greater freedom (comparable with that of teletex) in the representation of the body part of the message.

At the moment intra-company electronic mail is well established (at least in large organisations) using a number of proprietary systems, but the ability for such systems to intercommunicate is limited. In the UK there are also a number of value added services, AT&T's Mail and Instant Mail (incorporating Easylink) BT's Telecom Gold, Comtext International One-to-One, Datalinx,General Electric Information Systems' (GEIS) Quik-Comm, Istel's Comet, Mercury's Multimessage and Sprint International's Sprint Mail, which provide mail box services with varying degrees of specialisation in terms of their customer base.

These value added services did not intercommunicate in 1988 but began to do so in 1989 by using the X.400 standard. With the widespread adoption of International Standards, particularly the X.400 standards for message handling system (MHS), the situation should change to the point where the majority of VADS and in-house message systems can intercommunicate freely. When this happens the potential for growth in public electronic mail will be enhanced enormously and it is an underlying assumption in what follows that this will be so. It seems likely too that many service providers will switch from proprietary protocols to X.400 for intra-network mail as well as providing an external X.400 interface. BT for example provide their

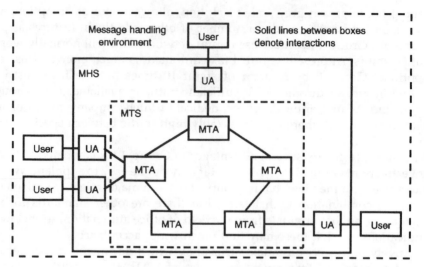

Fig. 17.1 X.400 functional model

Gold 400 service which interconnects with several tens of other services nationally and internationally.

In the USA there are over 3 million E-mail customers connected to LAN-based services with cc:Mail and Consumers Software as the major suppliers. The LAN suppliers are providing serious competition to the public E-mail providers which include a number of the companies mentioned above.

On receipt electronic mail has to be viewed on a terminal and because that terminal may be switched off or in use for something else, the incoming mail must be stored in a 'mail box'. A mail box is like a poste restante address; the mail has to be collected by the recipients interrogating their mail boxes regularly. This of course is done electronically. The mail box may either reside in the system of the service provider or the service provider may be asked to provide direct delivery to a terminal or mail box in the recipient's equipment. With some proprietary systems mail box delivery is the only option but there are X.400 standard protocols to cover mail box working and direct delivery and it is probable that all large users and eventually some small users will adopt the direct delivery form of working. However, this distinction, while it has some bearing on cost (you have to pay to fetch information from and have it stored in a mail box) does not impact the basic issue of being attached to the network and can be ignored in the following.

Figure 17.1 shows in schematic form the X.400 configuration. The Message Transfer Agent (MTA) performs the message routing and store and forward function. A service provider would have at least one MTA to serve all his customers. The User Agents (UA) are attached to a parent MTA and the user agent function might typically be contained within a PC or intelligent terminal. Assuming the UA is owned by an end user, it can be attached to a parent MTA belonging to a service provider or to a privately owned MTA. In the latter case the MTA has the option of communicating directly with other similarly situated MTAs without calling on the services of a third party's MTA. The availability of software that provides the MTA and UA functions within one PC at a reasonable cost will result in large users choosing to work directly with each other, thus saving the costs associated with the use of a third party service provider. Such users will still need a packet switching or other data network for basic conveyance, however.

The MTAs and UAs provide the basis of the X.400 system and various services (Message Transfer Services, MTS) can be run over them. The first application specific MTS to be defined by ISO is called the InterPersonal Message Service (IPMS) and this is essentially an electronic mail service as outlined above. Other services are defined or are being defined: for example, one specifically tailored to suit EDI as discussed in Section 17.5.3 (p.308) and other possibilities are mentioned subsequently. The addition of voice messaging to IPMS is covered by CCITT Recommendations F.440 and X.440 for service and protocols respectively.

Under the heading of Telematic Services, X.400 includes protocols for interworking IPMS with telex and teletex in both directions and with fax in the outgoing direction. The critical mass (or start-up) problem associated with such a new service is eased very significantly by this ability to interwork with an existing and substantial base of EM subscribers. There is also an interface to provide physical delivery; this would, for example, enable persons not on an electronic mail system to have their messages delivered to the nearest post office and to be forwarded by post from there.

The widespread acceptance of electronic mail should lead to falling costs and an increased ability to compete with fax and post. In the longer term mixed mode messages (e.g. a combination of textual and pictorial information) will be conveyed by X.400 and this could render fax as a 'service' in its own right functionally redundant.

Standardisation of such a mixed mode of operation is however in the very early stages of acceptance and the options under consideration are discussed in the next section.

The major advantage of electronic mail in a business environment where the PC and word processor are ubiquitous is that it provides a much more convenient form of message/document conveyance than any of the alternative services. There is no need to provide and address an envelope as a separate operation, there is no need for preparation on a separate terminal, there is no need to take documents to another part of the building for despatch and to monitor that despatch and there is no need to keep paper copies of documents by either corresponding party. Security (back-up) procedures can be made automatic as can acknowledgement and confirmation routines.

Document interchange and relational data base access

At a complete document level OSI standardisation is covered by Office Document Architecture (ODA) (ISO 8613) which specifies standard document types. But because of the widespread adoption of personal computers (PC), ODA has not received the acceptance that was anticipated and must be regarded as having a less than certain future. There are, however, a number of products available or planned.

At a more detailed level, and as a means for example of describing the documents that ODA might send, the Standard Generalized Markup Language (SGML) (ISO 8879) seems to be receiving wide acceptance. SGML provides a means of combining text, graphics, scanned images, audio messages, etc. within the one framework. The language describes the logical structure of documents, for example some particular word processor format or teletext could be described; part of a document might be a picture which could be described as being scanned using Group 4 fax standards or have a computer graphics description. The description is stored in a neutral format and there is freedom of choice in the way the information may be displayed or printed out. The use of this neutral format means that PCs or word processors using different proprietary word processing rules can communicate with each other through it and hence the need for ODA to describe a complete document layout is somewhat diminished. Nevertheless the translation of these specific formats into a neutral format and back again is by no means a trivial task, in terms of both software development and the computing resources required to handle it in real time, and it remains to be seen to what extent it is used in preference to standardisation at the document level. There are several SGML products available.

For conveyance of SGML documents by a telecommunications network there is a standard called Standard Document Interchange Format (SDIF) (ISO 9069) and for ODA there is a standard called Office Document Interchange Format (ODIF). The service used for conveyance in the interchange format might be based on MHS or FTAM (see Section 17.6.1) with the emphasis on MHS for short transactions and FTAM for long ones. ODIF and SDIF products are available.

Irrespective of the ability to convey graphical information and mixed mode information in general, there are very many of textual documents that could be conveyed using electronic mail but at present use the post, telex or facsimile.

Electronic data interchange

The term 'electronic mail' applies broadly to services that convey information between humans in a relatively free format. A large number of business transactions involve orders, bills, invoices, payments and receipts which have standard formats and are often computer generated by the sender and provide computer input for the recipient (being at the same time readable by humans). Such transactions come under the heading of Electronic Data Interchange (EDI) and a substantial traffic in such messages already exists on proprietary networks serving sets of business users with common interests. EDI over X.400 will allow such networks to intercommunicate and permit small businesses which are not otherwise catered for to join in.

The CCITT recommendations that covers X.400 operation are F.435 and X.435 which describe the service and protocols respectively for the EDI Messaging System (EDIMS). X.435 contains a new protocol called Pedi (similar to P2 in IPMS) and there are two basic message types which are communicated by the Pedi protocol called EDI Message (EDIM) and EDI Notification (EDIN). EDIM contains the basic EDI data suitably enveloped and EDIN is a means of acknowledging receipt of EDIMs.

Not all EDI transactions involve short messages, some involve bulk or file transfers and in this case FTAM (see Section 17.6.1) may provide a more suitable means of conveyance.

There are a number of emerging standards relating to the format of the actual EDI material itself, of which the EDIFACT is the International generic standard and probably of the greatest long-term importance. There various national standards and semi-proprietary protocols in everyday use, for example TRADACOMS and ODETTE. Details are rather beyond the current scope and further information may be found in the report referenced under Department of Trade and Industry (1981).

Under the heading of invoicing, i.e. bills, payments, statements, orders, and receipts, it is estimated that there were some 10^9 postal business-to-business transactions in 1988 with something of the order 10% of this amount being being sent by electronic means in 1988/89. Clearly, therefore, there is considerable scope for growth in this sector. Between three and four times as much material falling into the same category is sent between businesses and residences. However, the prospects of economically providing equipment in the home capable of handling EDI appears remote and it must be well into the next century before the domestic market can be exploited.

Transaction processing

In Transaction Processing (TP) one end of the connection is usually a human being operating a keyboard and the other end is a machine. The human operator expects to get an answer in a period of seconds so a store and forward mode of operation is inappropriate. One of the earliest applications of this kind which began development in the 1950s was for seat reservation systems for aircraft; nowadays applications are commonplace, with Electronic Point Of Sale (EPOS) terminals being one of the most widespread and likely to be the major source of traffic. Because of the specialist nature of most of the applications, there has been little standardisation at what corresponds to layers 4 and above in OSI terms; however, the use of X.25 packet networks is common as a means of basic transport and hence there is a fair measure of conformance to the lower OSI layers. Because many of the early transaction processing applications were provided by large organisations, such as airlines and banks, there has been a tendency for TP to run over private networks, and this too has meant that there has been less pressure on standardisation activities.

However, ISO did start work on what is known as OSI-TP in 1987 and by 1988 a fair measure of agreement had been reached on the principles involved. One of the major problems was the degree to which the OSI standard should relate to an IBM proprietary protocol known as LU6.2/APPC (APPC = Advanced Program-to-Program Communication). It seems to have been accepted that APPC should form the basis of the standard but there are significant differences in terms of excluded and additional features. In particular the OSI standard will contain the option of what is called Two Way Simultaneous (TWS) (or full duplex in communication parlance) communication as opposed to only the Two Way Alternate) (TWA) (or half duplex) protocol of APPC.

The availability of a standard should open up the market to competition and enable smaller organisations to be less dependent on large organisations for the provision of service. The ability to obtain immediate information on a customer's creditworthiness and to debit directly should facilitate purchasing transactions and in the longer term do away with the need for cheque books and a plethora of credit cards and perhaps in many instances the need to carry money. In 1989 penetration of EPOS terminals in the UK was estimated as being 15% and equal to the rest of Europe combined; clearly there is still considerable scope for growth. By the next century the number of transactions per day could well by the next century approach that of the number of telephone calls per day, but of course each transaction will have a relatively short holding time, so that the traffic measured in Erlangs is never likely to be as great as for telephony.

In the service reservation field OSI-TP offers the facility called 'provider supported transactions' in which a reservation is made conditional on the availability of other services. For example it is no use booking a flight from London to Nice if there are no suitable hotels available in Nice. OSI-TP will interrogate a number of data bases and only make reservations if all the specified conditions are met.

OSI-TP standardisation is not complete but demonstrations of its capability have been given in Japan; products were beginning to emerge in 1992.

Radio paging

Paging systems are treated in Section 1.2 and the intention in this section is to concentrate on the service aspects.

Radio paging started in 1956 with the first on-site system and has grown relatively slowly; there are currently some 200 000 users of on-site paging systems in the UK. Wide area paging systems, which were a later development, have grown more rapidly and towards the end of 1990 there were estimated to be some 670 000 users in the UK and 2 million in Europe as a whole with up to that time an annual growth rate of 25–30%, comparable with that of cellular radio.

Wide area paging system service providers fall into two main categories, those providing near (70%) national coverage and those who provide a service in a region or in several regions. In 1989 there were three service providers in the first category and some six in the second category.

Figure 17.2 attempts to summarise the features available to paging service subscribers using on-site and wide area systems. The mobile unit is called 'a pager' but is held by the called party or 'pagee'. It is a device that fits readily into the pocket and relies heavily on VLSI technology to provide small size and a wide range of features. Some pagers provide an option of non audible calling and others have the option of queuing messages for subsequent display by the pagee when he/she becomes free. Very often the purpose of the message is to ask the pagee to telephone for more information and some service providers have a voice messaging system that avoids the need for the calling party to stay by his telephone. Only on-site (460 MHz) systems provide a limited two-way voice capability.

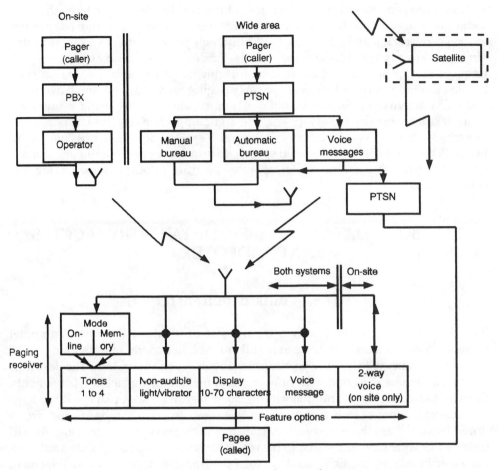

Fig. 17.2 Radiopaging system showing feature options

Some of the smaller service providers tend to offer special features, e.g. financial information, and other special features include time stamping of messages, numbering of messages and group calling.

A paging service using a satellite could provide Europe-wide coverage, and BT have been engaged in service trials in which the satellite aerial (operating in the 900 MHz band) is mounted on a vehicle roof. This would allow, for example, lorry drivers to be paged throughout Europe. Two proposed European paging systems called Euromessage for the shorter term and ERMES for the longer term are outlined in Section 12.2.2. Hutchison and Mercury launched satellite paging services in the UK in April 1992 in the hope that they would stimulate the paging market, which has tended to stagnate since 1990.

Compared with two-way radio telephone networks, paging networks are more efficient in their use of spectrum, and for the conveyance of short textual messages their protagonists claim that (owing to overload in cellular systems) the success rate is much higher. Clearly the mobile equipment is much cheaper and the ability to store messages makes the system less obtrusive. In the longer term there are some good arguments for convergence between cellular radio and paging networks/services. Racal–Vodophone have built their paging network using the infrastructure already established for cellular radio, while of course using the frequencies appropriate to each service. But this could be carried a stage further, because a cellular radio network could carry paging for a relatively small overhead. The GSM activity for European radio standardisation is proposing a both-way short textual message facility (using the signalling channel), which is a move towards such a concept. In future cellular radio systems it will be possible, by the use of TDMA and digitally encoded speech, to interleave speech and paging messages, thus removing the need for separate spectrum allocations for paging and cellular radio.

17.6 BULK DATA TRANSFER, DATA ENTRY/ACCESS AND VIDEOTEX

17.6.1 Bulk data transfer

Bulk data transfer usually involves the transfer of what are known in computer terms as 'files' between one computer and another. In general one is talking about the transfer of many kilobytes, perhaps megabytes, in one operation. The data flow is basically in one direction with only protocol messages, e.g. acknowledgements, flowing the other way. The majority of such transfers take place within large organisations which can if they wish set their own internal standards and procedures. Nevertheless there is a need for transfers between organisations, and the UK universities were among the first to recognise this need; they have made a substantial contribution to the OSI standard for File Transfer Access and Management (FTAM).

FTAM is designed to facilitate the transfer of computer files between non-collocated computers and to allow remote access to, and the management of, computer data bases. The first version was published in 1988 in the ISO 8751 series of standards with the emphasis on its transfer role. The addition of more material, particularly that covering filestore access and management protocols, has resulted in the ISO 10607 series published in 1992. Enhancements to the existing services to ISO 10607 are planned for completion in about 1994. There are a large number of FTAM products available with varying degrees of sophistication.

FTAM is not the only standard relevant to file transfer but it is the most advanced one and it seems to be recognised by telecommunication organisations as the basis for a potential value added service for the high volume information transfer. At the time of writing it is an open question as to who will provide such services and whether they will be aimed at specific market sectors, e.g. for EDI, or be widely available as with electronic mail. Certainly a number of British and European banks are interested in FTAM and have commissioned pilot studies, with one of the main applications being that of bulk EDI transfers.

A file transfer IPMS body part type has been defined for MHS which enables files that conform to FTAM at layer 7 to be conveyed using MHS protocols by electronic mail in store and forward mode rather than having to rely on a specialised FTAM service working in on-line mode. IPMS is a person-to-person application and is not ideal for FTAM and for this reason there is also work going on on a File Transfer Environment as another X.400 service distinct from IPMS.

The ability to send bulk data relatively cheaply, albeit less reliably, (e.g. via an ISDN interface over the PSTN) may in any case inhibit the exploitation of FTAM for some applications.

In the short to medium term it is not anticipated that bulk data traffic will be a substantial item as far as public networks are concerned, but in the longer term, i.e. in the next century, there may be an increase, as indicated in Section 17.2 above, in bulk distribution of what is currently printed material.

DFR (document filing and retrieval) covers similar ground to FTAM although the emphasis is more directed to storage and the structure of the document store than it is to communication. DFR fits into the distributed office applications model whereas FTAM is a transport service.

17.6.2 Remote data entry and remote data base access

Remote data entry includes the input of pure data and commands to carry out specific tasks such as batch processing. At the procedural level the ISO OSI Job Transfer and Manipulation (JTM) standardisation activity is relevant and was planned to reach fruition by 1992, at which time products should begin to emerge. Most of such activities are likely to be in house and therefore represent traffic on private networks and the need for any sort of public service appears distant.

Remote data access on the other hand is much more ubiquitous and most electronic mail services provide access to various proprietary data base services. There are also many library services, for example the Institution of Electrical Engineer's INFOSPEC service, which can use the PSTN or PSS networks for access. The response to an individual query is likely to result in the delivery of kilobytes of material, rather than megabytes, but as Table 17.2 shows, the resultant traffic is significant if not substantial. With the increasing archiving in computer readable form of vast amounts of material this is surely a growth area. It is anticipated that the searching criteria will become more sophisticated, leading perhaps to the bulk transfer of data from an archives to specialised searching engines.

Standardisation of (relational) data base access is likely to be based on the Standard Query Language (SQL) which is the subject of standardisation by the American National Standards Institute (ANSI). Standardisation of access must be distinguished from the transport protocol which might well be SDIF (see Section 17.5.3, p.308) and the service used for conveyance which might be based on FTAM or MHS.

The terminal that provides the means of data entry or access is the subject of standardisation in the context of the OSI Virtual Terminal (VT) protocols. However, there is such a large existing investment in proprietary systems that perform this function that the future of VT is somewhat questionable, although in 1992 there were some 21 suppliers with products available or in advance development. From the user's point of view there would be great benefit in not having to learn new sets of procedures; for example every time one goes to another library, catalogue access seems to involve terminals with a different set of rules.

17.6.3 Manufacturing and design protocols

The need to communicate manufacturing and design information has led to the development of two OSI based standards. Manufacturing Automation Protocol (MAP) covers the conveyance of manufacturing information either within or between organisations. Technical Office Protocol (TOP) is more concerned with design information. However, the distinction between manufacturing and design is not a rigid one and there is considerable functional overlap. MAP is intended for relatively short bursts of information for rapid delivery while TOP is intended for longer data files at low urgency. There is however a tendency for the two standards to converge, a situation that is recognised in version 3.0 of both of them.

In terms of the OSI 7-layer model MAP and TOP version 3.0 have their major differences at layers 1 and 7. At layer 1, TOP was originally based on the use of IEEE 802.3 (CSMA/CD or Ethernet) (see Section 16.3.2) while MAP used IEEE 802.4 (token bus). Now in version 3.0 both can now use IEEE 802.4 (token bus) and IEEE 802.5 (token ring) is an additional option for TOP. In 1992 MAP 3.0 was further extended so that MAP as well as TOP can use IEEE 802.3. At layer 7 both can use FTAM and directory services while TOP has the additional options of MHS, Virtual Terminal Protocol (VTP) and a special protocol called TOPP, while MAP has the option of a special messaging service called Manufacturing Message Service (MMS).

Both MAP and TOP were originally developed by large organisations (General Motors and Boeing respectively) for their own internal use but the fact that they have received international recognition means that they appeal widely to large organisations. They also represent a means by which large organisations can communicate with each other and with their suppliers. It seems rather unlikely however that MAP and TOP will ever become part of a public service but highly likely that VADS services will be provided, as EDI is, to special interest groups.

The General Motors Saturn Project has based the design of a new manufacturing plant around MAP and GM regard MAP as playing a key role in the plant's considerable success. The ability of all GM users and GM's suppliers to communicate over a common network with common protocols is one very important factor here.

17.6.4 Videotex

Videotex needs to be distinguished from teletext. Teletext, as provided by the BBC and ITV under the names Ceefax and Oracle respectively, is not under consideration here. The only public videotex service currently operated in the UK is the Prestel service of BT, although videotex is provided by a number of VADS operators in the context of proprietary systems. At the time when Prestel was introduced the main application was seen as being in the residential sector, but growth was much below prediction and recently the emphasis has moved towards business applications.

With videotex, users access a data base over (currently) the PSTN and the resultant information is displayed on a terminal, for example a PC. Another option for this terminal, which was the primary option in the early days of Prestel, is a conventional television set modified to include the additional interface and circuitry. Videotex is interactive; having consulted a data base the user can key in requests, for example to purchase a product that has been displayed on the TV screen. From the technical point of view there is a move towards improving the quality (colour and definition) of the displayed information using perhaps ISDN to provided the increased data rates that are implied in a new syntax called 'photo videotex'.

The conveyance of videotex over ISDN requires that lower layer protocols conform to OSI standards and the work on defining these standards is in a fairly advanced state.

Growth in videotex has been disappointingly slow in the UK: there were approximately 30 000 residential terminals and 90 000 business terminals in early 1990. The situation in France, where the Government subsidised the introduction of videotex primarily to provide an alternative to paper telephone directories, is more buoyant.

Videotex is in partial competition with a number of the other services listed above; its current strength is that it can provide a mixture of text and pictorial information on a TV screen which none of the other services currently do. In our view there is a future, particularly in the home, for a display that uses a television set, and there will be a convergence between a number of existing services and videotex. We are encouraged by a recent move to produce a combined standard for office terminal displays and HDTV.

17.7 BIBLIOGRAPHY

Department of Trade and Industry, (1981) *EDI Standards: a Guide for Existing and Prospective Users*, A Vanguard report, HMSO.

E. M Dickson and R. Bowers (1973) *The Video Telephone—Impact in a New Area of Communications*, Preager.

W. Hough (1970) Potential communication service market by the year 1990, *Computer*, Sept/Oct.

R. Johansen (1984) *Teleconferencing and Beyond—Communication in the Office of the Future*, McGraw-Hill Data Communication Book Series.

OSN: The Open Systems News Letter (Monthly) Technology Appraisals Ltd.

S. Roberts and A. Hay (1987) *Electronic Message Systems and Services—an International Handbook*, Commed Books.

S. Schindler and C. Heidebrcht (1990) Broadband technology with the DIDAMES project (RACE R1060), *Integrated Broadband Services and Networks*, IEE Conference Publication 329, October, pp 148–152.

A. Whyman (1972) Current developments in X.400—part 1, *Open Systems Communication— Data Transfer*, July, 1–8.

M. Wish (1975) User and non-user conceptions of picturephone service, *Proceedings of the 19th Annual Convention*, Human Factors Society.

18 INTELLIGENT NETWORKS

18.1 INTRODUCTION

The word 'intelligence' is used in electronics to describe the ability to access and process stored information. Thus an intelligent network is a network that uses stored information. The public fixed networks are already intelligent to some extent because they use stored information for number translation and routing, but the term 'intelligent network' is normally reserved for networks that have more intelligence (stored data and processing) than the current public switched networks, and that have some form of architectural separation between the basic switching, and the stored data and processing.

There are already two simple types of intelligent network in use in many countries: the networks that provide premium, freecall and other services, and the cellular radio networks. However a considerable amount of standardisation work is in progress to define intelligent networks, including their architecture and interfaces, in considerable detail. This work is taking place in both CCITT (Study Groups XI and XVIII) and ETSI (Technical subcommittee NA6).

Intelligent networks are also the subject of growing interest amongst regulators because of their potential for providing a wide range of new services and of separating the provision of services from the provision of the underlying physical network. Regulators are therefore concerned to ensure that there is a competitive supply of services by independent organisations, and also to avoid excessive domination from monopoly network providers.

18.2 INTELLIGENT NETWORK ARCHITECTURES

An intelligent network consists of a number of basic elements. The 'ordinary parts' of the network are the transmission and switching facilities that carry the call traffic. These facilities include a common channel signalling system (CCITT No. 7) that provides a sophisticated message carrying facility that is independent of the call traffic.

The additional parts that make the network intelligent are:

- additional processing associated with the basic switches to initiate, receive and transfer messages concerned with the handling of calls and the provision of special features;

- information resources (e.g. databases or message recorders) that are necessary for the provision of special features;

- service control facilities that control the provision of special features and services—the basic network and the information resources are used to provide these services.

Both the service control facilities and the information resources may be run by an organisation different from the one that runs the basic network.

Figure 18.1 shows a simple model of an intelligent network. The 'basic network' is formed of the switches and terminals that handle the main traffic. To these is added an intelligent peripheral that can be controlled by the intelligent part of the network in the context of basic traffic. An example is a recorded announcement facility which might be used in the interception of calls addressed to specific network numbers.

The Service Control Point handles single or multiple data base interrogations and controls routing for the intelligent network services. The Service Management System manages services, such as billing, authorisation and monitoring. The management terminal provides the man–machine interface through which the service provider controls all the intelligent network's services.

Figure 18.2 shows a more complex architectural model using the terminology that is being developed for intelligent networks by CCITT and ETSI. The basic communications network is made up of the connection control function (CCF), the service switching function and their interconnections. The call control agent function (CCAF) provides user access to the network service independently of underlying technologies.

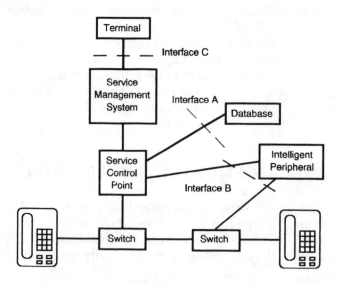

Fig. 18.1 Simple diagram of an intelligent network

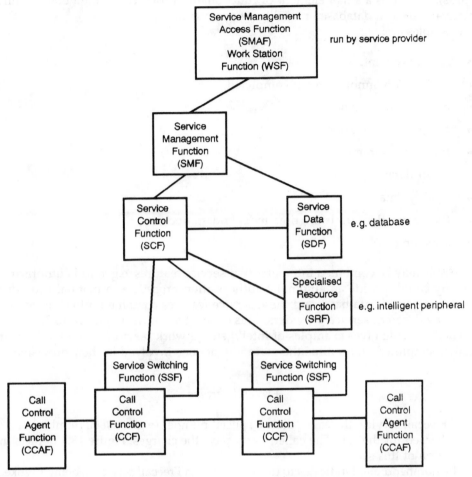

Fig. 18.2 Functional architecture of an intelligent network

The service control function (SCF) controls the basic facility for the provision of network services. The specialised data base function (SDF) and the specialised resource function (SRF) are specific to particular services and a multiplicity of these functions, and resources are likely to be available.

The service definition and management are carried out by the workstation function through the service management access function (SMAF) and service management function (SMF).

18.3 INTELLIGENT NETWORK SERVICES

Because of the potential to separate the supply of networks and services, a main focus of standardisation is the definition of Service Independent Building blocks

(SIBs). An SIB is a small element of functionality provided by the network infrastructure or by a database or intelligent peripheral.

Examples are:

- announcement;

- notification announcement completed;

- collect user information;

- charging information;

- traffic measurement;

- insert data;

- modify data;

- time dependent decision, time independent decision;

- screening

The SIBs may be combined into particular service features, e.g. a full authorisation facility could include an automatic authorisation procedure, a manual procedure and a customer data base, and these service features combined into different services by different service providers. This concept is shown in Figure 18.3.

The following gives examples of intelligent network services. The first three are rather simplified and outline the main interactions involved in their provision

Freecall service

In a Freecall service, the caller dials a special number (e.g. an 0800 number) and is not charged for the call. The called party pays the charges. Figure 18.4 depicts the provision of this service.

The number dialled indicates to the switch that a Freecall service is being invoked, but it does not give the switch sufficient information on where to route the call. This information is obtained by the service control point from the data base. At the end of the call, a call record is passed to the service management system for billing.

The Freecall service is defined, and in that sense controlled, from the terminal that is run by the Freecall service provider. Several different Freecall services, with different tariff arrangements, may be run on the same network; they would probably have different numbers, e.g. 0800, 0801, 0802, etc.

Also the services may be provided across multiple networks, i.e. the caller may be connected to public network A, the called party be connected to public network B, and the Freecall service be provided by Party C. Party C will be connected directly to either network A or network B but not necessarily to both.

Virtual Private Network (VPN) service

Although there is no common and precise definition of a Virtual Private Network (VPN) service, it is generally understood to be the provision by the public network

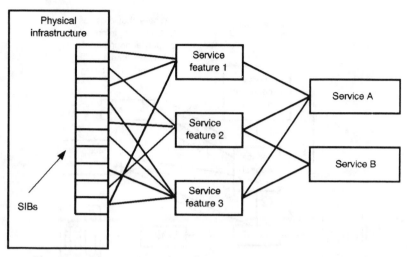

Fig. 18.3 Construction of intelligent network services from SIBs

of connections between different parts of a private network in a way that is more complex and provides more features than simple leased lines. For example the VPN service may assign circuits only when calls are in progress and may therefore provide savings by sharing transmission facilities between several private networks. It may also switch traffic at intermediate nodes so that calls which originate via the same 2.048 Mbit/s circuit arrive at different destinations. The tariff arrangements will be different from those for leased circuits and public networks. The VPN will normally work according to the numbering plan of the private network.

From the point of view of the intelligent network, the issues are the different numbering and billing arrangements and the requirement to provide customised

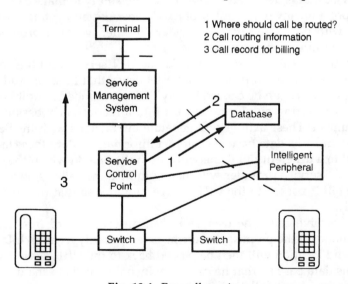

Fig. 18.4 Freecall service

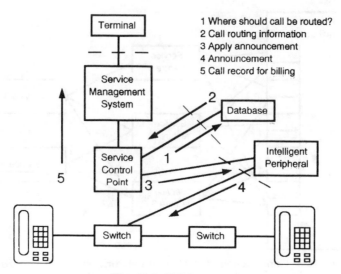

Fig. 18.5 VPN service

recorded announcements. The interactions for these features are shown in Figure 18.4. As in the above, the database is interrogated for routing information. The Service Control Point controls the call and arranges for a recorded message to be provided when appropriate. Call records are stored for billing.

Universal Personal Telecommunications (UPT)

UPT is a service that is designed to give a user the ability to make calls and receive calls at any point on any of a number of fixed networks, and yet have a contractual relationship with only one service provider. In other words it provides mobility and roaming within fixed networks.

In order to make a call, a user will invoke the UPT service and then complete an identification/authorisation process that establishes that he is an authorised customer. This procedure can be carried out by dialling numbers, but it is more likely to be done by inserting an electronic card into a reader and possibly entering a short PIN number. These actions will generate interactions with the Service Management System and a data base of customer information. Both these facilities may be connected to a different public network from the one on which the call is being made. The UPT service provider will then authorise the caller's network to carry the call. The bill is passed to the UPT service provider so that the customer is not billed directly.

In order to receive calls, the user will again insert his card into a terminal or special telephone on the premises where he is temporarily located. Information on his identity and location will then be passed back to the UPT service provider for storage on his data base so that he can route incoming calls to the user.

UPT is described more fully in Chapter 20.

Wholesaling and retailing of network services

At present basic telecommunication services are ordered from and provided by the operators of the networks Consequently the network operators need to deal directly with individual customers and the cost of the retailing element of their business is a significant part of their whole operation. With the introduction of intelligent networks, it would be possible for the records of each call to be passed to a third party retailer who would buy capacity in bulk from the network operator and look after the billing of the customer. The retailer could add value by tailoring the form of presentation and itemisation of the bills to the requirements of the accounting system of business customers. The retailer could also take orders for services from customers and negotiate the best terms for the provision of the services by the competing network operators.

Where there is competition for the provision of a particular service, the best choice may depend on the locations of the parties involved in a call or transaction and on the time of day; a third party could provide real time selection between basic service providers and also provide an aggregated itemised bill.

Given the natural monopoly element involved in the provision of basic telecommunication networks, the introduction of retailers could be an attractive way of introducing more competition into the administrative areas of service provision. For example, credit card companies who have highly developed billing systems could consider extending their services in this way.

Three-way services

A three-way service is a service where the network connects a customer not only to a service provider but also to a third party involved in the provision of that service. An example would be buying goods by credit card over the telephone. At present a customer communicates only with the seller, and the seller communicates separately with the credit card company to confirm that the account is good and arrange for payment. With an intelligent network, it would be possible for the network to manage a three-way communication with the supplier and the credit card company so that the whole transaction would be accomplished in one call. (See also in Section 17.5.3, p.309, under Transaction Processing.)

Emergency services

At present the controllers of emergency services normally have no information about persons to whom they are called. With intelligent networks it would be possible for the calling line identification of a caller to be used to interrogate a data base of key medical and other facts about victims and to send this information to the mobiles of the emergency services which are being sent to the incident, so that the crews arrive adequately briefed. The same or additional information could be sent to the hospital to which any injured would be taken.

Intelligent access and routing

A terminal equipment can choose routes over available networks that make best use of the options open to it. (The OSI model provides for this.) Thus routes can be chosen on the basis of such factors as time of day, tariffs and quality of service. The choice could be made by intelligence in the terminal or private network, by means of a third party value-added service or by use of a public network.

18.4 STANDARDS FOR INTELLIGENT NETWORKS

Intelligent network standards are being prepared in both CCITT (Study Groups XI and XVIII) and ETSI. In ETSI, the work on intelligent networks is being led by NA6, which is concentrating on defining an intelligent network architecture and defining intelligent network services. NA6 is liaising with a large number of other technical subcommittees within ETSI, the most important being the SPS subcommittees that will define the signalling protocols for use at particular interfaces. The work is at a comparatively early stage of development, although an industrial grouping based mainly in the USA has been formed to expedite the standards work.

The aim of ETSI's work is to define a set of standards that are:

- vendor and technology independent;

- network implementation independent;

- service implementation independent.

The work is planned in several phases with each phase producing standards for a particular capability set. The present work is the development of capability set 1. The current work on the architecture and the service definitions is likely to continue through 1992, and will then have to be followed by detailed work on protocols before there is an adequate definition of interfaces to enable the interconnection and joint provision of services. Thus it is likely that standardised intelligent networks are unlikely to become available before about 1995.

One key question is whether the market will wait that long. It is to be expected that the standardisation of intelligent networks will follow much the same pattern as the standardisation of data communications in OSI. Standards work on OSI proceeded in a rather laborious manner over many years while manufacturers served the market with proprietary solutions or mixtures of proprietary and standard solutions. Inevitably there will be a tension between the extremes of standards theoreticians who work in the abstract and seek universal solutions and users and suppliers who want solutions to immediate needs. It can be argued that such a tension is a healthy sign in a rapidly developing area of new technology, and that to impose complete standardisation would be to impede the market.

18.5 REGULATORY ASPECTS

The main issue is whether there should be a free competitive market for intelligent network services based on regulations to ensure the provision by the public network operators of the necessary interfaces and service elements and to enable third party organisations to offer services. This issue is closely analogous to the provision of value added services. This issue can be divided up into the following more specific issues:

- What interfaces and access to database information should public network operators provide to intelligent network service providers?

- Should the public network operators be allowed to offer intelligent network services themselves? If so, how can regulations ensure that they compete fairly with independent intelligent network service providers?

- Do interconnection requirements need to be defined to enable or oblige intelligent network services to be offered across multiple, and possibly competing, public networks?

- Should there be any obligations on the provision of access to particular information data bases.

Within Europe, the Commission has funded an extensive study to examine the relationship between intelligent networks and ONP (see Section 23.5). The study recommended that the Commission should promote competition by opening the following interfaces:

- access to databases run by network operators to enable other service providers to use those data bases (interface A in Figure 18.1);

- access to switches for the connection of intelligent peripherals to enable other service providers to supply information and processing resources for use by themselves and by others in the provision of services (interface B in Figure 18.1); and

- access to the service management system to enable organisations other than the network operator to provide services over the public network (interface C in Figure 18.1).

The study also recommended that the Commission should promote the provision of:

- Freephone,
- VPN and
- UPT

services through their standardisation, and through a memorandum of understanding on their provision and the implementation of the set of service independent building blocks needed for these services. The Commission is considering its response.

Within the USA interest in intelligent networks has been enhanced by the Federal Communication Commission's Computer III proceedings, whose Open Network Architecture requirement obliges telephone companies to unbundle their basic services, and whose Comparably Efficient Interconnection requirement obliges them to facilitate the connection of outside service providers to the network although the detailed implications of these requirements are not very clear. These developments provided an early stimulus to the Commission's Open Network Provision programme.

18.6 CONCLUSIONS

Public networks already incorporate a small degree of intelligence, and the further development of this intelligence and its standardisation will be a major feature of the nineties. Because of demand for more sophisticated services, there will be a real tension between the development of standards and the implementation of proprietary solutions.

The regulatory importance of intelligent networks has been recognised, but it will take a long time to develop a consistent and well thought out regulatory policy in such a complex area. The implementation of regulations will depend to a large extent on the availability of standards and therefore there may be a disincentive to public network operators and some manufacturers to support the standardisation process as fully as they would in other circumstances.

Notwithstanding these factors, advances in intelligent networks will lead to the provision of important new services for the user, the most notable being universal personal telephony.

18.7 BIBLIOGRAPHY

A. Beaty and E. S Albagli (1989) Intelligent networks. In R. Reardon (ed.), *Future Networks*, Blenheim OnLine.
P. Bloom and P. Miller (1987) Intelligent networks, *Telecommunications*, Jan 66–75.

19 CENTREX AND VIRTUAL PRIVATE NETWORKS

19.1 INTRODUCTION

Centrex and virtual private networks are covered briefly and together. Briefly because these are concepts that are not likely to generate new technology on their own and together because centrex has a role to play in the context of virtual private networks.

19.2 CENTREX

19.2.1 The function of centrex

Centrex is a service provided over a public network as an alternative to a subscriber having his own PBX or own PBXs and private network. With centrex, the public network provides the switching for calls between the telephones on the subscriber's site, or sites, as well as the switching for calls to other users of the public network. The tariffs for calls between the subscriber's own telephones are normally different in structure and value from those for calls to other subscribers, although there is no typical tariff structure for centrex in general.

Centrex services originated in the USA but after a period of popularity began to decline. However, there appears to be renewed interest in the service in the USA and an awakening of interest in the UK where a service from Mercury has been available since 1988.

A centrex service can be provided either by the same switch that provides the normal public network switching function or by a separate switch run by the public network operator. In either case the switch will be partitioned between centrex customers, with the switch ports assigned to any one customer in effect making a closed user group. The connection from the public network switch to the subscriber's premises will normally be a multiplexed circuit (e.g. one or more 2Mb/s circuits) rather than a collection of individual exchange lines. Consequently the subscriber needs to have at least a multiplexer on his own premises. Calls between telephones on the subscriber's premises may all pass through the public network switch or, if the equipment on the subscriber's premises includes a remote concentrator, they will be switched locally under control of the public network switch. These possibilities are shown in Figure 19.1.

When the centrex facility extends over more than one customer site, the network operator may also offer virtual private network service (see below). Depending on the geography of these sites the centrex service will involve one or more exchanges. If all the sites were in one town for example, then one centrex exchange would normally suffice.

A private exchange, e.g. one provided by a landlord and shared by business tenants, can be partitioned in the same way as a centrex, but whether such an arrangement should be called a centrex or partitioned PABX is debatable. Functionally the two are not very different.

19.2.2   Centrex technology

Centrex and public network switching technology are essentially the same; the resurgence of centrex in the USA owes much to the adoption of stored program control switches (e.g. AT&T ESS No. 5 or Northern Telecom's DMS system) as main network switches followed by their adaptation to provide the additional features needed to compete effectively with a PABX.

19.2.3   Centrex management

The key advantage of centrex, particularly to the medium sized business, is that the telecommunication network management function and the PABX operator function can (but don't have to) be handed over to the service provider, while centrex maintenance is always the responsibility of the service provider. Clearly the service provider must provide at least as good a maintenance and installation service as is provided by PABX suppliers or agencies. A major factor in the renewed interest in centrex in the USA was poor service on the part of certain PABX suppliers. A key factor in the growth of centrex in other countries will be the credibility of the network service providers in the fields of maintenance and general support.

19.2.4   The future for centrex

ISDN is already being offered as a centrex feature in the USA and it is expected that the availability and use of centrex will grow within Europe. Because public networks tend to lag behind private networks in the adoption of technical innovation and in particular because centrex is not able to compete very effectively with high speed LANs, there is little likelihood of centrex ever completely supplanting private networks in the call routing function. Centrex is most likely to be attractive to businesses with a number of separate small sites.

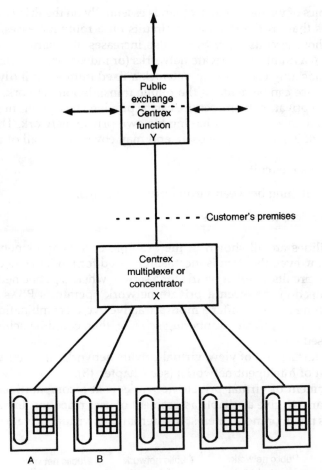

Note: Call from A to B will normally take route AXYXB, but may be switched at X under control of centrex processor at Y, in which case it will take route AXB.

Fig. 19.1 Simple centrex facility

Considering centrex in isolation, a relatively modest penetration of centrex into some specific niche markets seems to be the most likely outcome but the prospects in conjunction with a virtual private network are another matter and are discussed in the next section.

19.3 VIRTUAL PRIVATE NETWORKS

The word 'virtual' needs to be explained. It is used to describe something which appears to exist but does not in fact exist all the time. Thus a virtual private circuit appears to a subscriber to be a private circuit but the circuit is only established for as long as the subscriber actually uses it; when he is not using it the transmission capacity may be used by the provider of the virtual private circuit for other purposes.

The economics of virtual circuits depends essentially on the Erlang traffic formula which tells us that, as the number of circuits on a route increases, the efficiency with which those circuits carry traffic also increases. By sharing transmission capacity between a number of private networks (or indeed with the public network), rather than allocating each private network a fixed number of individual circuits, more efficient use can be made of the public transmission network.

Thus a virtual private network service can be provided by a public network operator as an alternative to a subscriber having his own private network. The term virtual private network is used rather loosely and may cover any or all of the following:

(a) virtual private circuits;

(b) tandem switching between virtual private circuits;

(c) centrex services to given sites.

These possibilities are all shown in the example of a complex network given in Figure 19.2. However, the term is increasingly used for tandem switching between virtual private circuits, in other words for a service where a public network operator provides connections between a private network operator's PBXs and by using tandem switching is able to offer a more attractive overall combination of tariff and features, including a private numbering scheme, than could be achieved by using dedicated leased circuits.

From a standards point of view, virtual private networks are seen as one particular application of intelligent networks (see Chapter 18).

The development of virtual private networks is at a comparatively early stage, and it is too early to give a reliable assessment of user reactions. Interestingly there seems to be as much or more interest in international virtual private networks than

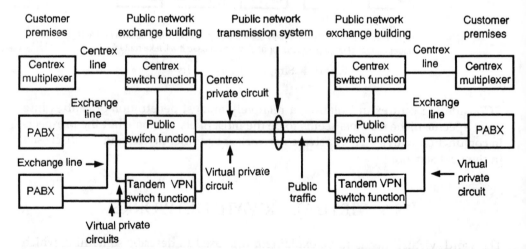

Note: This diagram shows separate public switching, centrex and VPC switching functions, but these functions may be fully integrated in the same switch

Fig. 19.2 Complex virtual private network

in national ones, possibly because the high tariffs for international leased circuits give a greater potential for cost saving.

The motivation for running a private network is to achieve cost savings and/or performance improvements (e.g. quicker call set up, better reliability, better transmission, or better facilities). In the future, as centrex and virtual private network facilities improve and benefit more from economies of scale, it is likely that they will be able at least to match the performance of traditional private networks, and if so there is likely to be a gradual migration to these new services.

One argument against this trend is that many private network operators have to have their own networks because either their businesses are so highly dependent on the special facilities of their networks that they cannot be replaced by services from others, or they cannot trust the public network operators with the provision of a facility which is so crucial to their business. Inevitably some facilities will always remain too highly specialised for others to provide and a hard core of private network use will continue indefinitely, but there is no reason why public network operators should not provide levels of service and reliability which will satisfy even the more demanding customers, nor why they should not give customers the level of control which they require over the network facilities. The crucial issue is one of public network operators establishing the necessary degree of confidence in their capability.

These sections have been written in terms of centrex and virtual private networks being provided by public network operators. If the regulations permit it, which they do in the UK and USA, there is no reason why these services should not be provided by other operators, in which case they would normally be called managed network services.

19.4 BIBLIOGRAPHY

B. Ralph (1989) Centrex and virtual private networks. In R. Reardon (ed.) *Future Networks*, Ch. 20, Blenheim OnLine.

20 UNIVERSAL PERSONAL TELECOMMUNICATIONS

20.1 INTRODUCTION

Universal Personal Telecommunications (UPT) is a new service that will enable a subscriber, who is identified by a personal number, to make and receive calls at any terminal. The service is currently being defined and is not expected to be offered before 1994–95 at the earliest, and then only in certain countries. A pan-European service is not expected until the late 1990s. The description given here is based on a draft of CCITT Recommendation F.851.

In both fixed services and first generation cellular services, the subscriber is associated with a particular terminal or line. In UPT this association is broken and the subscriber is identified by a new UPT number which may be used in a temporary association with any terminal, and thus provides personal (not terminal) mobility. All billing is carried out with respect to this UPT number, and so the subscriber is billed at his home address for all his calls, irrespective of where he makes them. The UPT service may be provided across multiple networks including networks of different types (e.g. fixed and mobile).

20.2 SERVICE DESCRIPTION

The basic aims and principles of UPT are to:

- provide personal mobility by enabling a UPT subscriber to make and receive calls from any terminal on a global basis (not just within the same network);

- charge on the basis of UPT identity rather than terminal identity;

- provide standardised access procedures across multiple networks;

- provide choice and flexibility with regard to the service features selected by a subscriber.

The concept of the service is as follows. The customer subscribes to the UPT service of a UPT service provider who allocates him a UPT number. The subscriber selects the features that he wishes to have, and this selection is stored in a personal user profile. He also selects the method of access and authentication that he wishes to use.

The subscriber's UPT number will normally be shown and identified clearly as such on his business card, which should also show his default or home location.

When the subscriber is away from his normal telephone, he may register his current location via any terminal of any network, and receive calls or make calls from that terminal. Registration involves an authentication procedure. Initially this procedure is likely to be limited to the entry of his UPT number and a secret personal identification number (PIN), using terminals with a tone dialling (DTMF) facility. However the wiping of a magnetic card in a card reader may also be used, and another alternative is to use a more secure authentication procedure with a small hand-held tone generator. In the longer term, more sophisticated interrogation/response systems may be used.

There are several options for registration at a local terminal. Registration may be set to enable incoming or outgoing calls or both, and registration may be indefinite or for a set period of time. Several UPT subscribers may be registered at the same terminal simultaneously. However, a facility will be provided to enable the owner of a terminal to reset or annul the registration, so that, for example, he will not receive incoming calls for a visitor who registered at his terminal but has left subsequently without cancelling the registration. Any new registration at any location will automatically cancel a previous registration by the same subscriber. When a UPT subscriber registers at a terminal, he may select whether or not he should give any identification (e.g. account number and PIN) when receiving incoming calls; identification could be useful if several people have access to the same terminal.

20.3 CHARGING AND BILLING

The charging and billing arrangements for UPT will need careful planning. The intention is that the subscriber will be billed only by his UPT service provider, not by the operators of the networks that he may use. All billing and charging will be based on the UPT number, which should be clearly recognisable so that callers to UPT number can be aware that they may be paying special tariffs. Callers to UPT numbers will be charged only for successful calls.

20.4 NUMBERING

The numbering arrangements for UPT will be described in CCITT Recommendation E.168. Currently this recommendation deals only with incall numbering, i.e. numbering for calls to a UPT subscriber; the identification arrangements for outcalls have yet to be made. The UPT number will conform to E.164, having a maximum length of 12 digits before time T and 15 thereafter (see Section 25.3). The draft of this recommendation contains three scenarios as shown in Figure 20.1.

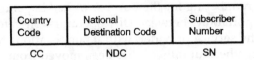

Country Code	National Destination Code	Subscriber Number
CC	NDC	SN

Scenario 1 - Home Related Scheme
    CC = Country Code
    NDC + SN = National Significant Number

Scenario 2 - Country Based Scheme
    CC = Country Code
    NDC = UPT Indicator or UPT/Service Provider Indicator
    SN = Subscriber Number

Scenario 3 - Global Scheme
    CC = UPT indicator
    NDC = Global or Country Identification (up to 3 digits)
    SN = Subscriber Number

Fig. 20.1 UPT numbering scenarios in E.168

Scenario 1—home related scheme

This scenario is designed for UPT services that are limited to a local area or home domain. The number does not itself contain any form of UPT identity.

Scenario 2—country based scheme

In this scenario, the national destination code contains a UPT indicator or a combination of a UPT indicator and a service provider indicator. The subscriber number may be structured to identify the particular node or data base where the subscriber's service profile is held. National dialling formats may be used.

Scenario 3—global scheme

In this scenario, the country code is used as the UPT indicator, with the national destination code being used either to give a global identity for the data base holding the subscriber's service profile, or to identify the country in which the subscriber is registered. In the former case, the numbering arrangements would have to be administered under a global scheme run by CCITT, in the latter case the subscriber's number could be administered nationally.

It is likely that different scenarios will be used in different countries and that there will be progression towards scenario 3 as a long term solution.

Within the UPT number, it will be necessary to identify the data base of the UPT service provider so that the correct data base can be interrogated. Were this information not available, a global data base would have to be interrogated and currently such a data base is considered to be impracticable. The need to identify the data base means that the UPT number will be specific to that UPT service or data base, and so will not be portable between UPT service providers. Thus although

UPT will offer personal numbering for the purpose of making and receiving calls, it is unlikely to provide portability between UPT operators. Therefore the problem of independence from the service provider, with its implications for competition, has not been solved altogether but rather it has been moved from the provision of the telecommunications network to the provision of the UPT service. One could therefore argue that the UPT number is not a true personal number so long as it contains information about the data base that holds the subscriber's information. If you change UPT service provider you will have to change your number.

20.5 SERVICE PHASES AND FEATURES

Within Europe, UPT is being planned in four phases:

Phase 1: Restricted short-term UPT service scenario

Phase 2: Basic UPT service scenario

Phase 3: Advanced UPT service scenario

The exact content and planned time scales for these phases are not yet defined. The features for UPT have been divided into two main sets: core features and additional features, but the provision of these features has yet to be related to the phases.
   The following features form the core of the UPT service:

- In call registration (registration for incoming calls)—A UPT subscriber may register at any terminal for incoming calls. When registered all incoming calls to the UPT subscriber will be presented to that terminal for the duration of the registration specified by the subscriber. A new registration for incoming calls will cancel a previous registration.

- Outgoing UPT call set up—A UPT subscriber may set up a single call using an authentication procedure. The set up will have to be repeated for subsequent calls.

- Outcall registration—A UPT subscriber may register for outgoing calls from any terminal. Once registered, all outgoing calls from that terminal will be charged to his account without further authentication unless he elects for such authentication.

- Allcall registration—Combination of incall and outcall registration.

- Linked registration—Same as Allcall registration except that the registration may be effected from a remote terminal, and is not cancelled by a subsequent incall registration from a different terminal.

- Profile interrogation—The subscriber may interrogate the status of his profile.

- Profile modification—The subscriber may modify his service profile.

- Secure answering of incoming calls—The subscriber may specify that authentication is necessary before incoming calls can be answered.

- Intended recipient identity presentation—Presentation on the alerting terminal of the identity of the intended recipient by UPT number or name.

- Follow-on facility—A follow-on procedure without further authentication.

- Specific announcements—A set of friendly UPT specific announcements.

The additional features are as follows:

- Multiple terminal registration.

- Call pick-up.

- Remote answering.

- Service personalisation (e.g. language for announcements and dialogues).

- Variable default incall registration.

- Access to user profiles by subscriber responsible for several users.

- Operator assisted services.

20.6 IMPLEMENTATION

The implementation of the UPT service is dependent on the implementation of an intelligent network architecture (see Section 18.2) where the translations between the subscriber's UPT number and the terminal to which his calls are to be routed are held in a data base. The design and management of these data bases will be similar in many respects to that of the cellular radio networks. The public network operators in Europe and the USA are in the process of adding intelligence to their networks, in some cases through overlay networks, in others through additional features for the digital local exchanges. These developments suggest that the service should commence, with some constraints on availability, in about 1994–95.

The implementation of the data bases and their management is only one aspect of the implementation of the service. The service will introduce a number of new and difficult issues in the areas of charging and billing, privacy, security and intrusion or disturbance in the use of other people's terminals, and the resolution of these issues may prove to be the determining factor for the successful implementation of the service on an international basis.

The principal elements in implementing the service are as follows:

(a) The UPT service provider needs to implement a data base to maintain a record of the translation between the UPT number and terminal number.

(b) The provision of outgoing calls requires an arrangement between the local operator and the UPT service provider for the authentication of the caller and the transfer of billing information and payments. This arrangement would not necessarily have to be implemented at the local exchange but could be implemented at an overlay exchange to which UPT signalling would be forwarded by the local exchange.

(c) The provision of a simplified incoming call facility could be made in any network if the subscriber used an ordinary call and tone dialling to register his location with his UPT service provider. However, in the longer term, registration will be local with signalling using CCITT No. 7 between the local digital exchange and the UPT service provider.

(d) Access to UPT numbers will be limited to networks that have the necessary routing and charging facilities, and will probably be limited to networks that can signal to the UPT service provider using CCITT No. 7.

In the light of these considerations, it seems probable that UPT will be introduced initially within individual public networks, or parts of public networks, and will be extended gradually across different public networks. It may be unrealistic to expect the widespread availability of UPT on an international basis much before the year 2000.

20.7 CONCLUSIONS

UPT is expected to be an important service from the late 1990s onwards. Both UPT and radio based mobile communications will provide mobility and in the longer term there will be an element of competition between them; however, the two services are fundamentally different in that a mobile service provides mobility between the terminal and the network whereas UPT provides mobility between subscriber and terminal. The subscriber who needs to communicate on the move needs a mobile service, but the subscriber who frequently works in different places will probably prefer UPT because it enables calls to his home number, which is fixed, to be redirected cost effectively to his current location. Some subscribers will need both services.

Both UPT and radio based mobile services depend on intelligent networks, so in the long term the relative costs depend on the cost of radio links compared with physical exchange lines.

From the point of view of competition, UPT decouples a person's number from the identity of the physical network provider, but, as has been pointed out in the chapter on numbering, because the UPT number is likely to be tied to the the identity of the UPT provider, the barrier to competition of such an association remains. Thus the problem is moved, not solved.

21 DIRECTORIES

21.1 THE FUNCTION AND SCOPE OF THE DIRECTORY

Directories are a necessity when the address information needed to establish a network connection is incomplete. Everyone is familiar with written telephone directories in which entries are in alphabetical name order but are of little use if you can remember the postal address you want but have forgotten or do not know the name of the person whom you wish to contact. Each service currently tends to have its own directory, there are telephone, telex and facsimile directories which are in book form and electronic mail directories which are on-line to service subscribers. Because it would be prohibitively expensive to provide every telephone customer with a complete set of UK telephone directories, there are directory enquiry services for telephony which use advanced computer technology for storing and retrieving information. The concern here however is with directories in a multiservice environment which are the subject of a fairly radical approach and one in which international standardisation is the driving force.

Within CCITT work has been proceeding for several years on the X.500 series of standards for the directory, the first version of which was published in 1987. In the context of the various CCITT and ISO numbering and naming and addressing standards, the Directory can be viewed globally as an integrated set of on-line directories. Logically there is this one global directory, but physically it is realised in separate parts with some duplication.

It is not the function of the Directory to provide explicit routing information, but routing information may be implicit in a directory number or implicit if more than one directory number is provided for an entry.

Directory development has taken place in the OSI context and therefore the emphasis has been on data services although if customers have the terminal equipment necessary for interrogating them they may also be used for voice services.

21.2 THE DIRECTORY IMPLEMENTATION

A directory is a data base that contains a large number of entries with each entry having various attributes, for example Personal Name and Organisation Name. If users specify a sub-set of a total directory entry (e.g. Personal Name) the local directory can be searched and the matching entry or entries will be returned to them. The X.500 set of standards has been influenced by and is compatible with the addressing structure of X.400. X.400 defines a number of alternative addressing structures but the interest here lies in a user friendly structure and one that corresponds fairly closely to a postal address. For example an entry that uniquely defines the intended recipient might be:

Address	Attribute
John White,	Personal Name
Production Manager,	Organisation Unit
The Bootle Bottle Company,	Organisation Name (or PRMD name)
Bootle,	Town Name
UK	Country Name

(*Note*. If (as will generally be the case) the Organisation Name is unique in the UK and the organisation unit name is unique within the organisation, then the town name is redundant in the example entry.)

Each line in the above example is an attribute, e.g. Personal Name is an attribute type and each entry can have associated with it any number of attribute types each with a value or values, for example the above entry could have associated with it an attribute of type 'Telephone Number' with a value for example 065 789999. Any attribute value that is unique among its peers (e.g. The Bootle Bottle company among organisation names) is called a distinguished value. There is no guarantee that in interrogating a directory the query is going to take exactly the form given above and there is a need for alternatives called multi-valued attributes. In the example given above, J. White and BBC might be entered for Personal Name and Organisation Name respectively. BBC as the Bootle Bottle Company's Organisation Name would not be unique and is therefore not acceptable as the Distinguished Value in their UK directory entry. Nevertheless both J. White and BBC are acceptable as multi-valued attributes for use in searching but the proper entry as tabulated above must be used for addressing. In this example the interrogant should not be surprised if the response he/she gets as the result of a search containing multi-valued attributes, e.g. BBC, includes details of someone working for the British Broadcasting Company.

Figure 21.1 attempts to put the above example in a more general context. Each entry in the Directory Information Tree (DIT) points to the items below it until the bottom of the tree is reached and the target information is found. At the object entry level there is no limit, as the figure indicates, to the number of attribute types that can be listed. The left-hand arm might be the USA while the right-hand arm might be the UK. The alias entry in the USA arm points to the object entry in the UK arm for what might be a an organisational unit of a multinational company. Thus with aliases there is only one entry that needs to be kept up-to-date.

Clearly the Directory entry contains the numbering information (see Chapter 25 for a discussion on routing and numbering issues) necessary to establish a connection but there is no need in the case of X.400 electronic mail (at least) for users to aware of this; the translation process can be made quite invisible to them and the

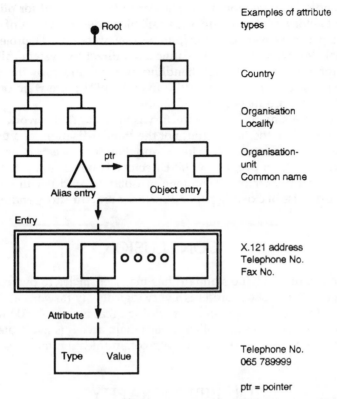

Examples of attribute types

Country

Organisation
Locality

Organisation-
unit
Common name

X.121 address
Telephone No.
Fax No.

Telephone No.
065 789999

ptr = pointer

Fig. 21.1 Structure of the directory information tree with examples of attribute types

routing information entered automatically into a message; however, the routing information could still be made visible on special request.

As indicated above, the intention is that the Directory be a global one and it is clearly impracticable that there should be one and one only global Directory; even producing national copies of a complete global Directory is rather impracticable and inefficient. The proposed solution to this problem is a distributed directory with the ability of directories to consult peer directories if they do not have the relevant entry. The distributed directory approach solves the problem of size but gives rise to other problems, e.g. of updating, particularly if aliasing is not used and the same information is contained in two directories, as is the case with many multinational organisations. The question of who pays for what in accessing the Directory is also likely to lead to complication and perhaps controversy.

The problems of directory security are by no means trivial because it is necessary at the same time to provide an easy updating mechanism and to prevent unauthorised and malicious changes from being made.

Enough has been said to indicate that the Directory represents an extremely complex software system, probably more complex than anything hitherto attempted either under the OSI label or more generally in the field of standardised communication protocols. It would be extremely optimistic to expect the more sophisticated directory features to be available much before the mid-1990s, although the complete and second version of the standard was scheduled for publication in 1992. In 1989 directory pilot software was available under the names of QUIPU and THORN; these packages were developed in two separate ESPRIT projects; QUIPU is public domain software. In 1990 BT launched a directory system called 'Cohort 500' which supports X.400 message handling systems and provides a company directory which can be attached to a PBX. In early 1992 there were over 20 companies with directory products under development.

The method of accessing the X.500 Directory is via a computer work station (PC) and work has started on the design (but not the standardisation) of a user friendly WIMP interface (Windows, Icons, Mouse and Pointers, see Section 10.2). This means that, while easy to use, the X.500 Directory is never going to be of much direct relevance to the user of a telephone in isolation, and it will be at least 20 years before the demise of existing paper directories and directory enquiry services.

21.3 CONCLUSIONS

The full realisation of the X.500 standard has many obstacles to overcome, nevertheless, we believe that X.500 represents a very logical way forward and that in the medium to long term it will be seen as providing great benefit. X.500 will become widely used for data services and all services where access is available through a suitable terminal such as a work station with communication facilities.

21.4 BIBLIOGRAPHY

M. Corby (ed.) (1992) *Telecomms Users Handbook 1992//3*, CommEd Books.
Directory services pilot project (1989) *Network News*, No. 24, July, 6–8.

Part 3

ADMINISTRATION

OVERVIEW

The opening up of public telecommunication networks to competition increases the importance of regulations because of the need to control the established dominant operator in order to give competitors a fair opportunity to establish a service and also because of the need to impose standards openly for terminal equipment and private networks. Numbering is of greater interest in a competitive situation where changing your supplier could mean changing your directory number. Competition leads to choice and a need on the part of the user to make careful tariff comparisons in order to obtain the best deal. All these topics are treated in the first four chapters of Part 3.

The increasing complexity and choice of networks and the need to establish whether faults lie in your network or someone else's means that network management must be put on a much more open footing. Network management and requirements for network security are the subjects of Chapter 16.

The emphasis is on principles rather than on ephemeral details; there is no point, for example, in providing specific tariff information for a situation that changes many times each year.

22 STANDARDS

22.1 INTRODUCTION

ISO, CCITT, ETSI, NET, ANSI and BSI; the world of standards is one of acronyms, and it is the purpose of this chapter to explain the connections between the various standards bodies and to list the standards that are important for telecommunications. Of course a number of standards have been discussed elsewhere in this book but the intent here is to list rather than discuss and to cast the net somewhat wider than the coverage given under the specific subject headings of the other chapters.

Individual standards might be listed under the standard body that promulgates them or under subject headings; it is felt that the latter is the more useful approach and is the one adopted in Section 22.3. However, in Section 22.4 an exception is made for the European CTRs and NETs, which are referenced in numerical order because this information does not seem to be readily available generally.

22.2 STANDARDS BODIES

Figure 22.1 shows in diagrammatic form some of the International European and national bodies that are concerned with major aspects of standard making for telecommunications.

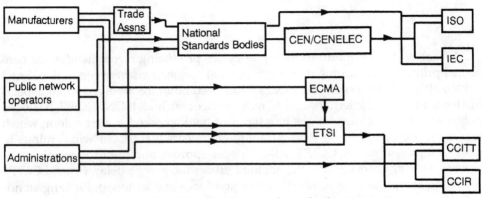

Fig. 22.1 Origination of standards

22.2.1 International

The two principal international standards making bodies at world level are the International Standards Organisation (ISO) and the International Telecommunications Union (ITU).

ISO

ISO has a very wide remit and its work on information technology involves it in data communications. In general ISO does not produce standards for voice communications although some of its work on local area network standards may be suitable for voice as well as data applications.

The membership of ISO is made up of national standards organisations rather than of individual companies. ISO publishes both International Standards numbered ISO ... and Draft International Standards numbered DIS .... Draft International Standards are produced where the subject matter is not sufficiently stable for a full international standard.

The International Electrotechnical Commission (IEC) is responsible for electrical and electronic standardisation of equipments and components, and it has a particular interest in safety matters. The IEC and ISO formed a Joint Technical Committee (JTCI) in 1988 to cooperate on the production of standards for information technology.

ITU

The ITU has five divisions of which the Consultative Committee for International Telegraph and Telephone (CCITT) and the Consultative Committee for Radio Communications (CCIR) produce recommendations on technical and operational matters and are the main international 'standards' bodies for telecommunications at world level. However, the recommendations produced are not standards in the true or purist sense because CCITT and CCIR are not standards bodies although their recommendations in many cases serve the function of standards.

CCITT

CCITT has worked hitherto in four year cycles producing recommendations concerned primarily with compatible international connections between national networks, although in recent years CCITT has become more involved in all aspects (national as well as international) of new services such as ISDN. CCITT has been publishing its recommendations in a series of book known by their colour, which was red in 1984 and blue in 1988. Although it appears that CCITT will continue to work in four year cycles, CCITT will no longer approve and publish recommendations every four years because this method causes too great a delay. Instead CCITT is making more use of approval by correspondence and is now publishing standards as soon as they are ready.

CCITT is controlled by national administrations. Traditionally the Post Telephone and Telegraph (PTT) organisation has been the national administration. This has normally been the government department that was responsible for providing public telecommunications. However, with the growing trend towards competition and privatisation, in many countries the provision of public telecommunications has passed to separate state owned or privately owned companies, and the role of the administration is now played by a new national regulatory body or small government department.

In addition to administrations, scientific and industrial organisations including public network operators and manufacturers may also be members of CCITT, but hitherto they have not been able to vote and so the organisation has been dominated by national administrations who in most cases were monopoly public network operators.

The constitution of CCITT and its working methods are now changing, and a report recommending changes will be considered at the Plenipotentiary conference at the end of 1992.

Work in CCITT is carried out by Study Groups with the following subject areas:

SG I	Service definitions
SG II	Network operations
SG III	Tariff principles
SG IV	Maintenance
SG V	Safety, protection and EMC
SG VI	Outside plant
SG VII	Dedicated networks
SG VIII	Terminals for telematic services
SG IX	Telegraphs
SG X	Software
SG XI	ISDN, network switching and signalling
SG XII	Transmission and performance
SG XV	Transmission systems and equipment
SG XVII	Data transmission
SG XVIII	ISDN and digital communications

CCIR

CCIR works in a way similar to that of CCITT and also produces recommendations. It has the following study groups:

- SG 1   Spectrum management techniques
- SG 4   Fixed satellite services
- SG 5   Radio wave propagation in non ionised media
- SG 6   Radio wave propagation in ionised media
- SG 7   Science services
- SG 8   Mobile, radio determination and amateur services
- SG 9   Fixed services
- SG 10   Broadcasting services—sound
- SG 11   Broadcasting services—television
- SG 12   Inter-service sharing and compatibility.

Cooperation between CCITT and ISO

The work programme of CCITT and ISO have overlapped in the areas of data communications and Open Systems Interconnection (OSI). For example both CCITT and ISO have produced similar sets of standards for message handling systems but the standards have contained significant differences in certain areas with the CCITT Recommendations reflecting the perspective of a monopoly in the supply of public services.

In order to provide coordination between CCITT and ISO, a Joint Technical Committee (JTC) has been formed.

International agencies

Relevant international agencies are the International Telecommunication Satellite Organisation (INTELSAT) and the International Maritime Satellite Organisation (INMARSAT) which are funded by interested parties to set standards, procure and operate satellite systems.

22.2.2 European

There are several European bodies involved in the preparation of standards that affect telecommunications, the main one being the European Telecommunications Standards Institute (ETSI). In general the European Commission has encouraged the formation and strengthening of European bodies with two objectives:

- to provide a coordinated European input to world bodies;

- to produce European standards in areas where world standards are either insufficiently advanced or contain too many options to form the basis for a common European market.

At present the European bodies are proving very effective in accelerating standardisation in Europe and they are providing a very strong input into world bodies. In the longer term everyone's aim is to have world standards, especially in telecommunications, and so in each particular service area the European standards could be seen as having only a transitional role. However, it remains to be seen whether world wide agreement can be reached when there is not a political impetus equivalent to that provided by the European Commission.

ETSI

ETSI was formed in 1988 as a result of discussions between the European Commission and the Conference of European Posts and Telecommunications (CEPT) which was a body of European administrations very similar in function to CCITT. The main aim of the formation of ETSI was to produce the standards needed for the creation of a common European market in telecommunications services and apparatus, which was one main objective of the Commission's policy set out in the 1987 Green Paper (see Section 23.5.1). However, membership of ETSI is not limited to EEC countries but is open to any country within geographical Europe.

The formation of ETSI involved a number of important innovations and features. The most important and the most radical was the inclusion of a number of different categories of member with much more equality between categories than is the case in CCITT. The five categories are:

- administrations and national standards bodies;
- public network operators;
- manufacturers;
- users;
- private service providers, research bodies, consultancies, partnerships etc.

Each member pays a fee according to a formula related to its turnover, and many of its decisions are based on voting by individual members, although the final approval of standards is based on weighted national voting with the national positions being decided in the national standards organisations. Thus users and manufacturers have much more direct influence than they have in other bodies.

The overall management of ETSI is provided by the General Assembly and the Technical Assembly with the latter having ultimate responsibility for all matters concerning the technical contents of standards and programmes. ETSI produces standards through a number of Technical Committees (TCs). Initially the technical committees of CEPT complete with their work programmes were moved across

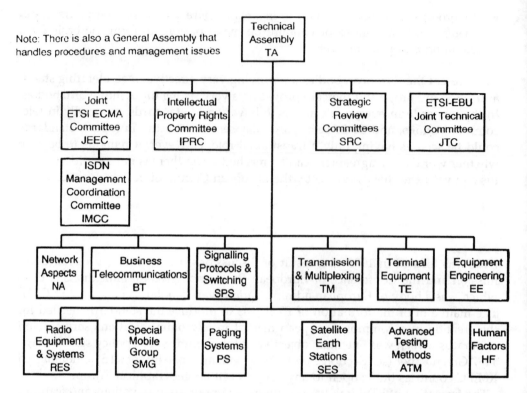

Note: There is also a General Assembly that
handles procedures and management issues

Fig. 22.2 ETSI technical committee structure

into ETSI, but subsequently new additional committees have been created to produce the structure shown in Figure 22.2. Each technical committee has several
Technical Subcommittees (STCs) which do the detailed technical work with the TC
functioning more as a management and advisory body.

Another important innovation is the use of project teams composed of paid
experts loaned from member organisations to produce drafts of standards. This
innovation is especially effective in speeding up the production of standards.

The standards produced by ETSI are known as European Telecommunications
Standards (ETS) or Interim European Telecommunications Standards (I-ETS), I-ETS
must be superseded by full ETSs within two years, or renewed, or withdrawn.

ETSI's work programme is composed of two parts: the main programme is
funded by membership subscriptions and revenue from sales, but there is an
additional voluntary part which can be funded by individual or small groups of
members. The European Commission is funding a considerable volume of work in
support of its regulatory programmes for common technical regulations and open
network provision through the voluntary programme.

A third innovation is that ETSI establishes special strategic review committees of
senior experts to examine its programmes and the need for standards in particular
areas. This procedure helps to ensure that ETSI is forward looking and responsive
to changes in telecommunications technology and markets.

CEN/CENELEC

The Comité Européen de Normalisation (CEN) is the European equivalent of ISO, and its sister organisation CENELEC handles electrical standardisation. Some confusion arose between ETSI and CENELEC over responsibility for private network standards because mandates for work in this area had been given by the Commission to CENELEC rather than ETSI, but this problem was resolved with an agreement that ETSI would be the leading body on all telecommunications standards whereas CENELEC would lead on subjects such as safety, EMC and building wiring.

The membership of CENELEC consists of national standards bodies only and so its work is much less accessible than that of ETSI.

CENELEC produces two types of standards:

- ENs, which must be published as national standards and may be used under European legislation.

- ENVs, the use of which is voluntary.

CENELEC used also to produce Harmonisation Documents, but they are no longer produced and existing ones will be replaced by ENs.

European Computer Manufacturers' Association (ECMA)

ECMA was formed in 1960 and has played a very influential role on the development of standards for data communications, especially OSI, and latterly in private network standards where it is developing the signalling system known as QSIG. The main objective of ECMA is to provide drafts to input to formal standards bodies.

The main ECMA committee is TC32 (Communications, networks and systems interconnection) which has four task Groups:

TG11	Computer Supported Telecommunications Applications
TG12	PTN Management
TG13	PTN Networking
TG14	PTN Signalling

Because there is an overlap between ETSI and ECMA in private network standards, the two organisations reached a coordination agreement in 1991 under which the allocation of work between ETSI and ECMA is handled at the top level by the Joint ETSI ECMA Committee (JEEC). ETSI is to focus on standards that involve interactions between public and private networks, and ECMA is to focus on standards that are concerned solely with the internal working of the private network. Information on work programmes and draft standards will be exchanged freely between the organisations and members of one organisation may participate in the activities of the other.

The standards produced by ECMA, once they are approved, will be passed to ETSI for an ETSI public enquiry and voting procedure, after which they will become European Telecommunications Standards.

22.2.3 National standards bodies and other bodies

Most countries have national standards bodies. Table 22.1 gives a list of some of the best-known ones.

Table 22.1 National standards bodies

Country	Body
France	AFNOR
Germany	DIN
Japan	Telecommunications Technical Committee
United Kingdom	BSI
United States	ANSI

Membership of these bodies is a national matter. In the UK individual companies may become members of BSI but attendance at the committees that draft the standards is limited to invited experts or representatives of trade associations and so the work is not as open or accessible as that of ETSI.

In addition to the formal national standards bodies, professional bodies can play an important role, the Institute of Electrical and Electronic Engineers in the USA being well known for its local area network standards.

22.2.4 Trade associations

Trade associations play a more direct role in setting standards in the USA than they do in Europe and the UK. The following trade associations in addition to ECMA (see above) have produced or are producing standards that are significant in their own right, and in most cases form the basis of international and national standards.

The USA's Electronic Industries Association (EIA) has produced important computer interface standards and jointly with the Telecommunication Industries Association (TIA) is active in the field of building wiring. The Exchange Carriers Standards Association produces standards for deregulated networks and is well known for its T1 committee, which develops standards for the interconnection of networks.

22.2.5 Proprietary specifications

Sometimes a proprietary specification becomes a *de facto* standard and there are a number of instances where for example IBM in particular has established interfaces and protocols that have been widely adopted throughout the industry. These *de facto* standards either get absorbed in national and international standards or they remain in competition with them for many years. Problems have arisen in the past because standards bodies have been followers rather than leaders of technological advances but there is increasing awareness of this problem which is to an appreciable extent is being redressed.

22.3 SPECIFIC STANDARDS

The purpose of this section is to provide a quick reference guide to the standards which are considered to be most relevant to the interests of readers but its scope is somewhat wider than the rest of this book in that it puts greater emphasis on the past. Standards are listed under subject and reference number but the formal title (which is often cumbersome) is replaced by a brief indication of content because this makes searching easier. An asterisk against a standard indicates that it receives more than a passing mention (explicitly or implicitly) elsewhere in this work (refer to index).

22.3.1 Coding

CCITT

*G.711	PCM coding
– 715	
*G.722	7 kHz audio coding at 64 kbit/s
*G.726	40, 32, 24, 16 kbit/s ADPCM
*G.727	5-, 4-, 3- and 2-bits sample embedded ADPCM
*G.763	Digital circuit multiplication equipment
*G.764	Packetised voice protocol
*H.261	Codec for visual and audio services
T.90	Group 4 facsimile encoding
*V.42bis	Textual data compression.

22.3.2 Component

More connectors are listed under the OSI physical layer heading in the IT section.

CCITT

K.12	Gas tubes for overvoltage protection

BS.

3041	Radio frequency connectors
3573	Polyfin–copper cables

4808	(Parts 4 & 5) LF cables and wire PVC insulated
6312	Plugs (Part 1) and sockets (Part 2) for line jack units
6513	Wideband cable
6558	Optical fibre and cable
9055	Surge arresters
9210	Radio frequency connectors
CECC 42201	Varistors

22.3.3 Equipment (terminals)

See also under ISDN

ETSI

NET 1	Data terminal equipment for circuit switched networks
NET 2	Data terminal equipment for packet switched networks
NET 4	Analogue interfacing to the PSTN
NET 10	GSM mobiles (three parts)

BS	Part	
6305		General requirements for PSTN
6317		Extension telephones
6320		Modems
6789	1	Particular apparatus—general requirements
6789	3.2	Auto calling and answering
6789	6	Series connected apparatus
6804		Social alarm systems
6833	1	Cordless attachments—general
6833	2	Apparatus using radio links

22.3.4 Interchange and access circuits

See also physical layer standards under the IT heading.

CCITT

*V.10	Unbalanced interchange circuit (5 V nominal)
*V.11	Balanced interchange circuit (5 V nominal)

V.24	Definitions for DTE–DCE interchange circuits
*V.28	Unbalanced interchange circuit (10 V nominal)
*V.35	Data transmission at 48 kbit/s (obsolete, see V.36, V.37)
*V.36	Data rates up to 72 kbit/s
*V.37	Data rates above 72 kbit/s
V.42	Error control protocols for modems
V.110	V. series interfaces in the ISDN context
*X.3	Packet assembly disassembly
X.20	DTE–DCE start/stop transmission on public data networks
X.20 *bis*	DTEs interfacing to asynchronous V-Series modems
X.21	DTE–DCE synchronous transmission on public data networks
X.21 *bis*	DTEs interfacing to synchronous V-series modems
*X.23	Packet switched network via PSTN
*X.28	Start/stop packet assembly disassembly
*X.29	Control of packet assembly disassembly
X.30	
*X.31	Packet mode terminals for ISDN.

BS	Part	
6328	1	Speech band circuits
6328	2	Base-band circuits
6328	3	DC private circuits
6328	4	SCVF signalling
6328	5	Wideband FDM
6328	7	X.21 *bis*
6328	8.1	G.703 (2048 & 8444 kbit/s)
6328	8.2	G.703 (64 kbit/s)
6638		DTE/DCE interface
6640		DTE/DCE (V.24 & X.24)
7326		Telex

22.3.5 ISDN

Most of the published material is from CCITT but there are a number of emerging European documents.

CCITT

*G.961	Transmission on metallic lines—basic rate
*I.120	Description of ISDN and BISDN
– 122	
*I.121	BISDN—audio visual frame structure
*I.130	Service definition
I.150	BISDN—ATM mode functional characteristics
I.211	BISDN—service aspects
*I.231	Bearer services—circuit switched mode
*I.232	Bearer services—packet switched mode
I.233	Frame mode—frame relaying (Pt 1) and switching (Pt 2)
*I.240	Teleservices
– 241	
I.250	Supplementary services—definition
*I.251	Supplementary services—number identification
*I.252	Supplementary services—call offering
*I.253	Supplementary services—call completion
*I.254	Supplementary services—multi-party
*I.255	Supplementary services—closed user groups, etc. (Pts 1 to 4)
*I.256	Supplementary services—charging
*I.257	Supplementary services—user signalling
I.361	B-ISDN ATM layer specification
I.362	B-ISDN adaptation layer—functional description
I.363	B-ISDN adaptation layer—specification
I.413	B-ISDN user network interface
*I.430	Layer 1—basic rate
*I.431	Layer 1—primary rate
I.432	BISDN—user network interface, physical layer
*I.440	Layer 2—general (also Q.920)
*I.441	Layer 2—details (also Q.921)
*I.451	Layer 3—protocols (also Q.931)
I.610	OAM principles of the B-ISDN access

ETSI

*NET 3   Basic rate access (ETS 300 104 & ETS 300 153)

*NET 5   Primary rate access (ETS 300 156)

 NET 7   Terminal adaptors at S/T reference point

See Tables 15.9 and 15.10 for an extended list of the ETS 300 series of ISDN bearer, tele and supplementary service standards.

ANSI

T1.606   Frame relay bearer service—service description

T1.606 Add Addendum to TI.606 on congestion management

T1.6ca   Frame relay bearer service—protocol description

T1.6fr   Frame relay bearer service—signalling specification

22.3.6 IT (services) in an OSI context

The following list of standards is broken down into the 7 OSI layers plus layer-independent standards; within each grouping they are in ISO number order with the equivalent CCITT and the British Standard given where appropriate and available. There is not always a one-to-one correspondence in document numbering, scope and content between the ISO and CCITT documents but the BS's in general correspond fairly closely to their ISO equivalent. Only the basic ISO document is given; in many cases this will be split into parts (e.g. 7498/3) and there may be addenda (e.g. 7498/AD2).

Layer-independent (generic) standards

ISO	CCITT	BSI	Subject
*7498	X.200	6568:88	Basic reference model
8509	X.210		Service conventions
8807			LOTOS—formal description technique
9074			ESTELLE—description technique
*9595	X.710		Common management information service
*9596	X.711		Common management information protocol
9646	X.290		Conformance testing
9798			Security techniques
9834			Registration procedures

ISO	CCITT	BSI	Subject
9843			Registration and information objects
10000			Taxonomy of profiles
10031			Distributed office applications
10040			Management overview
10164			Systems management (series)
10165			Structure of management information
*	X.800		Security architecture

Layer-7 application layer standards

ISO	CCITT	BSI	Brief title
*8505	X.400		Message handling system (MHS)
*8571		7090	File transfer access and management (FTAM)
*8613			Office documentation architecture (ODA)
*8649	X.217	7091	Association control service element (ACSE)
*8831			Job transfer and manipulation (JTM)
*8879			Graphical language (SGML)
*9040			Virtual terminal services (VT)
*9066	X.218		Reliable transfer service (RTS)
*9069			Documentation interchange (SDIF) protocols
*9545			Application layer structure
*9072	X.219		Remote operations service (ROSE)
*9594	X.500		Directories
*9804	X.237		Commitment, concurrency & recovery (CCRSE)
*10021	X.400		Message handling system (MOTIS) (EDI)
*10026			Distributed transaction processing (TP)
10035			Connectionless ACSE protocol
10163			Documentation—search and retrieve
*10166			Document filing and retrieval
*	X.435		Electronic data interchange using MHS
*10607			File transfer access and management (FTAM)

Layer-6 presentation layer standards

ISO	CCITT	BSI	Brief title
*8822	X.216	7093	Connection oriented presentation service
*8823	X.226	7094	Connection oriented presentation protocol
*8824	X.208	6962	Abstract syntax notation (ASN)
8825	X.208	6963	Encoding rules for ASN

Layer-5 session layer standards

ISO	CCITT	BSI	Brief title
*8326	X.215	6960	Session service definition
*8327	X.225	6961	Session protocol specification
9576			Connectionless mode session service

Layer-4 transport layer standards

ISO	CCITT	BSI	Brief title
*8072	X.214	7218	Transport service definition
*8073	X.224		Transport protocol specification
8602			Connectionless mode transport protocol

Layer-3 network layer standards

ISO	CCITT	BSI	Brief title
*8208	X.25		Packet level protocol for DTE
*8348	X.213	7220	Connection mode service
*8473		7235	Connectionless mode protocols
*8648		7221	Internal organisation of the network layer
*8878	X.223	7224	X.25 connection service
*8881			X.25 protocols on LANs
*9068			Connectionless mode service

ISO	CCITT	BSI	Brief title
9542			Routing exchange protocol
9574			Connection mode service by packet terminal
9575			Routing framework

Layer-2 data link layer standards

ISO	CCITT	BSI	Brief title
1155		4505 Pt3	Basic mode error detection
1177		4505 Pt2	Basic mode start/stop and synchronous transmission
1745		4505 Pt1	Basic mode digital data transmission
2111		4505 Pt4	Basic mode control procedure
2628		4505 Pt6	Basic mode control procedure complements
2629		4505 Pt7	Basic mode conversational transfers
3309		5397 Pt1	High level data link control (HDLC)
4335		5397 Pt2	High level data link control (HDLC)
7478	X.25/75	5397 Pt6	High level data link control (HDLC)
7776	X.25	5397 Pt7	High level data link control (HDLC)
7809		5397 Pt5	High level data link control (HDLC)
8471		5397 Pt8	High level data link control (HDLC)
8802			Logical link control for LANs
8885		5397 Pt9	High level data link control (HDLC)
8886	X.212		Data link service definition
9067		7249	Data communication automatic fault location
10171			List of protocols using HDLC transmission

Layer-2/1 link and physical layer standards

ISO	CCITT	BSI	Brief title
*8802/2			LAN logical link control
*8802/3			CSMA/CD
*8802/4			Token bus

ISO	CCITT	BSI	Brief title
*8802/5			Token ring
*8802/6			10 Mbit/s slotted ring
*8802/7			DTE for attachment to slotted ring
8867			Industrial asynchronous data link and physical layer

Layer-1 physical layer standards

ISO	CCITT	BSI	Brief title
2110		6623 Pt1	25 pin connector
2593		6623 Pt4	34 pin connector
4902			37 pin connector
*4903	X.21		15 pin connector
*7477	X.24	6640 Pt1	DTE physical connection (V.24)
7480		6638 Pt1	DTE/DCE interface signal quality
8480		6639	DTE/DCE 25 pin connector
*8481	X.24	6640 Pt2	DTE/DCE connection with DTE timing
8482		7248	Twisted pair multipoint
8877		7266	Connector for ISDN basic access
9160			Physical, layer interoperability requirements
10022	X.211		Physical layer service definition

22.3.7 Network (services/numbering)

CCITT

*E.163 Telephony numbering

*E.164 ISDN numbering—E.163 extension

*E.165 ISDN numbering—timetable

*E.166 ISDN numbering—interworking

*X.121 Data numbering

BS

*7521 Space convention for ADMD names

22.3.8 Radio

CCIR

Radio paging code No. 1 (POCSAG)

ETSI

(The GSM standards are listed in Table 12.1.)

* ETS 300 116 DECT test specification
*I-ETS 300 131 Interim CT2 standard
* ETS 300 133 EREMES (European paging system)
*I-ETS 300 168 Interim DSSR standard
* ETS 300 175 DECT system.

Radiocommunication agency (UK)

*MPT 1322 Angle modulated equipment for cordless telephone service (CT1)
*MPT 1327 A signalling system for trunked PMR
*MPT 1343 System interface specification PMR, band III, sub band 2
*MPT 1371 Angle modulated equipment fro cordless telephone service (CT1)
*MPT 1375 Common air interface (CT2)

22.3.9 Safety and protection

Safety

IEC	EN	BS	
65		415	Household equipment
	41003	6301	Telecommunication equipment
		6484	Power supplies
950	60950	7002	Information technology equipment

Protection

CCITT

K.11 Protection against overvoltage

22.3.10 Signalling

CCITT

Q.300 R1 multifrequency (North America)

_ 331

Q.400 R2 multifrequency (Europe)

_ 490

*Q.700 Introduction to signalling system No. 7

*Q.701 No. 7 message transfer part

_ 710

*Q.711 No. 7 signalling connection control part

_ 716

*Q.721 No. 7 telephone user part

_ 725

*Q.741 No. 7 Data user part

*Q.761 No. 7 integrated service user part

_ 767

*Q.771 No. 7 transaction capabilities application part

_ 775

*Q.920 ISDN layer 2—general

*Q.921 ISDN layer 2—details

*Q.931 ISDN layer 3

European (See also Tables 5.1 and 5.2)

*ENV 41004 Reference configuration—PBXs

*ENV 41006 Scenarios for interconnecting PBXs

*ENV 41007–1 Definition of terms for private networks

ECMA

*ECMA 141 Data link layer protocol between two PBXs

*ECMA 143 Layer 3 protocol between PBXs for circuit switched calls

British Telecom

*BTNR 188 Digital private network signalling system No. 1

*BTNR 189 Interworking between DPNSS1 and other signalling systems

22.3.11 Switching (network interfaces)

CCITT

*X.25 Packet switching—user interface

*X.75 Packet switching—network interface

BS Part

6450 1 PBXs—general

6450 2 PBXs—terminal stations

6450 3 PBXs—DDI

6450 6 Simple call routing apparatus

6450 7 In service observation

22.3.12 Transmission (including area networks)

Wide area networks

CCITT

*G.702 Plesiochronous digital hierarchy—bit rates

*G.703 Plesiochronous digital hierarchy—interfaces

*G.707 Synchronous digital hierarchy—bit rates

*G.708 Synchronous digital hierarchy—interfaces

*G.709 Synchronous digital hierarchy—multiplexing structure

*G.763 Digital circuit multiplication equipment

 M.1020 4-wire leased circuits (data or speech)

 M.1040 4 wire leased circuits (speech)

Metropolitan and local area networks

ISO	IEEE	BS	
*8802/3	8802.3		CSMA/CD (Ethernet)
*8802/4	8802.4		Token bus

ISO	IEEE	BS	
*8802/5	8802.5		Token ring
*8802/6	8802.6		Dual bus QPSX/DQDB
*8802/7		6531	Cambridge ring

22.3.13 Videophone and videotex

CCITT

F.300 Videotex service

H.100 Visual telephone systems

H.120 Codecs for videoconferencing

*H.261 Codec for visual and audio services

ETSI

ETS 300 072 Videotex terminal equipment

_ ETS 300 076

22.3.14 Wiring and installation

See also interchange circuits.

ISO	BS	
*	6701	Installation (Parts 1 & 2)
	6506	PBX installation
8492	7428	Multipoint interconnection of twisted pair

EIA/TIA

*568 Commercial wiring

*570 Residential wiring

22.4 CTRs AND NETS

The following is a list of CTRs; the CTR number replaces the NET number and the NET subject is given in the subsequent list.

CTR No.	Other ref.	Subject
1	NET 1	
2	NET 2	
3	NET 3	
4	NET 5	
5	NET 10	
6		DECT access for non-voice terminals
7		EREMES receive only access
8	NET 33	
9		GSM telephony terminal requirements
10		DECT telephony terminal requirements
11		DECT public network profiles
12	2020[*]	ONP leased lines 2048 kbit/s unstructured
13	2023[*]	ONP leased lines 2048 kbit/s structured
14	2026[*]	ONP leased lines 64 kbit/s
15	2028[*]	ONP analogue leased lines ordinary quality—2 wire
16	2030[*]	ONP analogue leased lines special quality—4 wire
17	2032[*]	ONP analogue leased lines ordinary quality—2 wire
18	2034[*]	ONP analogue leased lines special quality—4 wire

The following is a list of NETs against their recommendation ETS number and subject. (The formal titles are in most cases much longer.)

NET	ETS	Subject
1	T/TE 04–07	Approval of data terminal equipment using X.21
2	T/TE 04–06	Connection of data terminal equipment for X.25
3	T/TE 04–08	Part 1. Attachment—ISDN basic access
3	T/TE 04–22	Part 2. Attachment—ISDN basic access—layer 3
4	T/TE 04–13	Attachment of equipment to PSTN (superseded)
	T/TE 04–16	As above—current version
5	T/TE 04–24	Attachment—ISDN primary access
6	T/TE 04–37	Approval equipment for data networks (X.32)

[*] ETSI/BT reference

NET	ETS	Subject
7	T/TE 04–10	Attachment terminal adaptors at S/T reference point
10		Digital mobile access
11		Telephony characteristics of digital mobile equip.
20	T/TE 04–17	Basic approval—modems
21	T/TE 04–18	Approval—300 Bauds modems
22	T/TE 04–19	Approval—1200 Bauds modems
23	T/TE 04–20	Approval—2400 Bauds duplex modems
24	T/TE 04–21	Approval—1200 Bauds half duplex
25	T/TE 04–09	Approval—9600 or 4800 bauds modems
30		Facsimile—Group 3
31	T/TE 05–12	Facsimile—Group 4, class 1 on ISDN
32	T/TE 07–07	Attachment of teletex terminals
	T/TE 07–01	Annex to above on requirements
	T/TE 07–04	Annex to above on service intercommunication
	T/TE 07–05	Annex to above on testing
33	T/TE 10–06	Handset approval for ISDN telephony service

22.5 CONCLUSIONS

The number, complexity and diversity of standards is ever increasing and their importance is difficult to over-emphasise. We note particularly that standards bodies are leading rather than following events and how important it is for standard makers to be able to respond quickly to errors and omissions in their output as implementors get to grips with the problem of turning their standards into products.

The setting up of ETSI is changing the face of standardisation in Europe and must lead to European national standards institutions playing an increasingly subordinate role.

23 REGULATION AND COMPETITION

23.1 INTRODUCTION

Many books could be written on the subject of telecommunications regulation and competition. Changes are taking place with considerable speed in many countries, and any book would become quickly out of date. In this chapter we attempt simply to outline the background to these changes and to highlight some of the concepts and issues of greatest general interest.

23.2 HISTORICAL DEVELOPMENT

The 1980s was the decade when major changes in the regulation of telecommunications started. Although telecommunications began in many countries with private companies, these companies were either nationalised or became largely state controlled monopolies during the early part of the century.

During the 1970s, there was a renewed awareness in both North America and Europe of the growing importance of communications to the whole of the economy, particularly in view of the development of information technology and value added services.

A large number of separate local telecommunication companies had existed in the USA, but trunk and international communications had always been provided exclusively by AT&T. There was major concern that AT&T would be in too dominant a position if it was allowed to participate in computing and offer value added services as well as maintaining a monopoly on trunk services and ownership of many local services.

This concern led to important changes:

- the introduction of competition in trunk and international services;

- the introduction of unrestricted competition in value added services;

- the break-up or divestiture of AT&T into separate regional companies, and trunk and international carriers.

These changes, together with a political change from a Labour to a Conservative government, stimulated a major change of policy in the UK. The Acts of Parliament that gave the British Post Office a monopoly in telecommunications were repealed, British Telecom was created as a separate entity to the Post Office and the government took power to license new operators. Mercury was licensed to compete with British Telecom and the long process of liberalising the supply of customer apparatus began. The Office of Telecommunications was formed to fulfil the role of day-to-day regulation and was made formally separate from the Government. British Telecom was privatised with its shares being sold in two tranches.

The changes in the US and the UK stimulated considerable interest in Europe and in other countries. The European Commission began a programme to establish a European market in telecommunications apparatus and all services other than voice and basic infrastructure.

Australia and New Zealand both introduced competition and privatisation, with New Zealand being noted for the most deregulated regime in the world.

Japan began to introduce competition in the mid 1980s with the licensing of new national and international operators, and free competition in value added services and apparatus. Japan now has competition in all areas of telecommunications and, overall, has at least as high a level of competition as the UK and USA.

The following sections outline in more detail the major changes in the USA, the UK and Europe.

23.3 THE USA

23.3.1 Historical development

Up to the 1950s there was virtually no competition in the USA with all services being provided by the integrated AT&T Bell System and a number of independent local telecommunications companies. The Federal Communications Commission (FCC) existed but did not intervene to any great extent.

The first element of change came in 1959 when the FCC allowed the operation of private microwave links, and from that time on, the FCC took a more proactive approach and began to change the whole telecommunications industry.

The FCC began to re-examine its policy of allowing AT&T to prohibit the connection to its network of any apparatus not supplied by AT&T itself. This policy was based on claims that apparatus not made by AT&T might damage the network. This issue came to a head in the Carterfone case over the connection of an acoustic-inductive device to interconnect private two-way radios with the public network. The FCC ruled in 1968 that these devices could be connected to the AT&T network but that AT&T could charge different tariffs to reflect their costs in installing a protective device between the line and the Carterfone apparatus, despite the fact that there was no metallic connection to the network.

From 1966 to 1971 a five-year inquiry, known as Computer Inquiry I, was held into the issues raised by the interdependence of computing and communications.

The FCC ruled that all but the smallest common carriers could provide data processing services but only through separate subsidiaries.

Also in 1971, the FCC adopted a policy of allowing competition by permitting new operators (common carriers) in the areas of data communications and specialised services. This allowed new carriers to enter the microwave transmission area and Microwave Communications Inc. (MCI) began service in 1972.

In 1972 the policy of competition in specialised services led to the Open Skies policy for domestic satellite communications which allowed any company to establish a satellite communications network.

Between 1975 and 1977 major changes occurred in the area of customer apparatus. The FCC introduced a terminal equipment registration programme which allowed apparatus not supplied by AT&T to be connected electrically to the network provided that it satisfied technical standards for protection of the network from harm. This liberalised the market for all telecommunications apparatus.

In 1977, following a long investigation, the FCC decided that the Bell System operating companies should have more autonomy in the procurement of equipment and should no longer be obliged to procure from AT&T's subsidiary Western Electric. The FCC directed AT&T to establish a separate corporation to manage competitive procurements.

In 1979, AT&T and the specialised carriers reached an agreement which allowed the specialised carriers to interconnect to the Bell System provided that they paid an access charge.

In 1976, AT&T introduced an IBM compatible data terminal claiming that it was data communications equipment and not data processing equipment. This brought the FCC to focus on the inadequacies of the Computer Inquiry I decisions and so the FCC began Computer Inquiry II, which lasted until 1980. This second inquiry replaced the distinction between data communications and data processing with a distinction between basic and enhanced services; basic services were limited to the transport of information unchanged from source to destination. Basic services would continue to be regulated by the FCC but enhanced services would no longer be subject to regulation.

Computer Inquiry II also introduced two other major changes. Tariff controls were removed on the provision by AT&T of all customer premises apparatus that was not already installed and subject to state tariffs. However, in order to ensure that the regulated carriers did not cross subsidise their enhanced service or apparatus supply activities, they were permitted to provide enhanced services and apparatus only through a fully separated subsidiary. This ruling was modified subsequently in 1984 to allow regulated carriers to supply but not manufacture customer apparatus.

In 1984, a long running anti-trust case against AT&T by the Department of Justice accusing AT&T of abusing its monopoly privileges came to a conclusion in the Modified Final Judgement of Judge Green. This judgement split AT&T into several different parts and is known as the 'divestiture'. The 22 Bell Operating Companies were reorganised into seven separate Regional Bell Operating Companies (RBOCs):

- Ameritech

- Bell Atlantic

- Bell South
- Nynex
- Pacific Telesis
- Southwestern Bell
- US West.

AT&T itself was left as the main trunk carrier (Interexchange carrier) and international carrier. The areas covered by the RBOCs are shown in Figure 23.1. The judgement also prohibited AT&T and the RBOCs from offering enhanced services.

The RBOCs complained strongly against the prohibition on the offering of enhanced services except through separate subsidiaries because they considered it to be inefficient and to hinder innovation, and the FCC reviewed this issue in the Computer Inquiry III in 1986. Computer Inquiry III replaced the prohibition on the offering of enhanced services with two new requirements: Open Network Architecture (ONA) and Comparably Efficient Interconnection (CEI).

Open Network Architecture is an obligation on AT&T and the RBOCs to unbundle basic services and offer such unbundled services on the same terms to independent enhanced service providers as to their own operations. Once their ONA plans have been approved, as is the case for all the RBOCs, they are able to start to offer enhanced services.

Comparably Efficient Interconnection is an obligation to allow independent enhanced service providers the same facilities in connection to the regulated networks as the regulated networks enjoy themselves. Thus in theory independent

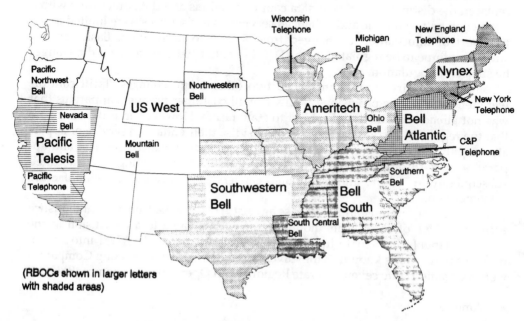

Fig. 23.1 Coverage of the RBOCs and BOCs

enhanced service providers should be able to site their equipment on the premises of AT&T and the RBOCs.

Neither ONA nor CEI have yet led to any great changes. ONA is the subject of disputes about the relative powers of state and federal bodies and the precise effects of the CEI concept have yet to be determined.

In 1989, the FCC decided to change from its traditional approach of regulating AT&T and the RBOCs through controlling their rate of return to one of adopting the UK approach of regulating prices. However, the FCC introduced a combination of price caps and novel price floors, the purpose of which is to prevent predatory pricing.

23.3.2 Overview of the US market structure

The domestic market in the USA is divided into two classes of carrier:

- Local Exchange Carrier (LEC)
- InterExchange Carrier (IEC).

Direct connections to the customer's premises are provided by the local exchange carriers who operate in a defined area. There are two types of local exchange carrier—the independent carriers and the 22 Bell Operating Companies (BOCs) which combined into the seven Regional Bell Operating Companies. The areas covered by the BOCs are called Local Access and Transport Areas (LATAs) and these areas may cross state boundaries. There is no direct local competition between conventional terrestrial switched networks, although there is competition from bypass technologies such as VSAT, teleports, and private microwave and optical fibre links. The local services are mainly subject to state regulation although interstate services within an LATA are subject to FCC regulation. The division of areas is shown schematically in Figure 23.2.

The Interexchange Carriers provide long distance communications between LATAs and are subject to FCC regulation. The Bell Operating Companies are not

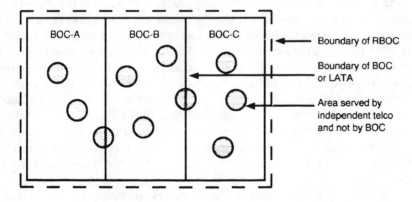

Fig. 23.2 Relationship between RBOC, BOC and independent telcos

allowed to offer services between LATAs. The Interexchange Carrier market is intensely competitive with a number of substantial operators, including:

- Allnet;
- AT&T;
- Metromedia Long Distance (ITT);
- MCI;
- US Sprint.

When a caller makes a trunk call he has to select his trunk carrier either by preselection or by dialling selection codes. All LECs with a market of 10 000 lines or more are obliged to offer equal access to all IECs. The IECs pay the LECs access charges for connection to the local exchanges.

There is also competition in the international market with the larger domestic IECs being the main players (e.g. AT&T, MCI, Sprint International, WorldCom).

23.4   THE UK

The first major step in liberalisation in the UK was the British Telecommunications Act of July 1981 which gave the Secretary of State for Trade and Industry power to license telecommunications operators, set apparatus standards and approve apparatus. The Act also separated the telecommunications and postal businesses of the British Post Office into British Telecom and the Post Office.

Early in 1982 Mercury, a consortium of Cable and Wireless, BP and Barclays Bank was licensed to run a competing network and began to construct a national network based initially on a figure of eight encompassing London, Bristol, Birmingham, Manchester and Leeds. Initially the Mercury network was built using fibre optic cables laid alongside railway tracks. The Mercury consortium later became a wholly owned subsidiary of Cable and Wireless.

The British Approvals Board for Telecommunications (BABT) was formed in March 1982 to oversee the approvals of apparatus. However, the growth in the market for apparatus developed only slowly because of the time needed to develop appropriate approval standards. Telephones were not liberalised until January 1985.

In June 1982, the Government announced that Racal-Vodafone and Cellnet (Cellnet is a consortium of British Telecom and Securicor) were to be licensed to operate competing analogue cellular radio networks based on the US TACS technology. Key features in the Government's policy were preventing BT from having too large a share holding in this new service, and insisting that all sales should take place through intermediary service providers who resell air time on the network. These measures were intended to stimulate competition and the cellular services enjoyed spectacular growth in traffic until they were affected by the recession at the end of the 1980s.

In October 1982, the first value added network services (VADS) licence was issued, opening a free competitive market in value added services.

In November 1983, a historic policy statement was made in the Commons which committed the Government to continuing its policy of liberalisation but undertook to preserve a duopoly of British Telecom and Mercury until November 1990 and not to allow simple resale of private circuits until July 1989. (Simple resale is the connection to the public switched network of both ends of a private circuit.) At about the same time, the Government announced the award of the first broadband cable licences, although cable operators were to be prevented from offering telephony until at least 1990 unless they did so in conjunction with BT or Mercury.

In April 1984, the Telecommunications Act was passed, creating the post of the Director General of Telecommunications and his organisation, OFTEL, as an independent regulatory body and assigning specific duties to the Secretary of State and the Director General. The Director General's independence is a notable feature. Although he is appointed by the Secretary of State, he is responsible to Parliament and not to the Government, and thus he has a measure of political independence. A major part of the job of OFTEL is the regulation of British Telecom, including the setting of price control formulae.

In August 1984, the first Branch Systems General Licence was issued permitting the operation of private networks connected to the public network. The first licence contained a very complex set of rules designed to prevent economic harm to the public network operators by restricting the routing of traffic within the private network. The purpose was to allow time for British Telecom to rebalance its tariffs by relating them more directly to costs. This licence has been modified several times subsequently to remove all the restrictions relating to the use of private circuits within the UK.

In November 1984, 51% of the shares of British Telecom were sold, thereby removing the company from Government control and enabling it to raise capital on the money markets.

A number of problems had arisen in the development of Mercury's services because Mercury was heavily dependent on British Telecom for the delivery of calls, for international connections and for providing indirect access to the Mercury network. The two companies were unable to reach satisfactory agreements and the matter was referred to OFTEL who issued a determination in October 1985 setting out the basis for interconnection.

In 1988, specialised satellite services were liberalised with six companies being allowed to operate uplinks. Bilateral arrangements were made with Japan and the USA to allow international value added and data services to operate over international leased circuits.

In 1989 four Telepoint licences were issued and a competition for licences to run Personal Communications Networks began. Subsequently PCN licences were promised to Mercury, Unitel and Microtel. The restrictions on simple resale were reviewed and removed permitting private network operators to make whatever use of national private circuits that they wish, including the running of voice and data services. However restrictions on the use of international private circuits for simple resale still remain.

In 1990 and early 1992, there was a major review of the duopoly policy and the development of telecommunications in general. The Government concluded that there was no case for continuing the duopoly policy and that in future it would consider issuing licences at any time to any applicant without running a competition unless scarce resources such as radio spectrum were involved. However, the Government indicated that it did not intend to licence new international carriers immediately and that it would permit simple resale on international private circuits only where the other country concerned had a similarly open market. A number of other measures were announced including plans to introduce standardised interfaces and arrangements for interconnection between competing operators and that OFTEL would become more active in managing the use of numbers and issue numbering conventions to define the use of numbers in the UK.

23.5 EUROPE

23.5.1 Introduction

Regulatory changes began to be considered in Europe in 1983 when the Council of Ministers formed a working group—the Senior Officials Group on Telecommunications (SOGT), chaired by the Commission to study the scope for Community action. In December of that year, the Council of Ministers agreed to pursue the following objectives:

- the creation of a Community market for terminal equipment through the creation of standards and the mutual recognition of testing;

- the improvement of the telecommunications infrastructure through projects of special interest and the development of broadband technology;

- the improvement of services in less favoured regions of the Community;

- coordination of action in international organisations such as CCITT.

In 1986 the STAR programme to improve access to advanced telecommunications services in less favoured regions began. A Council Recommendation (86/659/EEC) was also passed on the coordinated introduction of ISDN within the Community. This Recommendation led to the Memorandum Of Understanding (MOU) between public network operators on the introduction of ISDN.

A Council Directive (86/361/EEC) was passed on the initial stage of the mutual recognition of type approvals. This Directive obliged Member States to recognise and not to repeat approval tests carried out on apparatus in other countries where the tests were carried out against common European technical specifications. CEPT was given the task of preparing these specifications which were called NETs (Normes Européennes de Télécommunications). Because the analogue networks in Europe all differed from each other in various details, the programme for producing NETs focused on new digital networks and services, in particular ISDN and GSM.

In 1987 a Council recommendation (86/371/EEC) was made on the coordinated introduction of GSM, the pan European digital cellular system, and a Council Directive ensured the availability of the necessary frequency bands.

A wide ranging Green Paper on the future of telecommunications was issued in June 1987 arguing the case for greater liberalisation and competition. This was followed in February 1988 by a Commission publication proposing methods of implementing the development of a common market in services and equipment.

In March 1988, the European Telecommunications Standards Institute (ETSI) was created to develop the standards needed for the common European market and in particular the implementation of ISDN. The creation of ETSI was a radical move. Previously the main telecommunications standards and service description had been prepared within the Conference of European PTTs (CEPT) where only PTTs had a right to participate. In contrast ETSI was formed with a much broader base of members consisting of administrations, public network operators, users, manufacturers, private service providers and research bodies. Thus the important preparation of NETs passed from CEPT to ETSI and out of the exclusive control of the public network operators.

The important initiatives begun with the Green Paper developed into three main areas:

- competition in telecommunications services;
- the establishment of the internal market in services through Open Network Provision;
- the full mutual recognition of type approval.

These major activities are now described in more detail.

23.5.2 Competition in telecommunications services

Following the Green Paper there were protracted discussions on the extent to which competition in services should be introduced. Some countries such as the UK argued for competition in all services but others wanted to retain a monopoly in all forms of basic communications. Eventually a compromise was reached which resulted in the Commission Directive (90/388/EEC) under Article 90 of the Treaty of Rome.

This Directive obliges member states to withdraw all special or exclusive rights for the supply of telecommunications services except for voice telephony. Where services are subject to licence, member states have to follow objective, non-discriminatory and transparent procedures in the issue of the licences. As a special concession, member states are allowed to prohibit operators from providing simple packet or circuit switched data services by reselling leased line capacity until the end of 1992.

The second major area covered by the Directive is a requirement ensuring that from 1 July 1991 the granting of licences, the control of type approval and relevant standards, and the allocation of frequencies were put in the hands of a body independent of the telecommunications organisations. Thus separate national regulatory bodies were formed in each member state.

23.5.3 Open Network Provision

Open Network Provision (ONP) is a programme of regulation designed to ensure non-discriminatory access to and use of public telecommunications networks. The concept of ONP arose from concern to facilitate the development of an expanding and competitive market for value-added services, and from consideration of regulatory measures in the USA called Open Network Architecture and Comparably Efficient Interconnection (see Section 23.3.1).

The main concern about value added services is that public network operators are both monopoly providers of basic services and also providers of value added services. Competing value added service providers are dependent on basic services provided by the public network operators and therefore regulation is necessary to ensure that the public network operators do not use their monopoly to put these competing value added service providers at a disadvantage by failing to provide them with the basic services that they need.

The first proposal came in 1988 from the Analysis and Forecasting Group (GAP), which is a sub-group of the Senior Officials Group—Telecommunications, as part of discussions in connection with the Green Paper on telecommunications. A Framework Directive for ONP on the basis of Article 100A of the Treaty of Rome was passed in July 1990 at the same time as the Services Directive.

The ONP Framework Directive defines the areas to be covered by ONP as:

- leased lines;

- packet and circuit switched data services;

- ISDN;

- voice telephony;

- telex;

- mobile services;

- new forms of access such as data over voice and access to intelligent network functions;

- broadband networks.

The Directive addresses the preparation of ONP conditions for open and efficient access to the public networks and directs the Council to prepare specific directives to establish these conditions. The three conditions that may be harmonised are:

- technical interfaces;

- supply and usage conditions;

- tariff principles.

Supply and usage conditions cover provision time, quality of service including transmission, and maintenance. The tariff principles require tariffs to be cost based if the service is the subject of an exclusive or special right, and where possible to be unbundled, leaving the user to choose which elements he requires.

These ONP conditions have to be based on objective criteria, be transparent and published, and guarantee equality of access without discrimination. Access to the public network must not be restricted except for reasons based on the essential requirements of:

- security of network operations;

- maintenance of network integrity;

- interoperability of services in justified cases;

- protection of data, as appropriate.

It should be noted that these requirements are not exactly the same as the essential requirements in the Directive on type approvals (see Section 23.5.4). Thus the focus is on establishing a harmonised set of conditions for the use of ONP services.

The Framework Directive contains guidelines for implementation in the period up to the end of 1992. These guidelines propose:

- specific Directives for leased lines and voice telephony;

- implementation by January 1991 of harmonised technical interfaces and features for ISDN and packet switched data services;

- adoption by the Council by 1 July 1991 of a recommendation on the application of ONP to packet data services;

- adoption by 1 January 1992 of a similar recommendation on ISDN;

- examination of specific directives on packet switched data and ISDN to follow the recommendations.

The ONP programme has slipped somewhat but a directive on leased lines, and recommendations on ISDN and public packet switched data services were issued in 1992.

23.5.4 The full mutual recognition of type approval

The first stage of developing a European approvals system, to replace national approvals, was based on Directive 86/361/EEC passed on 24 July 1986. This directive requires Member States to recognise conformance tests by approved laboratories against common conformity specifications (NETs).

The second stage is based on Directive 91/263/EEC passed on 29 April 1991, which is a Council directive on the approximation of the laws of the Member States concerning telecommunications terminal equipment, including the mutual recognition of their conformity. This Directive requires Member States to implement a European-wide approval scheme with the same requirements being applied in all countries and terminal equipment that has been approved in one country having a right to connection in other countries without any further approval procedure. The standards to be used under this directive are called Common Technical Regulations (CTRs).

Directive 86/361/EEC was repealed by Directive 91/263/EEC with effect from 6 November 1992.

Under the second directive, terminal equipment that is intended for connection to the public telecommunications network has to satisfy the following essential requirements:

(a) user safety, in so far as this requirement is not covered by Directive 73/23/EEC;

(b) safety of employees of public network operators, in so far as this requirement is not covered by Directive 73/23/EEC;

(c) electromagnetic compatibility requirements in so far as they are specific to terminal equipment;

(d) protection of the public telecommunications network from harm;

(e) effective use of the radio frequency spectrum, where appropriate;

(f) interworking of terminal equipment with public telecommunications network equipment for the purpose of establishing, modifying, charging for, holding and clearing real or virtual connection; and

(g) interworking of terminal equipment via the public telecommunications network, in justified cases.

Under (g) a justified case service is a service which is either a reserved service under Community law or a service for which the Council has decided there should be Community wide availability. At present the only justified case service is voice telephony. It possible that there could be other justified case services created in future such as fax or packet switching but some countries such as the UK would be expected to oppose such moves.

National standards referred to in the Official Journal of the European Communities are used for requirements (a) and (b). Requirements (c) to (g) are to be set out in Common Technical Regulations (CTRs), and member states are directed not to impede the placing on the market of terminal equipment that conforms with the provisions of the directive.

There are three alternative procedures for complying with the directive:

• EC Type Examination and Conformity to Type, which relies on third party testing under control of the notified body for both design and production;

• EC Type Examination and Production Quality Assurance, which relies on third party testing for design but the manufacturer's declaration for production:

• full Quality Assurance, which relies on the manufacturer's quality assurance for both the design and production of the product.

23.6 DISCUSSION OF SPECIFIC ISSUES

23.6.1 Competition vs tariff regulation

The existence of a monopoly or dominant operator creates the need for some form of financial control to prevent that operator from making excessive profits to the detriment of customers. In the past, various measures have been used to adjust the profit level to a reasonable figure. This issue was the subject of considerable discussion within the UK with the conclusion that price controls would be introduced using the formula RPI-X, where RPI is the Retail Price Index, i.e. the change in the cost of living, and X is a factor determined by the regulator. This approach was chosen in order to give British Telecom maximum incentive to reduce its costs because such reductions would produce increased profits. Other approaches such as taxing excess profits were considered less desirable because they do not create the same incentive to reduce costs. However, in order to allow the incentives to work correctly, it is important that the formula is not reviewed too frequently.

The control of prices in a competitive regime, creates a terrible dilemma for the regulator. On the one hand he is obliged in the interests of the user to exercise a tight control on the prices of the dominant carrier, but on the other hand, if he wants to attract new entrants into the market and allow their businesses to grow profitable to fund further investments, he needs to allow the basic level of profitability to remain attractive to new investors.

23.6.2 Distinctions in service types

In both the USA and in Europe there have been distinctions in services types between service areas where there is no competition or restricted market entry, and areas where there is free competition. This distinction was drawn initially in the USA between data communications and data processing and subsequently changed to basic and enhanced services. In the UK the distinction initially was between value added and non-value added services with free competition in the value added area, with data services being added subsequently to the competitive area. Within Europe the distinction is built around the concept of a reserved service (i.e. voice telephony).

All the experience to date is that these distinctions are very difficult to define and apply in the complex world of digital telecommunications. Distinctions based on traffic type such as voice or data seem to be fundamentally flawed because networks mainly provide bearer services with the customer controlling whether he sends voice or data.

Distinctions based on the provision of physical transmission infrastructure seem to be the only durable ones.

23.6.3 Rebalancing of tariffs

Telecommunications tariffs traditionally have not been properly related to costs. International tariffs and trunk tariffs are traditionally high and provide a cross subsidy for local calls. Also exchange line rentals are low and subsidised by call charges although there are profound arguments as to whether the price relationship between exchange lines and call charges can be other than arbitrary.

When there is a single monopoly operator, the allocation of costs between different services is not of great importance overall, but when competition is introduced it is important that tariffs should be allowed to reflect costs in order to prevent artificially high profits which attract new competition in certain areas and result in an inefficiently large number of operators.

The general need for cost based tariffs raises difficult and sensitive political issues about the averaging of prices for example between rural and urban areas and between higher volume and lower volume users, as well as the basic provision of any service within very sparsely populated areas.

23.6.4 Privatisation

The privatisation, or transfer from state to private ownership, of the dominant carrier is a separate issue from the introduction of competition. The main attractions of privatisation are releasing the dominant operator from political interference and the slow decision making processes that almost inevitably accompany government control, and permitting him to make his own long-term plans on a commercial basis using commercially raised capital. A short-term attraction is the revenue from the sale of the shares.

23.6.5 Interconnection

When competition is introduced, it is necessary for the competing operators to interconnect their networks in order to provide a full range of connection possibilities to their customers. Where there are a number of operators of comparable size, interconnection is unlikely to be a major problem because it is in everyone's interests to resolve any problems as quickly as possible. However, when a competitor is being introduced into what was previously a monopoly market and has few direct connections to customers himself, the dominant carrier has every incentive to delay the establishment of connections. This situation can result in the need for detailed intervention by the regulator who can easily be drawn into an increasingly complex and technical dispute. The biggest danger to the new competitor is delay which prevents traffic and revenue growth and can cause dissatisfaction to his existing customers.

23.6.6 Quality of service

The quality, as well as the price, of the service provided by the public network operator is a major issue that is receiving increasing attention as the expectations of the customer rise. Quality of service covers a wide range of issues including:

- time to install exchange lines;
- frequency of faults;
- time to repair faults;
- probability of congestion;
- probability of calls being dropped;
- availability of working payphones;
- transmission quality;
- accuracy of billing; and
- provision of itemised bills.

Historically there was little published information on such parameters, but regulators have used their powers to require such information as a means of putting more pressure on the public network operators and effecting an improvement in services. When there is competition, the ability to publish comparative figures for the performance of the competing operators is a means of stimulating competition.

23.6.7 Apparatus supply

The liberalisation of the market for telecommunications apparatus has required the establishment of an approval system based on published technical standards, the clear delineation of the boundary of the public network with an arrangement such as a plug and socket which allows apparatus to be connected and disconnected, and a better definition of the technical characteristics of the service provided by the public network.

Thus the liberalisation of apparatus supply has led to a substantial increase in the writing of standards. In Europe, this effect has been enhanced further by the objective of harmonising the requirements in different countries.

One of the main issues is the extent to which the performance of apparatus should be subject to regulation. Most countries insist that the approvals system should ensure that apparatus is safe and does not harm the public network physically, but there are many difficult issues about indirect forms of harm such as bells that are too soft and cannot be heard and so lead to additional unsuccessful call attempts, and the extent to which apparatus performance should be included in approvals.

Whilst it is clearly necessary to have published standards and formal open approval processes, there has been a slight loss of flexibility in that it is less easy for a public network operator to tailor a service to the special needs of a particular customer because of the formal approval process.

23.6.8 Private networks

The process of liberalisation and the increasing availability of private circuits has created many new opportunities for companies to develop their own private networks. This has caused considerable discussion about the extent to which the quality of calls, particularly telephony calls, that pass to or from the public network should be regulated. The main technical issues for telephony are loudness ratings, delay, echo, and quantising distortion.

In the USA such issues have not been regulated. In the UK, OFTEL intended initially to introduce regulation on private network operators but later decided that such operators were themselves best placed to take these decisions taking into account any factors specific to their situation. There has been no noticeable degradation in quality as a result of this decision.

23.6.9 International communications

The structure, standards and regulations of telecommunications have been based on the division of the globe into separate nations. However, telecommunications is reducing the world to something of a global village and services and networks are becoming more international. Therefore if one looks to the future it seems increasingly likely that the national divisions and country-based approaches to numbering, transmission planning and tariffs will have to be replaced by a global approach that takes better account of the supra-national nature of many areas of telecommunications.

23.7 CONCLUSION

It is important to assess the benefits of competition and of changes in the approach to regulation, although it is far from easy to make rigorous conclusions that clearly relate causes and effects because of the enormous technical changes taking place in telecommunications and the interrelationships between regulation, competition and privatisation. Consequently it is hard to avoid a discussion about whether or not things are better than they would otherwise have been.

Within the UK there have been some substantial improvements in quality of service in areas such as;

- time to install exchange lines;

- time to repair faults;

- availability of working payphones;

- provision of partially itemised bills.

These have resulted from a combination of regulation and competition. However, there is still no obligation to provide an exchange line within a specified period, nor obligation to provide fully itemised bills even where there are digital local exchanges.

An area of considerable difficulty has been the quality of the cellular networks which became overloaded at the end of the 1980s resulting in some areas in unsatisfactorily high probabilities of failed call attempts and dropped calls, but this problem has largely been overcome by further expansion of the network infrastructure and a reduction in traffic as a result of the recession. Although the publication of comparative statistics on the performance of the competing cellular networks was considered, such statistics have not been published by the time of writing (early 1992).

Comparison of prices for a representative basket of services is a useful indicator of the effectiveness of telecommunications policies in different countries, although it is difficult to eliminate major factors that are peculiar to individual countries. Recent figures published by OFTEL indicate, surprisingly, that the UK residential and business service baskets are significantly more expensive than in France, where there has been no competition in public networks and a less competitive apparatus market, broadly equal to prices in Germany, and cheaper than Italy, especially for business. Prices in Europe for both apparatus and services tend to be higher than those in the USA.

It is difficult to make a fair assessment of these statistics. Most users in the UK consider that the current regime is much to be preferred to the old monopoly of the 1970s, and doubt whether BT would ever have undertaken major cost cutting exercises involving large scale redundancies had they not been subject to competitive and regulatory pressure on prices. Perhaps we must wait a few more years to make a better assessment.

23.8 BIBLIOGRAPHY

OFTEL *Updates*. Published frequently by the Office of Telecommunications, Export House, 50 Ludgate Hill, London, EC4M 7JJ.
US Code of Federal Regulations, title 47. Published annually.

24 TARIFF PRINCIPLES

24.1 INTRODUCTION

Charges for telecommunication services generally consist of three parts:

1. a connection charge;
2. a rental charge based on the cost of transmission plant terminal equipment, etc. that is rented rather than owned; and
3. a traffic or usage charge.

What matters most to the end-users is the cost of ownership of their telecommunication service and the tariff categories listed above are only a part, albeit a very significant part, of the total. Here we are not directly concerned with internal costs such as depreciation of telecommunication plant, switchboard operators, maintenance and communications management. Nor are we concerned with what any of the tariff charges are in pounds or ECUs or dollars but rather with their basis and with how charges are apportioned between the various parties involved.

First it is necessary perhaps to make the distinction between service and network providers as far as charges are concerned. A service provider provides a service such as cellular radio or packet data. A network provider provides a network over which a service or services runs, e.g. the PSTN. Of course in many cases, e.g. POTS, the service and network provider is the same organisation (or administration) although there may be separate profit centres within the one body. In any event the service provider is regarded as providing the billing interface and the network costs are a matter for negotiation between the network provider and the service provider. For connection and rental charges the billing interface is logically with the network provider, but the bill will still probably come from a service provider on a basis which is simple if there is only one service provider using the network connection but becomes complicated when more than one service provider uses the same network.

24.2 BILLING PATHS AND INTERNATIONAL TARIFF RECOMMENDATIONS

Figure 24.1 shows billing paths for a call. Customers interface with service providers at their end of the connection either directly or through a billing agent (the

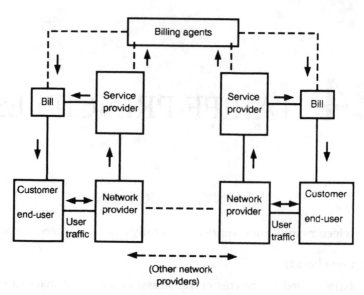

Fig. 24.1 Billing paths

dotted path in the figure). Any number of network providers can exist between the two end network providers and the negotiation of tariffs between these parties when they are within one country is the subject of commercial agreements that may be subject to national regulations but which currently are not covered by any general principles.

For service providers in different countries the CCITT D. series of recommendations give guidance that most countries adhere to. These recommendations include the following: D.1 to D.9 applying to leased circuits, D.10 to D.35 to data services, D.90 to D.93 to mobile services, D100 to D.155 to telephone service, and D.210 to D.251 to ISDN. The basic principles involved in all cases are:

- Per call charges are based on the utilisation of the international service rather than any charges for access to that service that are independent of utilisation.

- The basis on which charges are made should, as far as equipment constraints allow, be the same whichever end initiates a call (e.g. the number of meter pulses should be the same or the number of bytes of data that constitute a charge step should be the same).

- Within the context of a particular service, charges should not depend on the nature of the information conveyed.

- Generally the calling party administration or service provider is responsible for the total bill.

- Charges should be independent of the route used to connect a call.

What the recommendations do not say is how revenue is to be divided between the the two end administrations or service providers and any intermediate network providers. This appears to a matter for negotiation between the parties concerned. Although the basis for charging at each end of a connection is the same it does not follow that the charges will be the same; this of depends on exchange rates and many other factors.

24.3 THE BASIS FOR TRAFFIC CHARGES AND PAYING FOR CALLS

Some of the factors influencing the traffic-dependent aspects of call charging are given in Table 24.1. The trends are:

- For all services charges are becoming less dependent on distance, particularly when both ends of a connection are on the same network.

- For data and multiservice networks charges relate more directly to the amount of information conveyed rather than call duration. But on the other hand,

- for services with a high added value, e.g. information services, the call charge metering interval is becoming shorter and with stored program digital exchanges charges are based on the exact call time rather than fixed metering intervals.

- For telephony services there is a move towards charging for ringing with the caller paying, although it can be argued that if the called end is a business which is under-equipped or under-manned it should pay.

The fact that for some data services the called party has something to pay for unsolicited and unwanted material is increasing the pressure on regulators and the specifiers of facilities to provide means of preventing junk mail.

Table 24.1 Factors affecting traffic charges

Charge basis	Payer	Service	
		Telephony	Data
Fixed per call	S	0	2
Distance/route	S	3	1
No of service providers	S	3	3
Time of day/day of week	S	3	3
Duration	S	3	3
Quantity of information	S	0	2
Delivery	R	0	2
Storage	R	0	3
Call set-up	S	2	1
Ringing	S?	2	0
Multiple delivery	S	0	3

S = sender: R = recipient
0 = does not apply or very unlikely
1 = unlikely
2 = possible
3 = very probable

The widespread provision of freephone services means that increasingly networks are providing what amounts to an automatic reverse call charge facility.

Phone cards, telecommunication specific charge cards and general purpose credit cards, rather than cash, are becoming the accepted means of paying for calls from public kiosks and privately owned or privately rented payphones. The use of charge or credit cards where the customer is billed subsequently, rather than phone cards where the card is debited locally, gives rise to authentification problems and means that the checking facilities of an intelligent network (see Section 18.3) may be usefully invoked.

24.4   TARIFF SETTING BY NETWORK AND SERVICE PROVIDERS

When all telecommunication services were provided by one organisation there was no requirement for that organisation to apportion charges equitably between the services it provided or between different classes of customer within one service. In the UK, for example, it was long the practice for business telephony customers to subsidise residential ones, for international calls to subsidise national calls and for trunk calls to subsidise local calls. With deregulation and competition there is pressure for network and service providers to ensure that their accounts are kept in sufficient detail for services to be unbundled and, for example, for the balance between per-call charges and exchange line rental to reflect more precisely the true cost of each. As any one who has been involved in any detail of a cost apportionment exercise can tell you, this is far from being a precise science, and in particular, software, which represents an increasing proportion of the total system cost, may, because it can be common to a number of services and customer classes, be very difficult to treat in a non-arbitrary fashion. Nevertheless costs are being unbundled and in the UK and the USA, at least, charges reflect costs more closely than they used to do.

Some administrations are being heavily criticised for the way in which international call charges are used to subsidise national ones or, when that is not the case, for the very large profits that are made from international calls. This pressure is likely to lead to very significant reductions in international call charges during the 1990s. In 1992 a CCITT working party agreed in principle to bring international call charges into line with costs. Subject to ratification of this proposal there will be a five-year timetable during which developed countries will carry out the appropriate adjustments. Third world countries are to be allowed a longer period to make their changes.

Competition means rival networks and from the customer's point of view this is less than satisfactory if the networks are not interconnected with each other. The national regulator has a role to play here if the parties cannot agree on terms for interconnection. In the UK for example OFTEL had to make a determination when MCL was formed and needed to use BT's network to provide access to its own network nationally. The FCC has been involved similarly with trunk access problems in the USA. With increasing competition and network variety we anticipate a growth of activity under ONP in Europe and ONA in the USA with respect to both the technical and financial implications of network interconnection. (See further in Section 23.6.5.)

24.5 MINIMISING COSTS

Company communication management is concerned to minimise telecommunication costs and there are two main aspects of this:

1. choosing between PTOs as carriers, and
2. obtaining the best balance between leased and public switched network circuits for conveyance.

The principles involved in making judgements of this sort are basically accounting ones and are relatively straightforward; nevertheless personal experience indicates that the work is tedious, prone to error and needs frequent updating as costs and tariffs change. For information on and comparisons between tariffs in the UK, the Octagon Group's series of *Guides to Telecommunication Tariffs* are recommended (Octagon Group, 1988 onwards).

When a private network is a complex one, then a computer modelling program is a very useful tool and one that can be used both in the design and maintenance of networks. There are several proprietary products on sale in the UK and doubtless in other countries as well. While the principles are the same in all cases, every country needs its own tariff tables, maps (or the equivalent) of public network configurations, etc., and program vendors will generally provide a service for periodically updating this information. With such aids, maps of private networks can be constructed and costs against traffic levels, time of day, etc. can be calculated for various routes with, for example, the balance of traffic on the public switched network and leased circuits as a parameter. In the UK resale of private network capacity is allowed and gives another dimension to the equation.

The private operator, both for his own use and for resale, can, by limiting in geographical terms the sources and destinations that he is prepared to serve, use bandwidth compression techniques (e.g. ADPCM) to increase circuit capacity, whereas the public operator, because of universal service obligations, is unlikely to be able to compete in this way. Some of the available software design tools also allow the transmission performance of a network to be checked against regulatory rules or guidelines. (In the UK against the *Oftel Network Code of Practice*.)

24.6 CONCLUSIONS

The increasing number of service providers and telecommunication services means that the task of large users in minimising their costs is becoming increasingly complex and time consuming. The provision of tools to aid in this process is therefore opportune and there is scope for a considerable expansion in this type of activity.

24.7 BIBLIOGRAPHY

Octagon group, (1988 onwards) *Guides to Telecommunication Tariffs*, 8 Crinan Street, London N1 9SQ.

25 NUMBERING, NAMING AND ADDRESSING

25.1 INTRODUCTION

With the rapid development of telecommunications and the introduction of liberalisation and competition, the importance of numbering is beginning to be more fully recognised. Numbering is very important for three reasons. First the availability of numbers and the flexibility of numbering arrangements can have significant effects on competition between network operators. Secondly numbers can have commercial or social value and the economic and social consequences of a change in number can be considerable. Thirdly the structure and information content of numbers is a factor in the overall utility of telecommunication services.

Most people think of their telephone directory number as something which relates to them as a person, but in the case of the fixed networks it relates directly or indirectly to their exchange line, and in the case of mobile networks (e.g. cellular radio) it relates to the mobile terminal equipment. (*Note*: in the case of Strowger exchanges, for example, the directory number relates directly to the exchange line; in the case of crossbar and electronic exchanges there is normally a translation in the local exchange between the directory number and an equipment number. The equipment number bears a fixed relationship to the exchange line but is not known to the subscriber.)

Numbers may contain various information, e.g. on the geographical location of a user or on the network to which a user is connected. In the past there has been little flexibility with regard to this information because of the very limited intelligence of the number translating facilities of electromechanical exchanges, but with digital stored programme exchanges, there is much greater scope for providing numbering arrangements that are better suited to the customer's requirements. It is also becoming possible to provide arrangements that approximate to personal numbering, i.e. numbering where the customer can be reached on the same number even when at different locations or when using different services.

In this chapter, we will explore in more depth some of the fundamental characteristics of numbering and discuss the development of portable and personal numbering. The main emphasis is on telephone numbers, although data numbering and naming is also covered, but the wider topic of using alphanumeric addressing schemes in the context of data is covered in Chapter 21 on directories.

25.2 NUMBERING, ROUTING AND ADDRESSING

From a theoretical point of view a distinction should be drawn between numbering (or naming), addressing and routing. These distinctions have been drawn most clearly within the world of OSI which has considered the following questions:

- Who is the end user (person or process)?—end-user identification.

- Where is the user?—end-user address.

- How does the network route a message to the user?—routing information.

An end-user will have a single identification. This identification will be independent of the location of the user and the services used by the user. Thus it will not need to change but it may cease if the user ceases to exist. Thus ultimately there will be provision for only a single end-user identification on business cards.

Where the end-user is, is a different question. The address may be defined by reference to geography or networks. If a user is connected to several different types of network, there may be several different addresses (e.g. one for telephony, one for fax, one for data). The addresses may change due to a move. Most current numbering is addressing.

The fundamental role of a Directory Service, in the OSI sense, i.e. X.500 (see Chapter 21), is to provide a translation between a user identification (distinguished name) and the user's address (attribute type of particular value) for the service (attribute type of particular value) selected. However, the address does not tell the network how to reach the user.

To reach the user, routing information is needed. Routing information is conceptually separate from the address, and in principle a data base translation is needed to obtain routing information for a given address. Thus conceptually there is a three-layer model with data base (or directory) translations between each layer (see Figure 25.1).

Real communications systems are not implemented fully in conformance with these concepts. No true user identifications are in use, and X.500 Directory Services are not yet in full commercial operation, although some pilot projects have

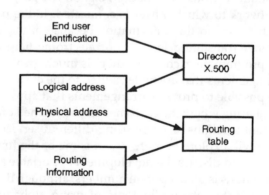

Fig. 25.1 Relationship between identification, addressing and routing information

commenced. Consequently current numbering is really addressing. When, in the distant future, these theoretical dreams reach fulfilment, users will not need to see addresses, which will be used only internally within networks, but in the meantime the main problems and issues are concerned with addressing.

Addresses are frequently structured to contain routing information. Addresses may be physical, logical or a mixture of both. The term physical address is used when routing information is explicit within the address, and the term logical address is used when routing information is not explicit and a data base interrogation is needed to find a physical address. For example, a number may begin with a country code and network identifier: this is physical information. The remainder of the number may take a logical or a physical form, or a combination of both. In a logical form, the number would give no explicit routing, and a data base would have to be interrogated to find out where to route the call. In a physical form, the middle part of the number would identify the node to which the terminal is attached, and the remainder of the number would identify the terminal.

Many services such as the PSTN use a combination of physical and logical addresses with physical addresses for nodes, and logical addresses for terminals or exchange lines served by nodes. (In cellular radio systems, a mobile number is a logical address that is translated by a location register into a physical address (the mobile station roaming number).)

As mentioned above, almost all current numbering and naming is really addressing, and the main issues are concerned with the division between the logical and physical parts of the address. The majority of the issues result from the conflicting objectives of giving the user the features that he requires such as portability and information on the tariff (e.g. freephone or premium rate) or service type (e.g. cellular) and the cost of implementing the numbering scheme, which is mainly concerned with implementing and maintaining the necessary data bases.

25.3 NUMBERING STANDARDS

The CCITT Recommendations define an international framework for numbering which ensures that the numbering arrangements of countries are compatible. Within this section we describe the main features of the Recommendations for telephony, data and ISDN.

Telephony numbering: CCITT Recommendation E.163

E.163 limits the length of an international telephone number to 12 digits and assigns country codes of 1–3 digits to individual countries, see Figure 25.2. The international number consists of the country code and the national significant number; it does not include any international dialling prefix. E.163 recommends the use of the international dialling prefix 00, although some countries still use other prefixes, e.g. 010 in UK. The national significant number does not include the trunk prefix which is to be defined nationally, but a single digit, preferably 0, is recommended.

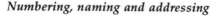

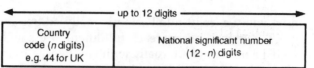

Fig. 25.2 International PSTN numbering (E.163)

Data network numbering: CCITT Recommendation X.121

X.121 limits the length of data network numbers to 14 digits and assigns country code values of 1–3 digits to individual countries (see Figure 25.3). These data country codes are different from the telephony country codes referred to above, and some countries have more than one data country code.

A very important difference between E.163 and X.121 is that the main structure of E.163 is based on a geographical division with country codes and area codes (with the implicit assumption of a single network within an area), whereas that of X.121 is based in practice on network identifiers called Data Network Identification Codes (DNICs) which consist of the first four digits in the number. These digits include the data country code and they are used for routing calls between networks. The data country code is not used for routing.

ISDN numbering: CCITT Recommendations E.164, E.165 and E.166

E.164 defines the extension and modification of E.163 for the ISDN era (see Figure 25.4). It uses the same country codes but extends the number length to a maximum of 15 digits, which must be capable of use from time *T*. Recommendation E.165 defines Time *T* as 2359 hours on 31 December 1996. The main modification necessary for networks will be extending the capacity of numbering registers.

E.164 also introduces several other concepts. The national significant number is divided into a national destination code and a subscriber's number. The national destination code may itself be divided into a destination network code and a trunk

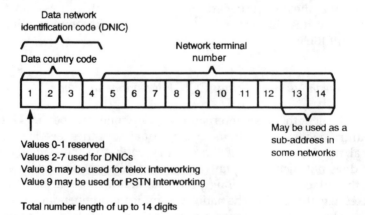

Fig. 25.3 International data numbering (X.121)

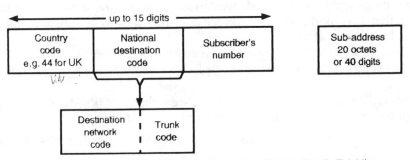

Fig. 25.4 International ISDN numbering (E.164, E.165, E.166)

code; the order and form of these codes are national matters. In view of these changes, E.164 recommends that after time T the originating network should be capable of analysing at least six digits (including the country code) for routing and charging purposes, whereas E.163 recommends only four or five digits. E.164 also introduces the use of a separate sub-address of up to 20 octets.

CCITT has recently decided to allocate only three digit country codes in future and to change existing country codes to three digits from time T.

E.166 (which from 1992 will be identical to X.122) defines numbering plan interworking in the ISDN era, or how a caller on one numbering plan, say E.164, will call a subscriber on X.121. E.166 outlines three solutions.

First a two-stage call set-up method can be used, whereby the caller first signals the number (in his own numbering plan) of an interworking function and then subsequently signals the number of the destination in the second numbering plan. This solution can be implemented now by means of a gateway.

The second solution provides for a single-stage call set-up by using escape codes digits at the beginning of the number that indicate the destination numbering plan and are followed by the number in the destination numbering plan. This second solution is planned for use until time T.

The third and long-term solution also planned for after time T provides single-stage call set-up by use of Numbering Plan Indicator (NPI) and Type Of Number (TON), e.g. local, trunk or international, fields in the call set-up packets in the CCITT No. 7 signalling system.

Message handling services, X.400, X.402

The naming arrangements for message handling services are defined in X.400 and X.402. Users of message handling systems are identified by Originator Recipient (O/R) names which are used to identify the senders and receivers of messages. An O/R name consists of a directory name (an end user identification for use with an X.500 Directory) or an O/R address or both. The reason for including the possibility of an O/R address is that not all users will have directory names during the early years of message handling services. However, because the directory name is more user friendly than the O/R address, and, because it is independent of the structure of the message handling system, the directory name is expected to become the primary means of identification in the future. If the O/R address is present, then it is used and the directory does not need to be consulted unless the O/R address is found to be invalid.

The O/R address tells the message transfer service where the user is. The O/R address is composed of a number of attributes and may take four different forms:

- mnemonic = country name + ADMD name + choice of (PRMD name, organisation name, organisational unit name, personal name);

- numeric = country name + ADMD name + numeric identifier + PRMD name (if required);

- postal = country name + ADMD name + postal address (formatted or unformatted);

- terminal = country name + ADMD name + network address (e.g. X.121 number).

The country name is a 2 character printable string or a 3 digit numeric string. The ADministration Management Domain (ADMD) and PRivate Management Domain (PRMD) names are 16 character or digit printable or numeric strings.

X.402 defines which attributes have to be included within an O/R address. The four types all require the country name and ADMD name to identify the ADMD, and at least one unique attribute, or combination of attributes, within the ADMD to identify the originator or recipient. The postal form of address is used when the message is to be delivered by a postal delivery service. The terminal form relies on the network address, i.e. an X.121 or E.164 number.

In the case of the mnemonic address form, the possible attributes, known as conditional attributes, are:

- PRivate Management Domain (PRMD) name;

- organisation name;

- organisational unit name; and

- personal name.

The exact requirements for these conditional attributes are not specified in the international standards and therefore need to be defined nationally or by the ADMD concerned. In network terms, the PRMD is the name of the private X.400 network that consists of one or more message transfer agents. The cost of implementing a message transfer agent is low and so most commercial users are likely to have their own PRMD. Thus the PRMD name equates normally to the company name. The organisation name and organisational unit name would normally be used for entities within a company.

25.4 USER REQUIREMENTS FOR NUMBERING

The following information is based on studies carried out by Ovum Ltd for the UK regulatory body, OFTEL, and for the European Commission. Users require:

- more consistent numbering and dialling schemes in different countries, including standard number lengths and presentations so that they can be sure that a number is complete (many business users would favour full national number dialling instead of separate trunk and local dialling);

- clear distinctions between geographic services, where the first part of the number identifies a particular geographical area, and individual non-geographic services (e.g. mobile, freephone, premium rate);

- harmonisation of prefixes for international and trunk calls, short numbers such as emergency services and directory enquiries;

- independent administration of numbering schemes and the facilitation of competition between different operators;

- portability of numbers between operators of non-geographic services;

- portability of numbers within a local area for numbers for geographic services, and, where competition exists, portability between operators within the local area;

- international (not country-based) numbering arrangements for services such as freephone.

25.5 IMPORTANT ISSUES FOR THE FUTURE

25.5.1 Competition in voice networks

For regulatory purposes a distinction is often made between voice and data services, but most networks are transparent to the information that they carry provided that it is present in the correct format. In practice telephony networks often carry data (e.g. fax) but data networks are usually confined to data alone.

Competition in voice networks has existed in the USA for trunk and international services, but not for local services, for over a decade. Competition was introduced in the UK in the early 1980s and is being introduced in Japan, Sweden, Australia, and New Zealand.

The introduction of competition creates some important numbering issues, including:

- The need for independent administration or control of numbering allocations to ensure that numbering is not used in an anti-competitive manner.

- The need to introduce methods by which subscribers can select particular trunk or international carriers. When there is no local competition and the local operator is independent of the trunk and international operators, selection can be left to the local operator, but when this is not the case, the subscriber will need either to instruct the local operator permanently about his choice of carrier, or to indicate his selection on a call-by-call basis using a prefix. This issue is further complicated by access security if the subscriber is billed by networks to which he is not connected directly: either a secret access code has to be used to prevent fraudulent access, or the local network has to pass on the calling line identity of the caller. Questions of equal or fair access arise if the numbering arrangements favour any particular trunk or international operator.

- The need to provide portability of numbers between competing operators in a local area so that subscribers can change from one operator to another without having to change their number. This portability requires the competing local operators to have access to common data bases for routing. If this portability is not provided, subscribers who wish to change operator but need to keep their existing numbers have to retain lines from the former operator for the delivery of incoming calls.

25.5.2 Competition in data networks

The competition issues for data networks are slightly different from those of voice networks because the numbering or naming is structured more in terms of the identity of the service provider rather than the location of the subscriber. Portability is again a major issue, but the solutions are different. Portability between service operators in X.121 could be provided only by allocating DNICs to subscriber networks but there are not enough DNICs to do this. In message handling services a 'space' convention has been introduced whereby the ADMD name is replaced by a single space in countries that operate the convention. To date only the UK has decided to operate the space convention and in a manner that is defined in BS 7521. (A number of countries consider it unworkable until problems relating to billing and non-delivery messages are solved.) Because the space convention absolves users from having to specify the ADMD, their records do not have to be changed when the ADMD is changed and this gives flexibility in making temporary or permanent changes to the service provider.

25.5.3 Information content

Numbers frequently carry implicit information about the geographical location of the number, the service to which the number applies, e.g. fixed or mobile, and the identity of the service provider.

Opinions are divided on the value of the information in a number. Some customers value the location information especially as it normally enables them to obtain some idea of the tariff level which will apply to a call. Also being able to find out where one is calling is of use when answering advertisements. Conversely some businesses would prefer to have numbers which do not contain location information because they may wish to offer services using the same number even if they move their location. Equally they may wish not to reveal their location.

The use of a number to identify a particular type of service, e.g. a mobile service or a premium service, is normally considered to be valuable because it indirectly provides information about the tariff, and subscribers can learn which numbers apply to services which carry particularly high tariffs.

The information about the provider of the service to which the number applies is generally of less value and can be a positive disadvantage if it means that a subscriber has to change his number if he wishes to change the supplier of his service. Such a restriction can greatly hinder the development of competition.

There has also been some interest in the possibility of providing tariff information within a number. Tariff information may be provided indirectly through the addressing and service information, which has been discussed above, but there seems to be little point and indeed many difficulties in providing direct tariff information because the numbers would then be dependent on the tariffs and would have to change when tariffs changed.

In this section we have been considering information which might be carried in a number in a way which can readily be extracted by a human. An alternative possibility is the provision of a data base or directory service which could be interrogated with a number and give the current tariff for calling that number and perhaps the location of the number.

25.5.4 Personal numbering and universal personal telephony

The types of numbering which we have considered so far have been mainly concerned with identifying exchange lines, although it is common practice to associate a directory number with a person. A radical new approach is to consider numbering people. This would mean that all the members of a household would have their own number instead of sharing the exchange line number.

To route a call to a person, the network would have to route the call to the equipment nearest to the person, whether it be mobile equipment such as a personal handset or fixed equipment such as a conventional telephone. To do this the network would have to keep a continually updated record of the location of the person so that it could translate the personal number to the appropriate mobile equipment number or exchange line number. The updating could be provided by messages sent by the person himself or by an automatic tracking system involving equipment attached to the person.

Mobile networks such as cellular radio maintain updated records of the location of mobile equipment and so the directory number of the mobile equipment approximates to a personal number. If cellular networks had the capability to translate between a personal number and a mobile equipment number, as GSM will, in case a user changes the mobile equipment which he is using, then that arrangement would give the closest possible approximation to true personal numbering.

Within fixed networks call diversion facilities can provide some of the benefits of personal numbering because calls can be diverted to a person's current location. However, call diversion normally diverts one exchange line to another (it does not translate a personal number to an exchange line number) and so all users of the exchange line that is diverted are affected simultaneously.

Both CCITT and ETSI (NA7) are studying the development of a 'personal' numbering service known as Universal Personal Telephony (UPT). Customers will be allocated their own personal UPT number. Wherever the subscriber travels within the area in which UPT is provided, he can register his presence with a fixed or mobile terminal and have incoming calls directed to that terminal. He can also make outgoing calls and have these calls charged to his UPT account. UPT is described in more detail in Chapter 20. CCITT is preparing a Recommendation E.168 on numbering for UPT. This recommendation specifies that the UPT number should conform to E.164 and outline three scenarios: local, national and international allocation and registration of the subscriber's number.

25.5.5 Calling line identity

ISDN offers the supplementary service of providing to the called party, the number (Calling Line Identity, CLI) of the caller. This information is potentially of considerable use, especially in a business context, where the number can be used to trigger the display of information about the caller, e.g. the state of his account, in order to facilitate the handling of queries or the placing of orders. However, the availability of CLI is of major concern as a civil rights issue and is likely to be the subject of privacy or data protection legislation in many countries. There is a growing consensus that callers should be able to suppress the presentation of CLI.

25.5.6 Administration of numbering schemes

Because of the development of competition, the administration of numbering schemes has become a major issue. In the past, numbering schemes have been administered by the monopoly PTT, but where competition is introduced there is a need to ensure that numbering schemes are structured to facilitate competition and that the allocation of numbers is fair and unbiased.

This issue has been considered in depth in the UK and is currently under detailed consideration within the European Commission. In the UK, the overall responsibility for numbering has been transferred from BT to OFTEL and OFTEL is preparing a set of rules called a numbering convention that operators will have to follow. By establishing a clear set of rules, a regulatory body can avoid becoming involved in day-to-day allocations which can be delegated to individual operators.

25.5.7 Regional and global numbering schemes

The European Commission is studying the case for proposing a pan-European numbering scheme that would use a new country code for Europe as a whole. This would be an optional alternative to national country codes. Such a scheme could be attractive for organisations or services which are pan-European and require numbers that are independent of any particular country. A prime example is a pan-European freephone service. However, although the concept is superficially attractive, the demand for and prospective benefits of a pan European scheme do not appear to be as great as might be expected, and where there is a clear demand, the real requirement is for a global rather than a European scheme, although a European scheme would be a good stepping stone towards a global scheme.

25.6 NUMBERING AND SWITCH TECHNOLOGY

In electromechanical Strowger exchanges the numbering arrangements were an integral part of the design and wiring of the exchange, and therefore there was little flexibility. In a Strowger local exchange the local numbers contained all the initial routing information. In stored programme controlled electronic exchanges there is much more flexibility because the exchange is able to translate between a subscriber's number and an internal equipment number used by the exchange. With the addition of data bases, and high speed signalling in an intelligent network architecture, a typical modern digital network is able to obtain routing information or make an additional translation from the number dialled to the number to be called. This ability can be used for call diversion type facilities in fixed networks, call routing in mobile networks, and numbering portability.

Although intelligent network architectures are able to provide whatever number translations are needed, the design and operation of these networks is far from trivial, and issues such as portability need to be considered carefully in terms of the costs of implementation. Although the costs of data base technology are falling steadily, the cost of the technology is not the only issue. The design and operation of the procedures for updating databases are equally important, and the administration may be a greater constraint than the technology in the future. These issues are, however, being tackled in the context of the directory (see Chapter 21).

Ironically, although portability is a solution to the problems of competition between networks, portable numbers would normally have to contain some information on the location or identity of the data base on which the subscriber's information is held. Thus the subscriber is tied through his number to the operator of the data base, and the problem of competition between network operator has been exchanged for the one of competition between data base operator.

25.7 NUMBERING CONNECTED NETWORKS

In this section we will consider the different ways in which the numbering of interconnected networks can be arranged to enable a call to be established from an apparatus or terminal on one network to an apparatus or terminal on another network.

A typical example is a call from a public network to a private network. For the following examples we will call the originating network 'network A' and the receiving network 'network B'. There are the possibilities outlined in the following four sub-sections (see Figure 25.5).

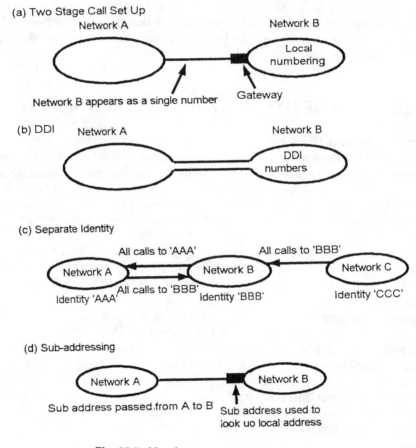

Fig. 25.5 Numbering connected networks.

Two-stage call set-up

In this case, Network B is identified in Network A as a single number. Calling this number takes the call across Network A to the connection point to Network B, and the call is then set up across network A—the first stage of the call set-up. To take the call across Network B a second stage of signalling has to take place across the established path through Network A to tell Network B which number in Network B the call is to be routed to. An example of this type of connection is dialling a single number for a private network and then asking the operator for a particular extension number.

Direct Dialling In (DDI)

With Direct Dialling In (DDI) all the apparatus or terminals on Network B are assigned numbers from the numbering plan of Network A. This enables single stage call set-up because Network B appears to be part of Network A. DDI is very popular for private networks which are connected to a single public network, but it is of limited use if Network B is to be connected to more than one other network because each apparatus or terminal on Network B would normally have to have a different DDI number from each network from which it could be accessed.

Separate network identity

We saw above that DDI provided a single stage call set up but had a disadvantage if Network B was to be connected to more than one other network. This disadvantage can be overcome if the interconnected networks are all part of one numbering scheme which gives each network its own identity, e.g. a particular value of the first few digits in the numbering space. Networks A and B would then be numbered as equals, and Network B's numbers would be recognisable and could be routed to directly from any other network because they would no longer appear to be part of Network A.

Sub-addressing

Sub-addressing is a powerful new technique for numbering and routing. Each network uses its own numbering and routing arrangements but provides a facility as part of the call set-up procedure to carry a number across the network. Thus in two-stage call set-up. discussed above, Network A would be able automatically to carry the number (i.e. the extension number) in Network B to Network B during call set-up. But this is only a simple example of sub-addressing.

The real power of the technique is that all apparatus or terminals of all types could be given unique numbers as part of a global scheme, and that each network involved in a call could translate the subaddress number to the number to be used in its own numbering plan for routing purposes and then pass the subaddress on to the next network. For data and within the OSI 7-layer model, the globally unique number is the Network Service Access Point (NSAP) address. For telephony, simple sub-addressing facilities are likely to become available in the near future as part of ISDN and as a result of the capabilities of the CCITT No. 7 Signalling System.

Of the four techniques described above, two-stage call set-up and sub-addresing are capable of handling the interconnection of different networks the numbering arrangements of which are entirely different, whereas DDI and separate network identity are capable only of handling the interconnection of networks of the same type.

25.8 THE FUTURE

In some respects it is particularly difficult to predict the future of numbering but in other respects it is easy. One can at least predict with confidence that there will be no reorganisation that will change every aspect of everyone's number: even radical changes will have to be introduced with the minimum immediate disruption.

Numbering is increasingly being recognised as a very important issue both for the convenience of the subscriber and the development of competition between service providers. Thus numbering is likely to become a major subject for debate within Europe and within CCITT.

Personal numbering is expected to be introduced in several countries by about 1995 through the UPT service based on intelligent networks.

In the long term, intelligent networks should provide a very flexible framework for numbering but the problems of the administration and updating of data bases will be the main constraints.

26 NETWORK MANAGEMENT

26.1 NETWORK MANAGEMENT IN GENERAL

26.1.1 Introduction

Management is commonly defined as making the most effective use of resources to achieve a given objective. In communications, the resources are the network and its various facilities, and the objective is the provision of a specified quality of service (QOS).

Network management therefore includes (but not exhaustively) the activities listed under the next five sub-headings. (*Note*: these activities are described in terms of a PTO network but very similar considerations apply to the management of a private network.)

Configuration management

- prevention of network overload by the management of transient traffic peaks using such techniques as alternative routing or refusal of originating traffic;

- the provisioning and configuration of a network's switching and transmission plant to meet changing user needs, ranging from peaks of traffic arising from major disasters through rearrangement of individual customer's equipment to long-term requirements for additional capacity;

- monitoring the installation and rearrangement of new customer equipment;

- the maintenance of equipment records and a network plan.

Performance monitoring

- the monitoring of network performance and the collection of network statistics which in turn may be used for configuration management.

Fault management

- the detection and reporting of faults, remedial action (e.g. by reconfiguration of the equipment or network) and then repair;

- the organisation of a maintenance service including monitoring of fault repair times.

Accounting management

- the collection, storing, processing and auditing of billing information.

Security management

- the management of access rights to various types of management information and control functions.

While these management functions have of course been recognised for a very long time, recent developments have depended on the use of computers to provide support and common services, e.g. to automate the collection of statistics and to provide a common and user friendly man–machine interface.

A maintenance service tends to be organised in terms of maintenance centres which are interconnected by telephone lines and data links. Such centres may be multifunction or specialised, depending on geographical factors and on how the owning administration wishes to organise itself. For example there may be a network maintenance centre covering the real time reporting of network performance over a relatively wide geographical region which is connected to a number of dependent repair centres which, because they involve the field repair staff in road travel, cover relatively small areas. The new generation of maintenance centres that have been introduced over the last decade or so depend upon the network using stored program controlled exchanges that act as collection points for fault reports and traffic statistics. Nevertheless manual input of fault reports is still required, particularly those reported by users, e.g. faulty telephone instruments. A maintenance centre may also act as a collection and forwarding point for passing call accounting information from a group of exchanges in the same region or district to a billing centre.

Management centres are configured and interconnected as an overlay on the network they serve. Their organisation is generally hierarchical, for example some centres will be manned during working hours only and fault reports will be passed to parent centres at other times. The network of centres involved in maintenance activities must be configured such that in the event of one centre failing the dependent exchanges which it serves can interwork, at least to some extent, with an alternative centre. Billing centres and centres that process network statistics are likely to be connected to and serve a large number of maintenance centres (see Figure 26.1).

Network management activities can be divided into long-term strategic and short-term tactical ones. The long-term strategic activities, specifically the provisioning of a network to allow for growth or decline in customer traffic, tend in practice to

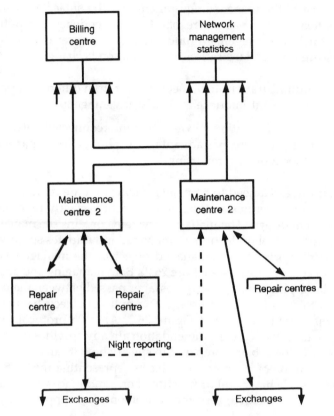

Fig. 26.1 Network management hierarchy

be centrally based and are divorced from shorter-term responsibilities. Strategic planning involves its own specialised computer-based support tools. A major input is of course the traffic statistics passed via regional network management centres. While network planning is important it is also a highly specialised subject and we shall concentrate on the shorter-term, closer to on-line, aspects.

On-line network management is important in both public and private networks. AT&T, BT, Mercury and most major telecommunication administrations have sophisticated network management centre designs and all use advanced network management technology to improve the flexibility with which combinations of exchange lines and private circuits are presented to their customers.

26.1.2 Network management standardisation

It is clearly desirable that the management functions listed in the introductory paragraph above should use a common set of procedures/protocols and perhaps a standard man–machine interface. Currently at least the man–machine interface is a matter for individual network providers but the protocols, particularly as they affect the interface to the managed equipment, are the subject of standardisation in the OSI context. In the past there has been a tendency for public network providers to regard the management function as their concern alone and there has been little pressure for standardisation. This situation is changing for two reasons:

1. The ability of administrations to buy equipment from different suppliers having a standard management interface is clearly advantageous.

2. The ability to pass information between two interconnected networks, whether private, public or a mixture, has advantages for both parties, particularly in the context of fault location and reporting.

In any context there is a need to manage systems that are remotely located with respect to management centres and this requirement is clearly reflected in the OSI management system design. The managing system contains the managing process which is under the control of the human manager and initiates some operation on, or receives event reports from, a managed object in the managed system either directly or via an agent process (see Figure 26.2). Both managing and agent processes are at layer-7, the application layer and messages passed between them are handled in a similar way to any other OSI message. Examples of objects being managed are identified on the right-hand side of Figure 26.2. An X.25 protocol machine is an example of a resource; the X.25 recommendation already specifies how it works and a new recommendation is being prepared to specify how it can be managed, which will lead to a definition of the managed objects representing the X.25 resource.

The OSI model is a behavioural or functional one and places no requirements on how systems are implemented; thus the equipment that supports say the bottom

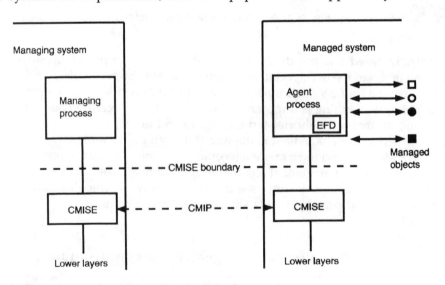

CMIP = Common management information protocol
CMISE = Common management information service element
EFD   = Event forwarding discriminator

Fig. 26.2 Architecture for systems management

three layers does not need to be physically partitioned to conform to those layers, and interfaces that are of interest for management may well be common to a number of layers or parts of layers. This situation is reflected in Figure 26.3, which shows two possible approaches:

1. The 7-layered structure of the OSI model means that there can be a conceptually identical management interface at each layer. Messages to and from layers can be passed up or down the hierarchy and between systems in a similar fashion to user messages except that their source/destination is not an end system but a layer or management/agent process somewhere in the network; objects which are managed in this way are called '(*n*)-layer managed objects' and can be managed from anywhere where there is the relevant capability.

2. Alternatively, as shown on the right-hand side of Figure 26.3, objects can be mapped directly onto the local management process when they are called 'systems

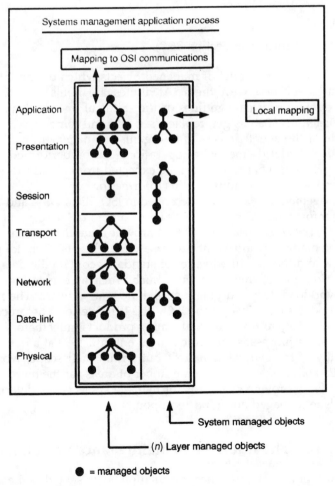

Fig. 26.3 Two methods of communicating management information

managed objects'. With this type of mapping remote management can take place only via the local systems management process.

For (*n*)-layer managed objects there is a limited interface at each layer over which exchanges with the managing/agent process take place which is quite distinct from and orthogonal to the telecommunications interface between the layers of the OSI model at that point. It seems likely that new designs of equipment will be based on (*n*)-layer managed objects while existing designs which pre-date the OSI management standard will have local management access.

The five major OSI management functional areas were defined in the introduction as:

1. configuration;

2. performance;

3. fault;

4. accounting;

5. security (treated further in Section 26.2).

There is an information model of managed objects which defines attributes and the operations associated with them. Attributes are items of information, for example the number of exchange lines or the status of one particular line. Operations involve actions such as get, replace, set, add and remove.

A major role of the agent process is to organise management communications. The information model defines what operations can be performed on a specific object or resource and what reports can be expected from it and the agent process can store, organise and communicate this information.

The ISO management framework is described in ISO/IEC 7498–4 and the Common Management Information Service (CMIS) and Common Management Information Protocol (CMIP) (ISO/IEC 9595 and 9596–1) provide a standard means of communicating between managing and agent processes. ISO/IEC 10040 contains a management overview and the ISO 10164 series of standards contain the detailed descriptions of the various management functions, such as alarms, events, logging, security, accounting, workload, test and diagnostics, time and software. The management information model is described in the ISO/IEC 10165 series. The process of standardisation is not likely to result in conforming products much before 1992–93 and the standardisation process is scheduled to go on until 1995 at a detailed level.

Independently of the general standards but within their scope, work (as exemplified for X.25 above) is going on in a number of areas for the bottom four layers of the OSI model to provide standards for managed objects, i.e. determining what it is that needs to managed and reported upon.

26.1.3 The interface to maintenance personnel

It is very important that maintenance information be presented to the maintenance personnel in a readily assimilable form and here WIMP (see Section 9.2) is a useful

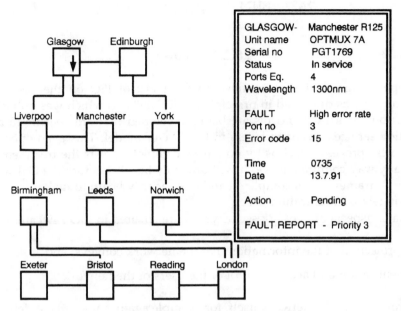

Fig. 26.4 Topological maintenance map

tool because it provides a topological menu with overlays and a mouse to pick items for which more detail is needed. Figure 26.4 shows a hypothetical topology menu in which details of a particular item of faulty equipment (an optical transmission link at the Glasgow end of the Glasgow to Manchester route) has been pulled down because of a high error rate.

26.1.4 Future prospects

The growing use of digital networks should increase the reliability of switching equipment to the point where its maintenance becomes a relatively trivial item. The emphasis on maintenance will then be on transmission equipment where damage due to extraneous causes is likely to become the major cause of failure rather than any basic unreliability of the transmission equipment as such. With the introduction of increasingly sophisticated systems for network management, especially in public networks, fault finding and repair times will decrease and the availability and reliability of services should improve quite noticeably. It remains to be seen whether the more sophisticated planning tools available to administrations will enable them in practice to solve the perennial problem of not providing adequate margins of spare capacity.

Increasing interaction between the management of private and public networks may be confidently expected with centrex and virtual private networks giving customers the option of handing over all or part of the management of their network to a service provider.

26.2 NETWORK SECURITY

26.2.1 Scope

The scope of this section is confined to the work of ISO on the OSI security architecture that is described in principle in ISO 7498 and which was published in 1988. Various security protocols are being developed that are specific to an OSI application service and/or to layers in the 7-layer model. This is an open-ended development process which is by no means complete, nor is the treatment given here in any way comprehensive. It will probably be 1993 before the OSI work on the security framework is completed and OSI security begin to make a significant impact on telecommunication users.

The major aspects of network security that are covered in ISO 7498 are:

- the protection of the information that the network conveys; and

- authentication and access control of the users of the network.

The protection of the network itself, for example against attempts to defraud the billing system, is not covered by the security model and is not covered in this section.

Figure 26.5 lists various 'security services' within the OSI 7-layer architecture and the security mechanisms that may be associated with them. For example, traffic flow confidentiality might be achieved by encryption, padding or routing control or some combination of these three options. Work is going on to specify the application-independent security protocols that are specific to a given layer of the model. In addition security procedures associated with applications such as X.400, directory, X.435 EDI, FTAM, transaction processing and data bases are being developed.

Some of the more general work on security techniques is still in the discussion stage but it seems likely that a protocol called End-to-End Security Protocol (EESP) will be adopted. This resides at the top of layer 3 and is concerned only with security protection defined by ISO 7498/2 to be above that layer. Both connection and connectionless modes are supported. The objectives of the protocol are to:

1. provide confidentiality and integrity so as to prevent interception and impersonation;

2. provide multiple security partitions within end systems;

3. control access to resources from outside and inside an end system;

4. provide the above features with various degrees of assurance.

There is also a need to establish international standards for the cryptographic key management protocols.

Security service	OSI layer							Security mechanism							
	1	2	3	4	5	6	7	Encipherment	Digital signatures	Access control	Data integrity	Authentication exchange	Traffic padding	Routing control	Notarisation
Peer entity authentication	N	N	Y	Y	N	N	Y	Y	Y	N	N	Y	N	N	N
Data origin authentication	N	N	Y	Y	N	N	Y	Y	Y	N	N	N	N	N	N
Access control	N	N	Y	Y	N	N	Y	N	N	Y	N	N	N	N	N
Connection confidentiality	Y	Y	Y	Y	N	Y	Y	Y	N	N	N	N	N	Y	N
Connectionless confidentiality	N	Y	Y	Y	N	Y	Y	Y	N	N	N	N	N	Y	N
Selective field confidentiality	N	N	N	N	N	Y	Y	Y	N	N	N	N	N	N	N
Traffic flow confidentiality	Y	N	Y	N	N	N	Y	Y	N	N	N	N	Y	Y	N
Connection integrity with recovery	N	N	N	Y	N	N	Y	Y	N	N	Y	N	N	N	N
Connection integrity without recovery	N	N	Y	Y	N	N	Y	Y	N	N	Y	N	N	N	N
Connection integrity selective field	N	N	N	N	N	N	Y	Y	N	N	Y	N	N	N	N
Connectionless integrity	N	N	Y	Y	N	N	Y	Y	Y	N	Y	N	N	N	N
Connectionless integrity selective field	N	N	N	N	N	N	Y	Y	Y	N	Y	N	N	N	N
Non-repudiation, origin	N	N	N	N	N	N	Y	N	Y	N	Y	N	N	N	Y
Non-repudiation, delivery	N	N	N	N	N	N	Y	N	Y	N	Y	N	N	N	Y

Y = yes   N = no

Fig. 26.5 OSI layers—security services and mechanisms

26.2.2 Protection of information

The concern here is basically with non-verbal transactions; the limitations of confidentiality on the telephone are well established and little fresh can be said about them except perhaps to note that with digital systems crossed lines are likely to become a less frequent occurrence. The major security mechanism is, and is likely to remain, some form of speech encryption.

In general the user of a messaging system would like confidentiality, assurance against wrongful delivery, proof of rightful delivery and proof that the message was sent by the stated originator. In the past no system has been able to provide this; the best that can be done is to encrypt the message and hope that if it then falls into the wrong hands it is unintelligible.

Confidentiality can be assured by encryption but a discussion of encryption methods is beyond the current scope, except to note that public key encryption seems to be becoming the main encryption method for non-military communication.

Within the the general scope of the OSI security architecture, X.400 (1988) extends the scope of message security by providing a number of additional protocols relating to delivery; and these include:

1. proof that the message has come from the identified originator (authentication);

2. proof of delivery;

3. proof of submission;

4. content integrity check;

5. message flow confidentiality—this is equivalent to double enveloping of mail, where only the external address information is unencrypted.

The protocols themselves are protected by the use of public key encryption, making it very difficult for a third party to emulate the terminal of a genuine participant.

26.2.3 Authentication and access rights

Authentication has already been mentioned above and it is essentially a means of trying to ensure that information comes from the source that it is claimed to come from. It might be based on automatic fingerprints or handwriting recognition, but is more likely to be based on passwords and keys.

Access rights are assigned in terms of the ability generally to send and/or receive via the transmission medium and to read information from or write information into a data base. Unless the system is closed and the terminals to it are secured against unauthorised access, the access mechanism will depend on some form of password. The success of hackers shows that the management of access by password is far from straightforward. The subject is however a rather specialised one and not central to the current scope.

26.2.4 The future

The legal position in the UK is that a contractual document sent by post has a status which the corresponding information sent by, say, telex or facsimile does not have. With the increasing confidence that can be placed in electronic conveyance, using for example the protocols developed in the X.400 context, there are good reasons for believing that electronic messaging can be regarded as becoming more fraud resistant than the mail. This changing situation has been recognised by the EC and work is in hand to establish a legal framework for electronically transmitted information. Clearly therefore the security aspects of OSI are becoming of major importance and one may anticipate an increasing emphasis generally on security in all telecommunication service specifications.

26.3 BIBLIOGRAPHY

European taxonomy of security standardisation (1991) *OSN: The Open System News Letter*, March, pp 12–18 (17 refs).

First set of OSI management standards now complete (1991) *OSN: The Open Systems News Letter*, October, pp 6–11.

J. Martin (1990) Management of facilities and services, *GPT Technology Symposium*, TMA Conference, Brighton.

26.5 BIBLIOGRAPHY

Part 4

SUMMARY AND THE FUTURE

27 THE FUTURE FOR TELECOMMUNICATIONS

27.1 INTRODUCTION

The purpose of this chapter is to summarise the trends and predictions about the next 10–20 years which we have developed in more detail within the body of the book, and to give an overview of the way in which we expect telecommunications in comparatively advanced countries to develop in the future. After providing this summary, we go on to explore in a little more detail what these changes may mean in practice for business and residential telecommunications users.

Predicting general trends is comparatively easy compared to making quantitative predictions about the precise sequence of events. Nevertheless wherever possible we dare to include our personal predictions about timescales so that the reader can form a clear idea of what will happen by when. This approach is undoubtedly risky because the more specific our predictions are, the more likely it becomes that we are proved wrong, but we believe that we owe it to our readers to be as specific as we can.

A matter of some concern is whether the information will be available to prove us right or wrong. A negative aspect of competition in public services, so far, has been the lack of provision of the statistical information that used to be provided for traffic carried by public network operators. Such information is now regarded as commercially confidential but is vital for the industry as a whole in all sorts of ways. We can but hope that some mechanism will be found for the information carriers to get together and produce some industry statistics just as trade associations do in a number of other fields.

27.2 SUMMARY OF TRENDS

27.2.1 VLSI

Silicon integrated circuits have been at the centre of the electronics and communications revolution for the last twenty years. We expect that silicon will continue to be the principal material for integrated circuits for at least the next decade, with other materials being used only in specialist applications.

We expect the reduction in circuit feature size to continue for the next decade but the rate of progress may decrease somewhat thereafter. The performance capability of integrated circuits should however continue to improve with rather less of a decrease in the rate of progress. We think that progress towards wafer scale integration will be slow because of the problems of reliability, yield and the general complexity of design and testing. In addition to these further improvements in basic digital circuitry, we expect that some of the most significant advances will be in the area of hybrid analogue-digital integrated circuits which will be used extensively in radio systems.

27.2.2 The man–machine interface

Terminal equipment will represent an increasing percentage of the total investment in telecommunications and the communications function of the terminal will be increasingly shared with other functions, particularly office procedures in business and entertainment at home. Portability and ease of use are key factors in improving terminal design. The gradual replacement of the cathode ray tube by flat panel, probably liquid crystal, displays will be a major factor in the quest for portability and the introduction of common channel signalling will, in conjunction with such a display, make the terminal more user friendly.

There will be a gradual evolution in terminal signalling from 10 ips (impulses per second) dialling to multi-frequency signalling and then to common channel signalling with ISDN resulting in faster call-set up and making it easier to use special services from an ordinary telephone. However, the large existing domestic investment in 10 ips instruments makes it unlikely that this form of signalling can become obsolete in the next decade except perhaps in a business environment.

27.2.3 Building wiring

One of the longer chapters in this book is on building wiring and this reflects the increasing importance of a subject that has been something of a Cinderella in the telecommunications world. The increasing added value of telecommunications for business and the large investment required to install building wiring that is multiservice and future proof means that structured wiring schemes are becoming the norm for large buildings.

Broadband services demand better transmission and crosstalk performance from the wiring and EMC is becoming an increasingly important issue as data rates become higher. This is leading to a major increase in standardisation activities and one can confidently anticipate that many new European, North American and International wiring standards will be published during the next 5 to 10 years.

While optical fibre has a role in building wiring, particularly in providing broadband backbone connections, twisted pair copper wire is far from being at the end of its service life and will continue to provide the horizontal or floor connections for many years to come. Some very significant improvements in the performance of multipair twisted pair cables have been achieved quite recently.

27.2.4 Line transmission systems

The developments in optical fibre communications which have taken place to date and which have already been of great importance, have used only a fraction of the inherent capability of optical fibres. We expect that there will be very substantial further progress during the next two decades which will take the maximum capacity of a fibre from the current 1–2 Gbit/s region into the 10–100 Gbit/s region through the use of wavelength division multiplexing, and higher bit rates at individual wavelengths as a result of coherent detection techniques. The 1500 nm band will soon be used as extensively as the 1300 nm band.

The distances achievable between repeaters will increase quite significantly, particularly as a result of coherent detection, which results in an improvement of up to about 20 dB in the signal-to-noise ratio of a receiver, and by about 1996 it should be possible to construct almost all of the links within all but the very largest countries without repeaters. Use of the soliton mode of transmission will mean that performance will not be dispersion limited and there will be less of a need to trade off a high bit rate against distance. By about 2000 it should be possible to provide trans-oceanic submarine cables without repeaters or underwater optical amplifiers.

Within the public networks we would expect fibre to have replaced almost all the trunk (or toll) and junction (or sideways) circuits between public exchanges by about 1995 and to be the standard means of serving all but the smallest business premises on 'greenfield' sites by about 1994. Where entertainment TV services are not provided, residential premises will however continue to be served by copper pairs probably for the whole of the decade. As a result of the continued use of copper there may well be further developments to exploit its capabilities and provide improved reliability. But by about 1995 it will be increasingly common for newly built residential areas to have the distribution cabinets in the street connected to the local exchange by fibre.

The transmission hierarchy of public networks, which is now almost exclusively plesiochronous, will change to being largely synchronous possibly by about 2000, enabling extensive use to be made of drop and insert multiplexers. At that time most of the traffic will be carried in synchronous mode but asynchronous mode traffic will be increasing rapidly.

Advanced multiplexing techniques will probably be introduced earlier in large private networks than in public networks.

27.2.5 Network access

ISDN access to public networks is becoming widely available in 1992 and this trend will accelerate. We expect the primary rate services to become very popular with the larger business customers, who are the purchasers of new PBXs. The take-up of basic rate services will be slower because smaller businesses tend not to have enough traffic of a type that would benefit from the extra features and cannot therefore justify the extra equipment and rental costs associated with ISDN. By 1997 we would expect basic rate services to start to become the norm for small businesses, but in the case of the residential market we expect a market penetration figure of only a few per cent. The availability and popularity of 64 kbit/s video will be a major factor here.

In any case, because of the increasing tendency for people to work at home, the distinction between business and residential usage is becoming increasingly blurred and it may be difficult to obtain meaningful statistics. Generally the amount of traffic from a home does not justify a second exchange line and currently the main reason for having one may be the need for the second line to be charged to a separate business account. It is convenient but not essential to have a fax terminal on a separate line and the widespread adoption of fax in the home could increase the demand for two lines, although the choice of ISDN or a second analogue line will depend on the installation and rental levels. It would be a great advantage to ISDN if the installation and rental charges did not exceed twice those of an analogue line and if provision were made for specific calls to be billed on designated accounts, thus avoiding the need for physical separation of equipment to achieve separate accounting.

27.2.6 Switching

In comparatively advanced countries throughout the world electromechanical exchanges are fast disappearing or have completely disappeared. The situation in the UK is behind North American and on a par with, or slightly ahead of, the rest of Europe but the UK is taken as an exemplifier of the general trend.

For the UK, the replacement of electromechanical exchanges with digital stored programme control electronic exchanges will continue as rapidly as possible so that by about 1993 all customers who require the extra facilities of digital exchanges will be able to be connected to digital exchanges or to analogue electronic exchanges with similar features. However, it will not be until about 1994 that the last Strowger exchange is taken out of service and until about 1996 for the last Crossbar exchange. Electronic analogue exchanges will be replaced by about 2000. The elimination of Strowger Group Switching Centres in 1991 means that the number of very noisy lines should be reduced quite significantly and only a very small proportion of subscribers will experience very noisy lines or poor quality transmission after the end of 1991. Most of the remaining noise is likely to be due to crosstalk from decadic dialling, which should not be a serious cause of annoyance and in any case will decrease as the older telephones are replaced by ones with DTMF signalling.

The current generation of UK digital exchanges (e.g. System X, System Y and DMS) should be expected to have a life of at least 15 years during which there will be significant feature enhancements (e.g. for ISDN) and an extension of their operational role. The two main extensions of role in the nearer term are Centrex/Virtual Private Network Services and Intelligent Network services. Centrex/Virtual

Private Network services will be provided in the main by enhancements through new software and perhaps some new hardware to existing exchange designs, but Intelligent Network services, while involving some modifications to existing exchanges, will also entail the addition of specialised exchanges and centres in the form of an overlay.

The next fundamental change to exchange design is expected to be the development of exchanges which will accommodate fast packet switching in the Asynchronous Transfer Mode (ATM). From the user's point of view the network will appear to be an integrated multi-service one. However, the network will contain the two transmission modes, STM and ATM. From the point of view of switching it remains to be seen whether STM and ATM switching nodes will be fully collocated and whether the switch fabric will be integrated or separated for the two types of transmission. The development of this next generation of public exchanges is expected to take about a decade and so it will not be until 2000–2005 at the earliest before they begin to appear as a major alternative to existing digital exchanges such as System X. There is increasing evidence of interest in broadband services and new exchange designs must provide a smooth means of transition during their introduction.

27.2.7 Fixed network costs and topologies

The developments in transmission will lead to steady and rapid reductions in transmission costs, especially as existing fibres can be used for higher and higher bit rates by replacing the transmitter, receiver and repeater (if needed) equipment. Switching costs will also decrease steadily but not as rapidly. These changes will lead to further convergence of long distance and local tariffs, and some observers have suggested that by the late 1990s trunk tariffs may be only 10–20% higher than local ones and transatlantic tariffs perhaps only 30–40% higher than local ones. Changes such as these will make the world much more of a global village.

The change in the balance of costs between transmission and switching will produce a trend towards fewer larger exchanges but is unlikely to produce radical changes in the topology of existing public trunk networks.

However, in the local area changes in topology are likely to occur for a number of reasons:

(a) The increased use of remote concentrators and intelligent multiplexers and the conveyance of their traffic on optical fibres.

(b) The much reduced floor area required for modern electronic equipment compared with electromechanical, making it possible to house in one building several existing exchanges that currently are geographically separate.

(c) The use of ring- or bus-like structures in the form of MANs connected to LANs. Initially at least these are likely to be confined to new development areas or areas of very high population density.

(d) Pressure to open up competition in the local area.

It is too early to say which of these factors is going to have the most influence and what that influence will be in detail. The pressure to open up competition could for example result in:

- competition in terms of distribution media, e.g. radio against land-line; or

- The provision of a broadband transmission medium, or distribution grid, by a network provider which would be available to any licensed service provider for the running of a service.

So far, there are two competing approaches to providing multiservices (telephony and broadband including TV) in the local area; the first is exemplified by the TPON/BPON pioneered by BT, which is based on the use of existing ducts, etc., and the second involves the extension of cable TV networks to become multiservice. Hitherto cable systems have developed more slowly than expected but in the future whether multiservice networks develop more rapidly from the existing telephony network or from cable systems will depend on commercial and regulatory decisions which have been taken, at least in part, in the UK but remain to be taken for the EC and USA.

27.2.8 Private networks

In the past the principal rationale for private networks has been to save money in a situation where on the public network trunk (toll) calls were subsidising local calls. However, with the rebalancing and convergence of trunk and local tariffs the potential savings from private networks are reducing significantly. The rationale for private networks in the future will increasingly be the relationship of the network and its special features to the main commercial activity of the company.

Within private networks there is likely to be increasing use of more sophisticated network configuration control and some increased use of low bit rate encoding although the justification for the latter may be more the result of the tariff structure of the providers of private circuits rather than a reflection of the true and decreasing cost of transmission.

During the next decade, private networks will also come under increased competition from centrex and virtual private networks provided by public operators and other managed network service providers. After the inevitable initial teething problems such services should be user friendly and reliable and provide a wide range of features. When they have become well established, possibly about 1996, the choice between owning your own private network and using a private network service will depend very much on quality of service in the widest sense of the term and on subjective and possibly rather arbitrary judgements about whether a business can trust its communications to a service provider. Therefore it is likely that both private network ownership and private network services will each have a significant proportion of the market in the long term but it is not possible to say which will have the major share; too many intangible subjective factors are involved.

The introduction of new features and technology has always tended to take place earlier in private networks than in public networks because PABX manufacturers can effect changes more quickly than public exchange manufacturers and because private networks are smaller than public networks. This phenomenon is likely to continue and therefore the convergence of voice and data on LANs and MANs may become quite significant in private networks by about 1995, which could be up to a decade before there are similar changes in public networks through fast packet switching based on the ATM.

27.2.9 Cable TV systems

The future of larger cable TV systems is uncertain, but recent signs are that investors are becoming more bullish again. From the point of view of television delivery they will have to compete with off-air broadcasting, satellite master antenna systems (i.e. a mini cable system with very limited facilities) and millimetre wave distribution systems as well as the rental of videos. Undoubtedly there is a high potential demand for television but the future of cable systems will depend to a large extent on the availability of attractive programmes, which have hitherto generally considered to be in short supply, but this situation could well improve in the future. To some extent the development of satellite broadcasting is helping the development of cable services by providing additional channels for distribution. One development which could really help cable could be high definition television. If it caught the public's imagination as some people expect, the extra bandwidth could perhaps be provided by cable systems based on optical fibres more cheaply than by competitive means of delivery.

From the point of view of services other than entertainment TV, we are not aware of forecasts of significant demand from residential customers for services which cannot be delivered over the PSTN/ISDN. The introduction of technology which would lead to cost savings by providing both TV and telephony is some way off and faces the problem of needing to achieve high volume production to bring the costs down. Any operator who wants to establish a business in local telephony would have to target business areas first, where economies of scope from also offering TV would not be available because the demand for entertainment TV in business areas would be low. All these factors make the growth of cable more difficult, and in the meanwhile radio-based technologies will establish themselves further as a means of delivering a wider choice of TV. Nevertheless we would expect cable systems to be successful at least in some areas.

A possible way of achieving more competition for telephony in the local loop would be if cable companies could use cable as a 'backbone' and use microwave or millimetre wave distribution for TV and CT2/DECT for telephony. This solution would avoid the expense of complex switched star systems and the expense of laying cables to individual premises.

27.2.10 Satellites

Satellite systems are most unlikely to achieve anything like the reductions in cost which are occurring in optical fibre transmission and they suffer from the problem of the 250 ms delay, which is an annoyance for telephony. Their future role will therefore be confined to services such as broadcasting and wide area mobile communications which exploit the satellite's natural characteristics, and significant growth is to be expected in these areas. Data broadcasting to VSATs is included in this category.

Satellites will continue to be used for international and intercontinental fixed services where either there is a strategic need for diversity or where the routes are thin and do not justify cables, but in general their use for fixed services will decrease. In developing countries they may be used as a means of providing links much more quickly than is possible with cable.

27.2.11 Mobile radio

On a medium-term view mobile radio is currently the area of most rapid growth in telecommunications. Despite some short-term reductions in traffic as a result of the recession, substantial further growth is to be expected lasting for at least the next decade. The market for mobile radio divides into four main areas:

1. message paging for short one-way messages;

2. private mobile radio for high volume closed user group type communications;

3. cellular radio for public mobile telephony;

4. CT2/DECT systems for cordless telephones, cordless PBXs and Telepoint services with less extensive coverage than cellular radio.

The public services, paging and cellular radio started initially as separate services, but in the future there is likely to be some convergence between them, especially so far as the user is concerned. The dominant public services are message paging and cellular radio and further growth is expected in both areas. GSM with its digital technology is set to replace analogue cellular and will provide pan-European coverage. New low-cost GSM services will be introduced to enable GSM to compete directly with PCN. In countries where analogue cellular networks are well established, there may be problems in giving sufficient incentive to customers to transfer from analogue cellular to GSM or PCN, but the main stimulus may be congestion on the analogue networks.

There should be further growth in message paging and PMR but both services may eventually lose market share when the price of cellular services drops significantly.

Within the UK, Telepoint has so far been a failure, although it is the subject of considerable interest in France. Telepoint may have been launched too early in the UK and it may be unrealistic to expect Telepoint to be much more than an added facility for customers who have bought CT2 handsets for use with cordless PBXs or as a domestic cordless telephone.

The key technological developments which will continue to drive these advanced mobile services are:

- low bit rate coding techniques for voice to give high spectrum efficiency;

- signal processing techniques to handle multipath propagation and fading, such as equalisation, and error detection and correction with interleaving;

- VLSI to allow compact handset design;

- dynamic channel assignment techniques to reduce the need for large margins in *a priori* planning.

The use of low bit rate techniques, for example in GSM, will lead to increased levels of quantisation distortion, but compared to fixed networks that use A-law coding there would appear to be some scope for further improvement in coding algorithms, although it is not necessarily easy to upgrade a coding system which is established in the market place. Therefore developments may have to await the introduction of a new generation of systems such as UMTS.

In many respects it is difficult to predict the future for UMTS. From a technical point of view, it will represent further convergence between the different mobile systems of earlier generations together with broadband capabilities. From a commercial point of view, it will need to provide a distinct improvement over existing systems in terms of cost or facilities for which there is demand.

At present mobile services such as cellular radio are expensive compared to the PSTN. As the services expand, as technology develops and as competition becomes more intense, the tariffs should reduce substantially and by the late 1990s might be possibly only 20% or less above PSTN tariffs. Radio could become a feasible alternative to line transmission in the local loop especially if there are further economies through combination with millimetre wave TV distribution.

27.2.12 Complexity and standardisation

There is no doubt that the the world of telecommunications is becoming more and more complex for the equipment designer, for the network designer and for the end-user. The increasing need for equipment to interwork puts enormous pressure on standardisation bodies whose task is made the more difficult by the same issues of complexity. To a large extent the rate of progress in a number of fields is primarily determined by the time it takes to reach agreement on standards, making it less likely that a revolutionary new idea can radically upset our predictions.

Because standardisation is such a slow process there is a tendency for 'industry' or *de facto* standards to emerge which, while giving immediate satisfaction in the market place tend to build up trouble for the future. We firmly believe that the ISO 7-layer model will receive widespread acceptance and will provide a sound framework which should help in the future to prevent excessive proliferation of design.

On the issue of complexity it is not possible to be as optimistic, there is as we have seen still a long way to go on the subject of sound software and VLSI design. On the question of understanding what the network could do and how to go about doing it, the user is very much in need of expert advice and only those companies who are in telecommunications in a pretty big way can afford their own expertise.

One can already see an increase in the number of companies specialising in the giving of advice and one can confidently predict a growth in such consultancies. Because complexity involves substantial development expenditure and the life cycle of products gets shorter as technology in general and VLSI technology in particular advances, the size of company needed to provide a 'total systems' capability gets bigger and bigger, leading to company mergers and a need to seek world rather than national markets. At some point this process must become self-limiting, either because of physical limits or because the sheer complexity of the problem exceeds human intellectual capability. We agree that the rate of progress should fall off at some time but at the component (e.g. VLSI) level at least there is not a lot of quantitative evidence that it is happening. There is however a considerable difference between what technology can achieve and the effect it has on the lives of the users of it and this is a theme that is continued implicitly and explicitly in the next two sections.

27.3 THE WORK ENVIRONMENT

For the purpose of the following discussion we shall regard the working environment as consisting of either an office (at home or at a business), a factory, a shop, a farm or an itinerant situation. The itinerant situation includes commercial representatives, repair technicians, and normally sedentary workers moving between one business appointment and another. We look about fifteen years ahead to gauge what the changing effect of communications on people's lives will be.

Before looking ahead, it is a salutary experience to look back fifteen years to see what the corresponding changes have been since 1975. Perhaps the most obvious changes are the following:

- push button telephones nearly everywhere (not necessarily using multi-frequency signalling however);
- offices full of PCs, word processors and computer terminals;
- fax in addition to telex machines in the general office or communications centre, and a few fax machines in individual offices;
- some telephones with display features;
- some PCs with modems attached for data communication;
- many more data terminals in accounts offices, particularly in the context of Electronic Data Interchange (EDI);
- many more car phones, personal mobile phones and pagers;
- quite a lot of portable PCs;
- electronic Point of Sale terminals in large retailers;
- some very large organisations adopting an automated, paperless design-production process and the adoption of computer aided design tools generally throughout the engineering design industry.

The major event that was forecast 15 years ago that has not happened is the paperless office. In fact the ease with which computers generate output has if anything made the situation worse.

Looking ahead fifteen years it is possible to make the following forecasts with fairly high confidence.

In the office (see Figure 27.1)

- Integrated communication and office terminals will be the norm; these terminals will have an excellent display capability and the ability to despatch electronic mail and mixed media documents by means of simple keyboard commands.

- Telex will virtually have disappeared and fax as a distinct service will be obsolescent by having become part of a mixed media service and by having been partially replaced by computer graphics.

- Voice communication will be provided by cordless pocket phones and PBXs will keep track of staff as they move around the building. This is in addition to the more secure wired voice facilities provided under the first item above.

- The move towards a paperless office will have had some success; facilities for moving documents between terminals and accessing the data bases that have replaced filing systems will be commonplace.

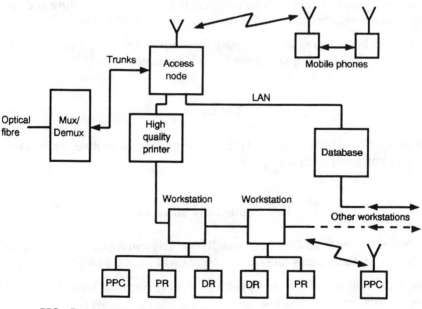

PPC = Portable personal computer
PR = Personal printer
DR = Document reader

Fig. 27.1 An example of office communications

- Portable PCs will be a very common means of transporting documents where today we would use a briefcase. PCN access will be an option on such PCs.

- The number of people working at home will have increased significantly. In 1991 BT estimated that there were about 0.5 million full time 'telecommuters' with another 1.5 million part time. We think however that the Henley Centre for Forecasting's estimate of 10 million by the year 2000 is too high.

In manufacturing and distribution

- In what will generally be a highly automated production and distribution environment, most operators will be equipped with portable communication terminals and paper work will be very much the exception. Automated warehouses will predominate.

- The engineering design process and its interface to production will be entirely automated in all large organisations. The data communications traffic between and within sites in this context will have become significant.

In retailing

- Electronic point-of-sale terminals with automatic stock recording and re-ordering facilities will be provided in the majority of retail outlets.

- There will be an increase in shopping from home involving the ability of sellers to provide a visual display of their goods to customers.

On farms

- Farm machinery, e.g. tractors, will be equipped with mobile communications, mainly voice but with the option of data.

In a mobile work situation

- Mobile telephones and/or pagers will be nearly universal and the provision of both-way data facilities on such terminals will be commonplace.

- The PCN will be widely accepted and pocket terminal equipment will have become unobtrusive and low cost.

- Portable PCs will be designed to be attached to mobile data systems so that for example reports can be filed from staff in the field and pro forma documents downloaded to sales staff for completion in the presence of the customer.

Network infrastructure

- The ability to locate an individual automatically whereever he/she may be will have improved very significantly owing to improved address (number) translation facilities and personal numbering.

- For call accounting, call itemisation will be optional for all services.

- Telephones and work stations will contain credit or direct debit card readers that will automatically put the cost of a call to the appropriate account.

There are two fairly major innovations that it is not possible to be too positive about, these are video telephony and voice input. We think that video telephone will have been accepted by 2005, particularly in the context of the interactive discussion of documents placed in front of the camera, but it is not possible to make concrete predictions about the extent of its acceptance. It is highly probable that low rate coding techniques will make a 64 kbit/s bandwidth acceptable for some applications but whether to the exclusion of higher bit rates is less clear. Voice input is another item that has continually been ten to fifteen years off since the 1970s and we remain unconvinced that the performance of speech recognition equipment in general will be good enough to have received widespread adoption within our timescale. One corollary of this is the general need to acquire keyboard skills, a need that the education system in general fails to recognise.

27.4 THE HOME ENVIRONMENT

27.4.1 Introduction

As a means of providing some light relief at the end of the book the impact of communications in the home environment is illustrated by a day in the life of an imaginary fairly prosperous family in an imaginary town in the Midlands. To put the situation in context it is perhaps advisable to say a few words about the sort of equipment configuration that might by then exist in the home.

Work is already under way under the auspices of an EC Eureka project to define a European standard for the interconnection of domestic appliances by a common bus network. The resultant system is called an Interactive Home System (IHS). The IHS has four service classes that can be briefly summarised as follows:

- *Class 1* Basic command and control for heating, lighting, air conditioning, etc., and as a control means for higher service classes. It works in a packet mode at a low data rate over the electricity main.

- *Class 2* Data services with three sub-classes for different speeds: viz, low speed up to 1200 bits/s, medium speed at 64 kbit/s and high speed at around 2 Mbit/s.

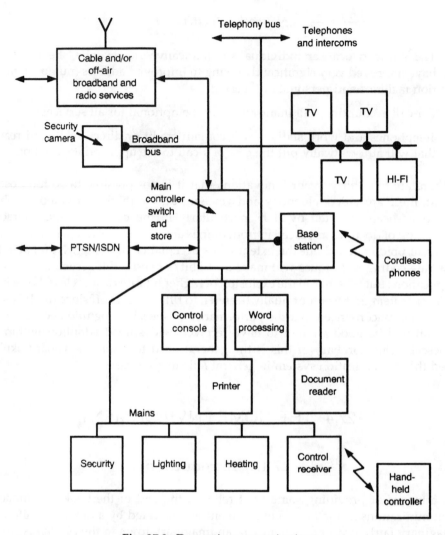

Fig. 27.2 Domestic communications

- *Class 3* All voice services (telephone quality) whether implemented in analogue or digital form.

- *Class 4* Audio/visual services including off-air and cable TV, locally generated video and high quality audio.

Figure 27.2 shows such a system in diagrammatic form.

There is clearly scope for some form of overall computer control for such a system and perhaps for common centralised storage on magnetic or optical disk of the data associated with the various entertainment and information services available in the home.

It is against this background that the following scenario is drawn

27.4.2 A day in the life of—*circa* 2009

George White is a financial analyst who commutes to work by train, his wife, Pamela, is a nurse who works part time at the local hospital. Samantha and James White are their teenage children, both still at school. The Whites live in the small market town of Farnston.

04.30 The printer in George's study prints out a financial analysis based on London and New York stock exchange closing prices.

05.00 Pamela's nursing news bulletin is delivered and stored ready for print out in colour. A delivery indication is displayed on the central control console in the hall.

06.45 The centralised alarm system wakes Mr and Mrs White according to its weekly schedule.

07.15 George starts to read his financial analysis over breakfast and leaves at 07.30 to walk to the station.

07.45 George's train is late (as usual) and he rings the office using his pocket phone, leaving a recorded message for his secretary to delay the 09.00 meeting until 09.15.

07.50 Samantha and James want to hear the latest No. 1 hit while having breakfast; they switch on the TV and key in the number of the local dial-a-disc service. The TV set displays the top twenty tunes and they select No. 1. The TV switches off and the tune is played on the White's Hi Fi system.

08.10 The White children leave for school. There has been an incident recently in Farnston of a man molesting a young girl and all the girls at Samantha's school have been issued with a miniature alarm caller which on the press of a button causes an alarm to be sent over the PCN to the local police and identifies the base station receiving the call. The alarm continues to broadcast a radio frequency signal for direction finding purposes which enables the police to pinpoint the incident. (Fortunately the alarm is not needed today.)

08.30 Pamela has had her nurse's bulletin printed out and is scanning the situations vacant column when there is a video phone-call (using the ISDN and 64 kbit/s encoding) from her friend Penelope Green. The two ladies chat for a few minute and agree to meet for coffee at 11.00.

08.50 While Pamela is having a bath the phone rings. Fortunately she has a cordless (DECT) telephone in the bathroom and picks it up to find that the caller has keyed a wrong number. (It is now very rare for the system to get the number wrong; system errors are confined to human errors in setting up the directory data base.)

09.15 Pamela makes a video phone call to the fishmongers in Farnston and inspects and chooses four plaice which she agrees to pick up later. The fishmonger and Pamela establish a three-party data connection to Pamela's bank's computer and the cost of the fish (30 Eurodollars) is transferred to the fishmonger's account.

09.30 Pamela types a letter to an old school friend on one of the family's PCs; she has mislaid her friend's address and using the same terminal consults a directory service to find the E-mail address information. This is inserted automatically in the address field of her letter and a colour photograph of her two children is inserted in the document reader. The letter and the photograph are despatched as a mixed mode document through the E-mail system.

11.00 Pamela and Penelope meet at Tanya's Tea Rooms. During the discussion they agree that it would be nice to to go to a matinee in London on Pamela's day off next week to see a revival of the musical *Cats*. On Penelope's mobile phone they contact a ticket agency, book two tickets which are debited directly to their individual accounts. Later the tickets are printed out on their terminals at home.

11.30 Pamela drives home, picking up the fish on the way. She gets home and prepares the evening meal, has a light lunch and leaves for the hospital. On the way there she realise that she meant to turn the thermostat on the central heating down. She stops the car and makes a telemetry call to reset it.

16.30 Samantha and James arrive home from school and settle down to do their homework. Samantha realises that she has left her French text book behind and she sets up a call to an information service where the book is available on-line. She obtains a copy of the relevant two pages on the family's printer. (George will not be pleased when he next looks at his itemised communications bill!)

17.30 Pamela arrives home and starts to prepare the meal. George rings from the train to say it's running 15 minutes late.

18.30 George comes home and as the family settle down to their meal the phone rings. James goes to answer it, sees from the calling number display that it's his Aunt Tracy Jenkins (who is a marathon talker) and switches on the telephone answering machine. (He knows that she won't leave a message but will ring back later.)

19.15 The meal over and the dishes stacked, George goes to his study to watch a special relay of a learned society meeting from Copenhagen where his boss is giving a paper. The rest of the family watch *Coronation Street* and then, because James wants to watch a science fiction film (pre-recorded earlier in the day) and the two ladies want to watch a repeat of an old Victoria Wood sit-com, James retires to his bedroom to watch on his own set. Meanwhile George's meeting has finished and he switches over to a news bulletin; an item appears in which a friend of the family is featured. George enters the item's key code on his remote controller and it is later copied to the White's video recorder to be available for replay to the whole family.

22.00   George remembers that he has to go to Manchester by train tomorrow and to save time in the morning he calls a travel agency and has a ticket printed at home and his personal business account appropriately debited.

02.00   The PTO's routing equipment detects an incipient fault on the White's land line termination equipment and switches service to the PCN pending a repair the next day.

27.5   EPILOGUE

Since the last century the emphasis has shifted from moving goods about towards the easy conveyance of information. Telecommunications is a key technology in the information revolution just as the railways and canals were in the industrial revolution. The need to move people to work on a daily basis is being called into question as urban transport, particularly in London, becomes increasingly slow and unpleasant. Telecommunications can help to provide a solution to this problem too but the extent to which it will do so remains to be seen. The issue of working at home well illustrates the point that in the end it is the human acceptance of what is possible that counts and not what the technology can do.

Undoubtedly we are going to see a continuing and substantial growth in mobile communications but the effect on people's lives is not so easy to gauge. Will it become normal for passengers to gossip with friends over a mobile telephone on a bus or train journey? Who could have predicted the use of personal stereos on public transport even a few years before it happened? Will information services spring up directing travellers to avoid road congestion or rail cancellations or will such services be self defeating? We can be confident that whatever happens we shall be in for some surprises.

In 1979 the late Chris Evans made a number of bold predictions: for example, that by the mid-1980s we would be working a 30 hour week, taking six weeks' annual holiday and retiring at 50 or 55. The trend to shorter hours is there, but it is one based on the human acceptance of new practices, which takes a time of the order of the working life span and is not determined by what technology might be able to do. He also predicted that the police would be equipped with pocket computers that would contain a set of criminal records. Again the trend is there but the manner of its execution, access to a remote data base by means of improved mobile communications rather than carrying the data base around with you, is rather different. More pointedly he predicted the decline of the communist system, attributing it to improved prosperity in the west brought about by the microprocessor and improved global communications. We have not attempted to be so bold; it remains to be seen whether our more circumscribed approach is too cautious or not cautious enough.

27.6   BIBLIOGRAPHY

C. Evans (1982) *The Mighty Micro: The Impact of the Computer Revolution* (2nd edn), Gollancz.

J.R Stern and R. Wood (1989) The longer term future of the local network, *Br. Telecom Technology J.*, **7**, (2), 161–170.

J.A Tritton (1988) Interactive home systems (IHS)—an overview, Private *Switching Systems and Networks, IEE International Conference*, June, pp 195–200 (CPN 288).

INDEX

Note: In many cases entries are identified by their acronyms rather than being spelt out—see page xxiii for a list of acronyms